T0281031

Lecture Notes in Computer Science   14500

Advanced Research in Computing and Software Science
Subline of Lecture Notes in Computer Science

More information about this series at https://link.springer.com/bookseries/558

Rayna Dimitrova · Ori Lahav · Sebastian Wolff
Editors

Verification, Model Checking, and Abstract Interpretation

25th International Conference, VMCAI 2024
London, United Kingdom, January 15–16, 2024
Proceedings, Part II

Springer

Editors
Rayna Dimitrova 🆔
CISPA Helmholtz Center for Information
Security
Saarbrücken, Germany

Ori Lahav 🆔
Tel Aviv University
Tel Aviv, Israel

Sebastian Wolff 🆔
New York University
New York, NY, USA

ISSN 0302-9743   ISSN 1611-3349 (electronic)
Lecture Notes in Computer Science
ISBN 978-3-031-50520-1   ISBN 978-3-031-50521-8 (eBook)
https://doi.org/10.1007/978-3-031-50521-8

This Springer imprint is published by the registered company Springer Nature Switzerland AG
The registered company address is: Gewerbestrasse 11, 6330 Cham, Switzerland

Paper in this product is recyclable.

Preface

This volume contains the proceedings of VMCAI 2024, the 25th International Conference on Verification, Model Checking, and Abstract Interpretation. The VMCAI 2024 conference was co-located with the 51st ACM SIGPLAN Symposium on Principles of Programming Languages (POPL 2024), held at the Institution of Engineering and Technology in London, UK, during January 15–16, 2024.

VMCAI is a forum for researchers working in verification, model checking, and abstract interpretation. It attempts to facilitate interaction, cross-fertilization, and advancement of methods that combine these and related areas. The topics of the conference include program verification, model checking, abstract interpretation, program synthesis, static analysis, type systems, deductive methods, decision procedures, theorem proving, program certification, debugging techniques, program transformation, optimization, and hybrid and cyber-physical systems.

VMCAI 2024 received a total of 88 submissions, of which 74 went through the peer review process (14 were desk-rejected). After a rigorous single-blind review process, with each paper reviewed by at least three Program Committee (PC) members, followed by an extensive online discussion, the PC accepted 30 papers for publication in the proceedings and presentation at the conference. The main selection criteria were quality, relevance, and originality. Out of the 30 accepted papers, four are tool papers and one is a case study, while the rest are regular papers. In addition to the contributed papers, the conference program included two keynote talks: David Harel (Weizmann Institute of Science, Israel) and Hiroshi Unno (University of Tsukuba, Japan).

By now, artifact evaluation is a standard part of VMCAI. The artifact evaluation process complements the scientific impact of the conference by encouraging and rewarding the development of tools that allow for replication of scientific results as well as for shared infrastructure across the community. Authors of submitted papers were encouraged to submit an artifact to the VMCAI 2024 artifact evaluation committee (AEC). We also encouraged the authors to make their artifacts publicly and permanently available. All submitted artifacts were evaluated in parallel with the papers. We assigned two members of the AEC to each artifact and assessed it in two phases. First, the reviewers tested whether the artifacts were working, e.g., there were no corrupted or missing files and the evaluation did not crash on simple examples. For those artifacts that did not work, we sent the issues to the authors, for clarifications. In the second phase, the assessment phase, the reviewers aimed at reproducing any experiments or activities and evaluated the artifact based on the following questions:

1. Is the artifact consistent with the paper and are the results of the paper replicable through the artifact?
2. Is the artifact well documented and easy to use?
3. Is the artifact available?

We awarded a badge for each of these question to each artifact that answered it in a positive way. Of the 30 accepted papers, there were 14 submitted artifacts, all of which were awarded two or all three Artifact Evaluation Badges.

The VMCAI program would not have been possible without the efforts of many people. We thank the research community for submitting their results to VMCAI and for their participation in the conference. The members of the Program Committee, the Artifact Evaluation Committee, and the external reviewers worked tirelessly to select a strong program, offering constructive and helpful feedback to the authors in their reviews. The PC and the external reviewers contributed a total of 233 high-quality reviews to the review process. The VMCAI steering committee provided continued encouragement and advice. We warmly thank the keynote speakers for their participation and contributions to the program of VMCAI 2024. We also thank the general chair of POPL 2024, Philippa Gardner, and the organization team for their support. We thank the publication team at Springer for their support, and EasyChair for providing an excellent conference management system.

November 2023
Rayna Dimitrova
Ori Lahav
Sebastian Wolff

Organization

Program Committee Chairs

Rayna Dimitrova	CISPA Helmholtz Center for Information Security, Germany
Ori Lahav	Tel Aviv University, Israel

Artifact Evaluation Committee Chair

Sebastian Wolff	New York University, USA

Program Committee

Ezio Bartocci	TU Wien, Austria
Nathalie Bertrand	Inria, France
Emanuele De Angelis	IASI-CNR, Italy
Coen De Roover	Vrije Universiteit Brussel, Belgium
Jyotirmoy Deshmukh	University of Southern California, USA
Bruno Dutertre	Amazon Web Services, USA
Michael Emmi	Amazon Web Services, USA
Grigory Fedyukovich	Florida State University, USA
Nathanaël Fijalkow	CNRS, LaBRI, University of Bordeaux, France
Hadar Frenkel	CISPA Helmholtz Center for Information Security, Germany
Liana Hadarean	Amazon Web Services, USA
Jochen Hoenicke	Certora, Germany
Hossein Hojjat	Tehran Institute for Advanced Studies, Iran
Qinheping Hu	Amazon Web Services, USA
Marieke Huisman	University of Twente, The Netherlands
Amir Kafshdar Goharshady	Hong Kong University of Science and Technology, Hong Kong
Joost-Pieter Katoen	RWTH Aachen University, Germany
Daniela Kaufmann	TU Wien, Austria
Bettina Koenighofer	Graz University of Technology, Austria
Burcu Kulahcioglu Ozkan	Delft University of Technology, The Netherlands
Anna Lukina	Delft University of Technology, The Netherlands
Roland Meyer	TU Braunschweig, Germany
David Monniaux	CNRS / VERIMAG, France
Kedar Namjoshi	Nokia Bell Labs, USA
Jens Palsberg	University of California, Los Angeles, USA
Elizabeth Polgreen	University of Edinburgh, UK
Arjun Radhakrishna	Microsoft, USA

Robert Rand	University of Chicago, USA
Francesco Ranzato	University of Padova, Italy
Xavier Rival	INRIA / ENS Paris, France
Philipp Rümmer	University of Regensburg, Germany
Anne-Kathrin Schmuck	Max-Planck-Institute for Software Systems, Germany
Mihaela Sighireanu	ENS Paris-Saclay, France
Gagandeep Singh	VMware Research and UIUC, USA
Hazem Torfah	Chalmers University of Technology, Sweden
Zhen Zhang	Utah State University, USA
Lenore Zuck	University of Illinois Chicago, USA

Additional Reviewers

Armborst, Lukas
Balakrishnan, Anand
Ballarini, Paolo
Bardin, Sebastien
Beutner, Raven
Biere, Armin
Biktairov, Yuriy
Blicha, Martin
Boker, Udi
Boulanger, Frédéric
Cailler, Julie
Cano Córdoba, Filip
Chakraborty, Soham
Cheang, Kevin
Chen, Mingshuai
Chen, Yixuan
Chiari, Michele
Daniel, Lesly-Ann
Eberhart, Clovis
Esen, Zafer
Fleury, Mathias
Fluet, Matthew
Golia, Priyanka
Gruenke, Jan
Gupta, Priyanshu
Hamza, Ameer
Helouet, Loic
Huang, Wei
Klinkenberg, Lutz
Lammich, Peter
Larrauri, Alberto
Liang, Chencheng
Lopez-Miguel, Ignacio D.
Mainhardt, Ana

Maseli, René
Matricon, Théo
Meggendorfer, Tobias
Milanese, Marco
Mora, Federico
Morvan, Rémi
Mousavi, Mohammadreza
Mutluergil, Suha Orhun
Nayak, Satya Prakash
Parker, Dave
Paul, Sheryl
Piribauer, Jakob
Pranger, Stefan
Quatmann, Tim
Rappoport, Omer
Rath, Jakob
Refaeli, Idan
Riley, Daniel
Rubbens, Robert
Saglam, Irmak
Schafaschek, Germano
Shah, Ameesh
van der Wall, Sören
Vandenhove, Pierre
Viswanathan, Mahesh
Waldburger, Nicolas
Williams, Sam
Wolff, Sebastian
Xia, Yuan
Ying, Mingsheng
Zanella, Marco
Zavalia, Lucas
Zhao, Yiqi

Invited Keynote Talks

Two Projects on Human Interaction with AI

David Harel[1]

The Weizmann Institute of Science

A significant transition is under way, regarding the role computers will be playing in a wide spectrum of application areas. I will present two work-in-progress projects that attempt to shed new light on this transition.

The first we term *"The Human-or-Machine Issue"*. Turing's imitation game addresses the question of whether a machine can be labeled intelligent. We explore a related, yet quite different, challenge: in everyday interactions with an agent, how will knowing whether the agent is human or machine affect that interaction? In contrast to Turing's test, this is not a thought experiment, but is directly relevant to human behavior, human-machine interaction and also system development. Exploring the issue now is useful even if machines will end up not attempting to disguise themselves as humans, which might become the case because they cannot do so well enough, because doing so will not be that helpful, because machines exceed human capabilities, or because regulation forbids it.

In the second project, we propose a systematic programming methodology that consists of three main components: (1) a modular incremental specification approach (specifically, scenario-based programming); (2) a powerful, albeit error-prone, AI-based software development assistant; and (3) systematic iterative articulation of requirements and system properties, amid testing and verification. The preliminary results we have obtained show that one can indeed use an AI chatbot as an integral part of an interactive development method, during which one constantly verifies each new artifact contributed by the chatbot in the context of the evolving system.

While seemingly quite diverse, both projects have human-machine interaction at their heart, and bring in their wake common meta-concepts, such as trust and validity. What, for example, are the effects of the presence or absence of trust in such interactions, and how does one ensure that the interaction contributes to achieving one's goals, even when the motives of the other party or the quality of its contributions are unclear?

In addition, the kinds of interactions we discuss have a strong dynamic nature: multi-step interaction on requirements elicitation, for example, is not just a search for something that should have been specified in full earlier. Rather, it is often a constructive process of building new knowledge, acquiring new understanding, and demarcating explicit boundaries and exceptions that are often absent from even the most rigorous definitions.

---

[1] Research joint with Assaf Marron, Guy Katz and Smadar Szekely.

Automating Relational Verification of Infinite-State Programs

Hiroshi Unno

University of Tsukuba

Hyperproperties are properties that relate multiple execution traces of one or more programs. A notable subclass, known as k-safety, is capable of expressing practically important properties like program equivalence and non-interference. Furthermore, hyperproperties beyond k-safety, including generalized non-interference (GNI) and co-termination, have significant applications in security.

Automating verification of hyperproperties is challenging, as it involves finding an appropriate alignment of multiple execution traces for successful verification. Merely reasoning about each program copy's executions separately, or analyzing states from running multiple copies in lock-step, can sometimes lead to failure. Therefore, it necessitates synthesizing a scheduler that dictates when and which program copies move, to ensure an appropriate alignment of multiple traces. With this alignment, synthesizing relational invariants maintained by the aligned states enables us to prove k-safety. Additionally, verifying GNI and co-termination requires synthesis of Skolem functions and ranking functions, respectively.

In this talk, I will explain and compare two approaches for automating relational verification. The first approach is constraint-based, reducing the synthesis problem into a constraint solving problem within a class that extends Constrained Horn Clauses (CHCs). The second approach is based on proof search within an inductive proof system for a first-order fixpoint logic modulo theories.

Contents – Part II

Program and System Verification

Runtime Verification

Security and Privacy

Contents – Part I

SAT, SMT, and Automated Reasoning

Concurrency

Petrification: Software Model Checking for Programs with Dynamic Thread Management

Matthias Heizmann$^{(\boxtimes)}$, Dominik Klumpp , Lars Nitzke,
and Frank Schüssele

University of Freiburg, Freiburg Im Breisgau, Germany
{heizmann,klumpp,schuessf}@informatik.uni-freiburg.de,
lars.nitzke@mailfence.com

Abstract. We address the verification problem for concurrent program that dynamically create (fork) new threads or destroy (join) existing threads. We present a reduction to the verification problem for concurrent programs with a fixed number of threads. More precisely, we present *petrification*, a transformation from programs with dynamic thread management to an existing, Petri net-based formalism for programs with a fixed number of threads. Our approach is implemented in a software model checking tool for C programs that use the *pthreads* API.

Keywords: Concurrency · Fork-Join · Verification · Petri Nets · pthreads

1 Introduction

We address the verification problem for concurrent programs with *dynamic thread management*. Such programs start with a single main thread, and create (*fork*) and destroy(*join*) additional concurrently executing threads. The number of threads changes during execution and may depend on input data. As an example, consider the program in Fig. 1. In a loop, the `main` thread creates new worker threads and assigns them the thread ID `i`, i.e., the current iteration. In all iterations but the first, the `main` thread also *joins* the thread with ID `i-1`, i.e., the thread created in the previous iteration. The `join` statement blocks until that thread terminates, and destroys the terminated thread. Dynamic thread management in this style is widely used in practice. For

```
1  c := 0; i := 0;
2  while (true) {
3      fork i w();
4      if (i > 0) {
5          join i-1;
6      }
7      i := i + 1;
8  }
```

(a) The initial `main` thread

```
1  c := c + i;
2  assert c <= 2 * i;
3  c := c - i;
```

(b) The worker thread `w`

Fig. 1. Program with dynamic thread management.

instance, the threads extension of the POSIX standard [1] specifies such an API (commonly known as *pthreads*), as do Java [24] and the .NET framework [22].

© The Author(s), under exclusive license to Springer Nature Switzerland AG 2024
R. Dimitrova et al. (Eds.): VMCAI 2024, LNCS 14500, pp. 3–25, 2024.
https://doi.org/10.1007/978-3-031-50521-8_1

Automated verification of programs with dynamic thread management is challenging. The control flow of the program (which code will be executed in what order) is not immediately apparent from the syntax. In fact, it is already an undecidable problem to determine statically *how many* and *which* threads will be created at runtime, and *when* or *if* a thread will be joined. For instance, for the program in Fig. 1, a control flow analysis must determine how many instances of the worker thread w can be active at the same time, and consider all interleavings of their actions. Furthermore, the analysis has to figure out that the thread joined in line 5 is always the thread created in the previous iteration, hence the program's control flow does not include interleavings in which the joined thread executes another action after the join statement. This difficulty of determining the control flow presents a challenge for many *software model checking* techniques, which typically require a representation of a program's control flow as input.

Given a description of the control flow, software model checkers reason about the program's data to construct an over-approximation of the reachable states. This allows them to either prove the correctness of the program (if the over-approximation does not admit a specification violation) or find bugs. Here, we encounter a second challenge for programs with dynamic thread management: Threads may have thread-local variables, and program states must assign a value to x_t for each thread t and each local variable x. Hence, if the number of threads created at runtime is unknown, then so is the program's state space. This makes it challenging for software model checkers to construct over-approximations of the reachable states that are precise enough to show correctness.

In this work, we present an approach for the automated verification of programs with dynamic thread management, which overcomes these challenges. Our focus is on programs where the number of thread that can be active at the same time is bounded. We call the maximum number of threads (with the same thread template) that can be active at the same time the *thread width* of the program. Programs with a bounded thread width represent a sweet spot for automated verification: Firstly, programs with a bounded thread width are still tractable for software model checking. Secondly, the number of active threads is often small in practice because programs are most efficient if the number of active threads is at most the number of CPU cores. This is in contrast to many other discrete characteristics of executions, which can grow unboundedly depending on input data (e.g. length of executions, and consequently, the overall number of threads) or are not under the programmer's control (e.g. number of context switches).

In order to apply software model checking techniques to the verification of programs with dynamic thread management, and overcome the challenges laid out above, we reduce the verification problem for a program with dynamic thread management to a series of verification problems with a fixed number of threads. The reduction is *sound*: If our approach concludes that a program with dynamic thread management is correct, then *all* executions of the program are indeed correct. To ensure soundness for all programs, our approach attempts to identify the given program's thread width (by trying out different values) and then *ver-*

ifies that the program indeed has a certain thread width. We again reduce this verification problem to a verification problem with a fixed number of threads. Note that the thread width is not supplied by the user of the analysis. Instead, the thread width is a property of the input program, derived from the program's semantics. Our approach is *complete* for programs with a bounded thread width, and for incorrect programs.

The key technical step in our approach is *petrification*, a transformation from programs with dynamic thread management to *Petri programs* [9] with a fixed number of threads. Petrification separates the control flow of the program (including thread management) from data aspects, and imposes a given *thread limit* β on the number of active threads. The control flow of a Petri program is encoded as a Petri net, while the manipulation of data (i.e., variable values) is expressed by labeling transitions of the Petri net with assignments and guards. The resulting Petri program faithfully represents executions of the original program where at most β threads are active at the same time.

We define two separate specifications for the petrified program: a *safety specification* which expresses correctness of all executions encoded by the Petri program, and a *bound specification*, expressing that the program's thread width is bounded by β. Both specifications can be verified independently, using an existing algorithm for the verification of Petri programs [9]. We thereby reduce the existence of a bound on the program's thread width to a safety specification for Petri programs; checking this specification does not require any changes to the underlying verification algorithm. We present a verification algorithm for programs with dynamic thread management, that repeatedly petrifies a given program (with different thread limits β) and invokes a Petri program verification algorithm [9] to check the safety and bound specifications. We investigate several variants of this verification algorithm and their differences.

Our approach is implemented in the program analysis and verification framework ULTIMATE. The implementation verifies concurrent C programs using the POSIX threads (*pthreads*) standard for multi-threading. Our approach is practicable and can be successfully used for automated verification.

Example. As an illustration of our approach, consider again the program in Fig. 1. The worker threads each perform the same computation involving the global variables c and i, and assert that $c \leq 2 \cdot i$. The verification problem consists of showing that this condition always holds when any worker thread executes the **assert** statement.

As the first verification step, we *petrify* the program. If we chose a thread limit $\beta = 2$ – i.e., at most two worker threads can run at the same time –, petrification produces the Petri program in Fig. 2. We see the main thread (in the middle, in red), as well as two instances of the worker thread (at the top resp. at the bottom, in blue resp. purple). Transitions (black vertical bars) encode the possible control flow, and are annotated with assignments and guards. The thread ID is tracked through specially-introduced variables id_1^w and id_2^w. The

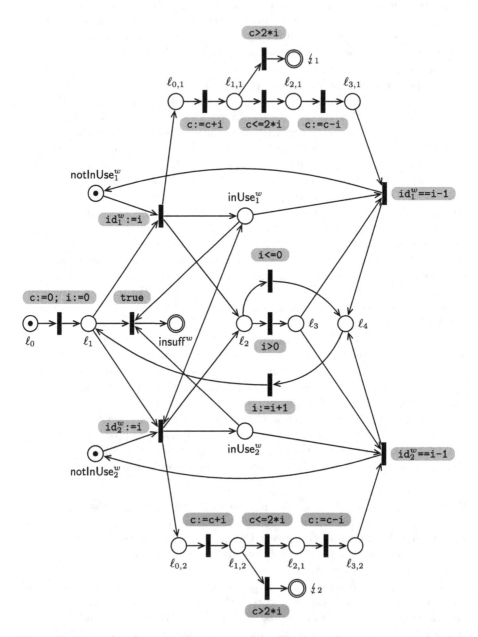

Fig. 2. Petri net for the example program from Fig. 1, with two instances for the worker thread.

transitions labeled $\boxed{\mathtt{id}_k^w\,\mathtt{:=i}}$ correspond to the `fork` statement, whereas the transitions labeled $\boxed{\mathtt{id}_k^w\,\mathtt{==i-1}}$ correspond to the `join` statement.

We employ the Petri program verification algorithm [9] to show that the petrified program satisfies both its bound and safety specification. Intuitively, this means that no run (or firing sequence) of the Petri net, that also obeys the assignments and guards of the transitions (in a sense made formal below), can reach the place insuff^w (for the bound specification) resp. the place $\frac{1}{2}_1$ or $\frac{1}{2}_2$ (for the safety specification). Intuitively, the former holds because at the start of any iteration, only the worker thread created in the previous iteration is active. The latter then follows, because all worker threads created in even earlier iterations must have terminated, and their overall effect was to either leave `c` unchanged or decrease it (if `i` was incremented in between the two assignments to `c`). If $n \in \{0, 1, 2\}$ active worker threads have executed the first but not the second assignment, we have $\mathsf{c} \le n \cdot \mathsf{i}$. We conclude that the program in Fig. 1 is correct.

Related Work. One approach for the analysis of concurrent programs found in the literature is *bounded verification*. In bounded verification, the user of an analysis supplies some bound, and the analysis then considers only those behaviours of the program below the given bound. For instance, one may only analyze executions up to a certain length [13,15,21], with at most a given number of threads [23], or a with a limited number of context switches [19,26]. In each case, the analysis only covers a fragment of the program behaviours. For programs that exceed the imposed limit, such analyses are often successful in finding bugs, but cannot soundly conclude that the program is correct. By contrast, we present an approach that uses a form of bounded verification as a sub-procedure (the bound is given by the thread limit β), and yet gives full correctness guarantees.

Our approach shares some similarities with works that combine bounded verification with the search for a *cutoff point* [7,8,11,30]: A finite bound such that correctness of the fragment of program behaviours up to the cutoff point also implies correctness of all other program behaviours. In contrast to such works, the satisfaction of a certain thread width bound by a program is wholly independent of the verified property, the logic in which it is expressed, or how it is proven. However, given a suitable cutoff point for the thread width bound, one may employ petrification to produce and subsequently verify the corresponding bounded instance.

There are many works concerned with the verification of *parametrized programs*, i.e., program with an unbounded number of active threads. For instance, in *thread-modular verification*, the idea is to find a modular proof that allows for generalization to any number of threads. For instance, thread-modular proofs at level k [18] generalize the non-interference condition of Owicki and Gries [25]. A thread-modular proof for a program with k threads establishes correctness of the program for any number of threads. Other forms of compositional verification, such as rely-guarantee reasoning, follow a similar approach. A challenge in this setting is that such proof methods are often incomplete, and while there is some work in automating the search for such a proof (e.g. the thread-modular abstract interpreter GOBLINT [27]), software model checking based on compositional proofs is not yet as mature. An additional challenge is that many such works are based on the setting of parametric programs, where an arbitrary number (not controlled by the program) of threads execute concurrently (i.e., all threads start at the same time). Supporting dynamic thread management often requires further work [28].

On a technical level, the work most closely related to our approach [4] describes the implementation of the software model checker CPACHECKER [5]. Upfront, this implementation creates a fixed number of procedure clones. It then performs a specialized configurable program analysis (CPA), which treats the cloned procedures as threads. In contrast, we describe a modular approach that separates the reduction to programs with a fixed number of threads from the analysis, and allows for different verification algorithms to be applied. Our approach is grounded in a formal semantics of a concurrent programming language, which allows for theoretical analysis.

Contributions. To summarize, our contributions are as follows:

- We identify the class of programs with a *bounded thread width* as a sweet spot for software model checking, both relevant and tractable.
- We present an approach for the automated verification of programs with dynamic thread management, by reduction to a series of verification problems with a fixed number of threads. The approach is sound, and it is complete for programs with bounded thread width as well as incorrect programs.
- The key technical contribution behind this reduction is *petrification*, a construction that captures the control flow of programs with dynamic thread management (up to a given thread limit β) as a Petri program [9] with a fixed number of threads.
- We implemented our approach in the program verification framework ULTIMATE and showed that it is successful in practice.

Roadmap. Section 2 presents the syntax and semantics of CONC, a simple language with dynamic thread management, and defines the verification problem. In Sect. 3, we quickly recap the essential notions of Petri programs and then introduce petrification, our transformation from CONC to Petri programs. We present the overall verification algorithms based on petrification in Sect. 4. Section 5 discusses the implementation of our approach in a verification tool for C programs, and evaluates the practical feasibility. We conclude in Sect. 6.

2   A Language for Dynamic Thread Creation

In this section, we introduce CONC, a simple imperative language for concurrent programs. CONC captures the essence of dynamic thread management. We define syntax and semantics of CONC, as well as the corresponding verification problem.

Syntax. CONC programs consist of a number of *thread templates*, which have a unique name, and a body specifying the code to be executed by a thread. We denote the finite set of all possible thread template names by **Templates**. Further, let **Var** be a set of program variable names. We assume as given a language of expressions over variables in **Var**, e.g., the expressions defined by the SMT-LIB standard [2]. Let x range over variables, ϑ range over thread templates, e_{int} range over integer-valued expressions, e_{bool} range over boolean-valued expressions, and e range over both integer- and boolean-valued expressions. The syntax of CONC commands is defined by the following grammar:

$$
\begin{aligned}
C ::=\ &x{:}{=}e \mid \texttt{assume } e_{\text{bool}} \mid \texttt{assert } e_{\text{bool}} \\
\mid\ &\texttt{if } (e_{\text{bool}}) \;\{\; C \;\} \texttt{ else } \{\; C \;\} \mid \texttt{while } (e_{\text{bool}}) \;\{\; C \;\} \\
\mid\ &\texttt{fork } e_{\text{int}}\ \vartheta() \mid \texttt{join } e_{\text{int}} \\
\mid\ &C; C
\end{aligned}
$$

The set of all commands is denoted by **Command**. The set of *atomic statements*, **AtomicStmt**, is the set of all assignments and **assume** commands. We call atomic statements as well as **fork** and **join** commands *simple statements*.

The key feature of CONC are the **fork** and **join** commands for dynamic thread management. The statement **fork** $e\ \vartheta()$ creates a new thread, whose code is given by the thread template ϑ. The expression e is evaluated, and its current value is used as the *thread ID* of the newly created thread. All forked threads run concurrently, and their computation steps can be arbitrarily interleaved (i.e., we consider a sequential consistency semantics). When a thread executes **join** e, the execution blocks until some other thread, whose thread ID equals the value of e, has terminated. The terminated thread is destroyed.

This style of dynamic thread management is inspired by the POSIX threads (or *pthreads*) API [1], in which a call to **pthread_create** creates a new thread, with the thread template given by a function pointer. In contrast to our **fork** command, **pthread_create** also computes and returns a thread ID. In CONC programs, the computation of a (unique) thread ID can be implemented separately, and the resulting thread ID can be used by a **fork** command. However, CONC also allows multiple threads to share the same thread ID. Generally, there is a tradeoff regarding uniqueness of thread IDs: One can enforce uniqueness of thread IDs on the semantic level, and possibly design a type system that ensures this uniqueness. This complicates the definition of the language, and slightly limits the expressiveness of the language. On the other hand, allowing thread IDs to be non-unique leads to a simpler and more expressive language. Though it slightly complicates some proofs in this paper, we choose the latter path.

The semantics of the other, not concurrency-related commands, is standard: An **assume** commands blocks execution if the given expression evaluates to *false*, otherwise it has no effect. An **assert** command *fails* if the given expression evaluates to *false*. The intuition behind the assignment, if-then-else and while commands, as well as the sequential composition (;) of commands is as expected.

We use the special values Ω to represent a command that has successfully terminated, and $\frac{1}{4}$ for a command that has failed (due to a violated **assert**). For convenience, we extend the sequential composition by setting $C; \Omega{:}{=}C$.

A program is given by a tuple $\mathcal{P} = (body, \texttt{main}, \textbf{Globals})$, consisting of a mapping $body$ that associates each thread template name ϑ with a command $body_\vartheta$, a thread template $\texttt{main} \in \textbf{Templates}$ for the main thread, and a set of global variables $\textbf{Globals}$. For each thread template $\vartheta \in \textbf{Templates}$, the command $body_\vartheta$ identifies the code executed by instances of this thread template. This command may refer to the variables in $\textbf{Globals}$, as well as to any other variables (which are implicitly assumed to be thread-local). We do not require variables to be declared, and uninitialized variables can have arbitrary values.

Semantics. We define the (small-step) semantics of CONC, in the style of structural operational semantics. To this end, we first introduce the notion of *local* and *global configurations* of a given CONC program $\mathcal{P} = (body, \texttt{main}, \textbf{Globals})$.

A *local configuration* is a quadruple $\langle X, \vartheta, t, s \rangle$ consisting of some X that is either a remainder program left to execute ($X \in \textbf{Command}$), or a special value to indicate termination (Ω) or failure ($\frac{i}{2}$), a thread template name ϑ, a thread ID $t \in \mathbb{Z} \cup \{\bot\}$, and a local state $s : \textbf{Var} \setminus \textbf{Globals} \to \mathbb{Z} \cup \{true, false\}$. The thread ID $\bot$ is used exclusively for the start thread. In general, thread IDs need not be unique. We keep track of the thread ID in the local configuration, in order to determine which threads can be joined when another thread executes a $\texttt{join}$ command. If multiple threads with the same ID are ready to be joined, one thread is chosen nondeterministically.

A *global configuration* is a pair (M, g) of a multiset M of local configurations, and a global state $g : \textbf{Globals} \to \mathbb{Z} \cup \{true, false\}$. We use a multiset to reflect the fact that several running threads could have the same local configuration. A global configuration (M, g) is *initial* if $M = \langle\!\langle body_{\texttt{main}}, \texttt{main}, \bot, s \rangle\!\rangle$ for any local state s; the global state g is also arbitrary. (The symbols $\langle\!\langle \ldots \rangle\!\rangle$ denote a multiset containing the listed elements.)

Figure 3 defines the small-step structural operational semantics of our language as a transition relation $(M, g) \xrightarrow{st} (M', g')$ over global configurations $(M, g), (M', g')$ and simple statements st. We assume here that, given a mapping $\tilde{s} : \textbf{Var} \to \mathbb{Z} \cup \{true, false\}$, we can evaluate an expression e to some value $[\![e]\!]^{\tilde{s}} \in \mathbb{Z} \cup \{true, false\}$. In particular, if s is a local state and g is a global state, we can set $\tilde{s} := s \cup g$. Given the semantic transition relation, we define:

Definition 2.1. (Execution). An *execution* is a sequence of global configurations and statements $(M_0, g_0) \xrightarrow{st_1} \ldots \xrightarrow{st_n} (M_n, g_n)$ where (M_0, g_0) is initial.

Definition 2.2. (Correctness). An execution $(M_0, g_0) \xrightarrow{st_1} \ldots \xrightarrow{st_n} (M_n, g_n)$ is *erroneous* if $\frac{i}{2}$ occurs in any local configuration of any M_i. A CONC program $\mathcal{P}$ is *correct* if there does not exist any erroneous execution of $\mathcal{P}$.

Definition 2.3. (Thread Width). We say that the *thread width of the execution* $(M_0, g_0) \xrightarrow{st_1} \ldots \xrightarrow{st_n} (M_n, g_n)$ is the maximum number $\beta \in \mathbb{N}$ such that some M_i contains β local configurations with the same thread template ϑ. The *thread width of the program* $\mathcal{P}$ is the supremum over the thread widths of all executions of $\mathcal{P}$.

$$\frac{[\![e]\!]^{s \cup g} = \mathit{true}}{\wr \langle \mathtt{assume}\ e; X, \vartheta, t, s \rangle \int, g \xrightarrow{\mathtt{assume}\ e} \wr \langle X, \vartheta, t, s \rangle \int, g} \ (\text{Assume})$$

$$\frac{x \in \mathbf{Globals}}{\wr \langle x\!:=\!e; X, \vartheta, t, s \rangle \int, g \xrightarrow{x:=e} \wr \langle X, \vartheta, t, s \rangle \int, g[x \mapsto [\![e]\!]^{s \cup g}]} \ (\text{AssignGlobal})$$

$$\frac{x \notin \mathbf{Globals}}{\wr \langle x\!:=\!e; X, \vartheta, t, s \rangle \int, g \xrightarrow{x:=e} \wr \langle X, \vartheta, t, s[x \mapsto [\![e]\!]^{s \cup g}] \rangle \int, g} \ (\text{AssignLocal})$$

$$\frac{[\![e]\!]^{s \cup g} = \mathit{true}}{\wr \langle \mathtt{assert}\ e; X, \vartheta, t, s \rangle \int, g \xrightarrow{\mathtt{assume}\ e} \wr \langle X, \vartheta, t, s \rangle \int, s} \ (\text{Assert1})$$

$$\frac{[\![e]\!]^{s \cup g} = \mathit{false}}{\wr \langle \mathtt{assert}\ e; X, \vartheta, t, s \rangle \int, g \xrightarrow{\mathtt{assume}\ !e} \wr \langle \mbox{\textbf{\textlightning}}, \vartheta, t, s \rangle \int, g} \ (\text{Assert2})$$

$$\frac{[\![e]\!]^{s \cup g} = \mathit{true}}{\wr \langle \mathtt{if}\ (e)\ \{\, C_1 \,\}\, \mathtt{else}\, \{\, C_2 \,\}; X, \vartheta, t, s \rangle \int, g \xrightarrow{\mathtt{assume}\ e} \wr \langle C_1; X, \vartheta, t, s \rangle \int, g} \ (\text{Ite1})$$

$$\frac{[\![e]\!]^{s \cup g} = \mathit{false}}{\wr \langle \mathtt{if}\ (e)\ \{\, C_1 \,\}\, \mathtt{else}\, \{\, C_2 \,\}; X, \vartheta, t, s \rangle \int, g \xrightarrow{\mathtt{assume}\ !e} \wr \langle C_2; X, \vartheta, t, s \rangle \int, g} \ (\text{Ite2})$$

$$\frac{[\![e]\!]^{s \cup g} = \mathit{true}}{\wr \langle \mathtt{while}\ (e)\ \{\, C \,\}; X, \vartheta, t, s \rangle \int, g \xrightarrow{\mathtt{assume}\ e} \wr \langle C; \mathtt{while}\ (e)\ \{\, C \,\}; X, \vartheta, t, s \rangle \int, g} \ (\text{While1})$$

$$\frac{[\![e]\!]^{s \cup g} = \mathit{false}}{\wr \langle \mathtt{while}\ (e)\ \{\, C \,\}; X, \vartheta, t, s \rangle \int, g \xrightarrow{\mathtt{assume}\ !e} \wr \langle X, \vartheta, t, s \rangle \int, g} \ (\text{While2})$$

$$\frac{}{\wr \langle \mathtt{fork}\ e\ \vartheta'(); X, \vartheta, t, s \rangle \int, g \xrightarrow{\mathtt{fork}\ e\ \vartheta'()} \wr \langle X, \vartheta, t, s \rangle, \langle \mathit{body}_{\vartheta'}, \vartheta', [\![e]\!]^{s \cup g}, s' \rangle \int, g} \ (\text{Fork})$$

$$\frac{}{\wr \langle \mathtt{join}\ e; X, \vartheta, t, s \rangle, \langle \Omega, \vartheta', [\![e]\!]^{s \cup g}, s' \rangle \int, g \xrightarrow{\mathtt{join}\ e} \wr \langle X, \vartheta, t, s \rangle \int, g} \ (\text{Join})$$

$$\frac{M_1, g \xrightarrow{st} M_1', g'}{M_1 \uplus M_2, g \xrightarrow{st} M_1' \uplus M_2, g'} \ (\text{Frame})$$

Fig. 3. The definition of the small-step semantic transition relation. Assume that $C, C_1, C_2 \in \mathbf{Command}$, $X \in \mathbf{Command} \cup \{\Omega\}$, $\vartheta, \vartheta' \in \mathbf{Templates}$, $t \in \mathbb{Z} \cup \{\bot\}$, s, s' are local states, g, g' are global states, and e is an expression.

The thread width plays a crucial role in our verification approach. Note that the thread width of an execution is always a natural number. However, the thread width of a program might be infinite.

Example 2.4. The thread width of the program in Fig. 1 is $\beta = 2$. Although the program might unboundedly often execute a `fork` command, there are always at most two worker threads active at the same time.

3   Petrification

In this section, we describe a process called *petrification*, which transforms a CONC program $\mathcal{P}$ into a representation suitable for verification algorithms. Specifically, we build on the formalism of [9], and transform $\mathcal{P}$ into a so-called *Petri program*. The resulting Petri program can be verified using the algorithm presented in [9]. The petrified program captures the control flow of $\mathcal{P}$, and maintains the concurrent nature of the program (as opposed to, say, an interleaving model). Petrification is parametrized in an upper limit on the number of threads. In Sect. 4, we show how an iterative verification algorithm can manipulate this parameter in order to determine a (semantic) upper bound on the number of active threads (i.e. thread width), if it exists.

3.1   Petri Programs

Before presenting our construction, we give a brief recap on Petri programs.

Definition 3.1. (Petri Programs). A Petri program is given by a 5-tuple $\mathcal{N} = (P, T, F, m_{\text{init}}, \lambda)$, where P is a finite set of *places*, T is a finite set of *transitions* with $P \cap T = \emptyset$, $F \subseteq (P \times T) \cup (T \times P)$ is a *flow relation*, $m_{\text{init}} : P \to \mathbb{N}$ is an *initial marking*, and $\lambda : T \to \textbf{AtomicStmt}$ is a *labeling* of transitions.

Equipped with Petri net semantics, a Petri program defines a set of *traces* (i.e., sequences of atomic statements) that describe possible program behaviours. Formally, we define a *marking* as a map $m : P \to \mathbb{N}$ that assigns a token count to each place. With $m \rhd_{\lambda(t)} m'$ we denote that transition $t \in T$ with the label $\lambda(t)$ can be *fired* from marking m, i.e., all predecessor places have a token – formally, $m(p) > 0$ for all p with $(p, t) \in F$ –, and the firing of t results in the marking m' – formally, $m'(p) = m(p) - \chi_F(\langle p, t \rangle) + \chi_F(\langle t, p \rangle)$, where χ_F is the characteristic function of F. A *firing sequence* is a sequence $m_0 \rhd_{\lambda(t_1)} m_1 \rhd_{\lambda(t_2)} \cdots \rhd_{\lambda(t_n)} m_n$, where $m_0 = m_{init}$ is the initial marking. We say that a marking m is reachable iff there exists a firing sequence $m_0 \rhd_{\lambda(t_1)} m_1 \rhd_{\lambda(t_2)} \cdots \rhd_{\lambda(t_n)} m_n$ with $m_n = m$. Analogously to [9], we only consider Petri programs that are *1-safe*, i.e., where all reachable markings have at most one token per place. We thus identify reachable markings $m : P \to \{0, 1\}$ with sets of places.

Unlike CONC programs, Petri programs do not have a notion of global and local variables. The semantics are formulated over (global) program states $\sigma \in$

State, i.e., mappings from variables to their (boolean or integer) values. Each atomic statement st has a semantic transition relation $[\![st]\!] \subseteq \mathbf{State} \times \mathbf{State}$:

$$[\![\mathtt{x\!:=\!e}]\!] := \{\, (\sigma, \sigma') \mid \sigma' = \sigma[x \mapsto [\![e]\!]^\sigma] \,\}$$
$$[\![\mathtt{assume\ e}]\!] := \{\, (\sigma, \sigma') \mid \sigma = \sigma' \wedge [\![e]\!]^\sigma = true \,\}$$

A *specification* for a Petri program is a set of "bad" places $\mathcal{S}$, i.e., places that should not be reached by an execution of the Petri program.

Definition 3.2. (Satisfaction). A *counterexample* to a specification $\mathcal{S}$ consists of a firing sequence $m_0 \rhd_{\lambda(t_1)} \cdots \rhd_{\lambda(t_n)} m_n$ and a sequence of states $\sigma_0, \ldots, \sigma_n \in \mathbf{State}$, such that $m_n \cap \mathcal{S} \neq \emptyset$ (i.e., a bad place is reached), and we have $(\sigma_{i-1}, \sigma_i) \in [\![\lambda(t_i)]\!]$ for all $i \in \{1, \ldots, n\}$ (i.e., the trace corresponding to the firing sequence can actually be executed).

The Petri program $\mathcal{N}$ *satisfies* the specification $\mathcal{S}$, denoted $\mathcal{N} \models \mathcal{S}$, if there does not exist a counterexample to $\mathcal{S}$ in $\mathcal{N}$.

3.2 Transformation to Petri Programs

In this section, we present the transformation that we call *petrification*. The input for this construction consists of a CONC program $\mathcal{P} = (body, \mathtt{main}, \mathbf{Globals})$ and a *thread limit* $\beta \in \mathbb{N}$. The constructed Petri program represents all executions of the program $\mathcal{P}$ where, at any time, at most β threads with the same template are active, i.e., executions whose thread width is at most β.

First, recall that, unlike CONC programs, Petri programs do not have a notion of thread-local variables; all variables are global. We thus rename variables of the program $\mathcal{P}$ to ensure uniqueness. Let $\mathbf{Var}$ be all the variables mentioned in $\mathcal{P}$. We define the set *instantiated variables* as follows:

$$\mathbf{Var}_{\mathrm{inst}} := \{\, x_k^\vartheta \mid x \in \mathbf{Var} \setminus \mathbf{Globals}, k \in \{\bot, 1, \ldots, \beta\}, \vartheta \in \mathbf{Templates} \,\}$$
$$\cup\, \mathbf{Globals} \cup \{\, \mathrm{id}_k^\vartheta \mid k \in \{1, \ldots, \beta\}, \vartheta \in \mathbf{Templates} \,\}$$

For each local variable x, we define instantiated variables x_k^ϑ, where ϑ is a thread template, and $k \in \{\bot, 1, \ldots, \beta\}$ is a unique *instance ID*. These IDs range from 1 to β, with the special $\bot$ for the $\mathtt{main}$ thread that is initially active. Global variables are not instantiated. We also introduce variables id_k^ϑ, which keep track of the (non-unique) thread IDs used by $\mathtt{fork}$ and $\mathtt{join}$ statements. We extend the idea of instantiation to expressions and atomic statements: e_k^ϑ denotes the expression derived by replacing every local variable $x \in \mathbf{Var} \setminus \mathbf{Globals}$ in e with x_k^ϑ. Similarly, the instantiated statement $[st]_{\vartheta,k}$ is derived by replacing every local variable $x \in \mathbf{Var} \setminus \mathbf{Globals}$ in the atomic statement st with x_k^ϑ.

We introduce a formalism to capture the program control flow. First, let us define the set of *control locations* $\mathbb{L}_\beta^\mathcal{P}$:

$$\mathbb{L}_\beta^\mathcal{P} = (\mathbf{Command} \cup \{\Omega, \lightning\}) \times \mathbf{Templates} \times \{\bot, 1, \ldots, \beta\}$$
$$\cup\, \{\, \mathrm{inUse}_k^\vartheta, \mathrm{notInUse}_k^\vartheta, \mathrm{insuff}^\vartheta \mid k \in \{1, \ldots, \beta\}, \vartheta \in \mathbf{Templates} \,\}$$

There are several types of control locations: *Program control locations* consist of a command to be executed, as well as a thread template and an instance ID that indicate to which thread the control location belongs. Additional control locations $\mathsf{inUse}_k^\vartheta$ and $\mathsf{notInUse}_k^\vartheta$ indicate if the thread with template ϑ and instance ID k is currently active or not. Finally, the control location $\mathsf{insuff}^\vartheta$ indicates that the thread limit β is insufficient for the program $\mathcal{P}$, specifically because more than β threads with template ϑ can be created.

We define the control flow as a ternary relation $\hookrightarrow$ between sets of control locations, (instantiated) simple statements, and sets of control locations. Specifically, let $\hookrightarrow$ be the smallest relation such that the following conditions hold:

1. Let us fix a command C, some $X \in \mathbf{Command} \cup \{\Omega, \lightning\}$, a thread template ϑ and a simple statement st. For every rule of the semantics definition (see Fig. 3) that has the form

$$\frac{\varphi}{\wr\langle C, \vartheta, t, s\rangle\wr, g \xrightarrow{st} \wr\langle X, \vartheta, t, s'\rangle\wr, g'} \; ,$$

where φ is only a side condition (i.e., φ does not refer to the semantic transition relation), it holds that

$$\langle C, \vartheta, k\rangle \xrightarrow{[st]_{\vartheta,k}} \langle X, \vartheta, k\rangle \tag{1}$$

for all instance IDs $k \in \{\bot, 1, \ldots, n\}$. In particular, this applies for the semantic rules (ASSUME), (ASSIGNGLOBAL), (ASSIGNLOCAL), (ASSERT1), (ASSERT2), (ITE1), (ITE2), (WHILE1) and (WHILE2).
2. The following holds for all $X \in \mathbf{Command} \cup \{\Omega\}$, $k \in \{\bot, 1, \ldots, \beta\}$ and $k' \in \{1, \ldots, \beta\}$:

$$\langle \texttt{fork}\ e\ \vartheta'();\, X, \vartheta, k\rangle, \mathsf{inUse}_1^{\vartheta'}, \ldots, \mathsf{inUse}_{k'-1}^{\vartheta'}, \mathsf{notInUse}_{k'}^{\vartheta'}$$
$$\xrightarrow{\mathsf{id}_{k'}^{\vartheta'} := e_k^\vartheta} \langle X, \vartheta, k\rangle, \langle body_{\vartheta'}, \vartheta', k'\rangle, \mathsf{inUse}_1^{\vartheta'}, \ldots, \mathsf{inUse}_{k'}^{\vartheta'} \tag{2}$$

$$\langle \texttt{fork}\ e\ \vartheta'();\, X, \vartheta, k\rangle, \mathsf{inUse}_1^{\vartheta'}, \ldots, \mathsf{inUse}_\beta^{\vartheta'} \xrightarrow{\texttt{assume true}} \mathsf{insuff}^{\vartheta'} \tag{3}$$

$$\langle \texttt{join}\ e;\, X, \vartheta, k\rangle, \langle \Omega, \vartheta', k'\rangle, \mathsf{inUse}_{k'}^{\vartheta'} \xrightarrow{\texttt{assume } \mathsf{id}_{k'}^{\vartheta'} == e_k^\vartheta} \langle X, \vartheta, k\rangle, \mathsf{notInUse}_{k'}^{\vartheta'} \tag{4}$$

Equation (1) captures the sequential control flow within a thread, in analogy to the CONC semantics. Equations (2) to (4) implement the dynamic thread management and the thread limit β. When a `fork` statement is executed, the newly created thread is assigned a currently inactive instance ID k'. Specifically, we always assign the minimal available instance ID. If all instance IDs are already active, control gets stuck in the control location $\mathsf{insuff}^{\vartheta'}$. When a `join` statement is executed, the joined thread must have terminated (the remainder program is Ω), and the corresponding instance ID k' is marked as inactive. The full definition of $\hookrightarrow$ can be found in the extended version of the paper [17]. We can now define petrification:

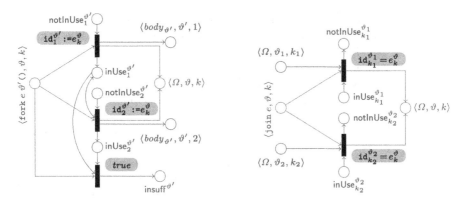

Fig. 4. Illustration of the transitions derived from Eqs. (2) to (3) (on the left) and from Eq. (4) (on the right), for $\beta = 2$.

Definition 3.3. (Petrified Program). We define the *petrified program* as the Petri program $PN_\beta(\mathcal{P}):=(P, T, F, m_{init}, \lambda)$, where

– the places are control locations: $P = \mathbb{L}_\beta^{\mathcal{P}}$,
– the transitions are triples defined by the control flow relation $\overset{.}{\hookrightarrow}$:

$$T = \{ \, (pred, st, succ) \mid pred \overset{st}{\hookrightarrow} succ \, \},$$

– the flow relation follows directly from the transition triples:

$$F = \{ \, (p, (pred, st, succ)) \mid p \in pred \, \} \cup \{ \, ((pred, st, succ), p') \mid p' \in succ \, \},$$

– the initial marking consists of those control locations that indicate that the `main` thread is in its initial location (i.e., it has to execute the complete body) and all other threads are not active

$$m_{init} = \{ \langle body_{\texttt{main}}, \texttt{main}, \bot \rangle \} \cup \{ \texttt{notInUse}_k^\vartheta \mid k \in \{1, \dots, \beta\}, \vartheta \in \textbf{Templates} \}$$

– and the labeling of a transition triple is given by its second component:

$$\lambda\big((pred, st, succ)\big) = st.$$

Note that $\mathbb{L}_\beta^{\mathcal{P}}$ has infinitely many elements, because there are infinitely many commands. However, only a finite number of control locations are actually relevant to the program. We omit certain unreachable places and transitions in $PN_\beta(\mathcal{P})$, such that it only has finitely many places.

Figure 4 illustrates the transitions created for fork and join commands in the petrified program (for $\beta = 2$). The three transitions on the left correspond to `fork` $e\ \vartheta()$. The uppermost transition creates the thread with instance ID 1. It can only be fired if the thread with instance ID 1 is not yet active (there is a token in $\texttt{notInUse}_1^\vartheta$). After firing the transition, the thread with instance ID 1 is active (there is a token in $\mathsf{inUse}_1^\vartheta$), its thread ID was stored in id_1^ϑ, and the command $body_\vartheta$ remains to be executed. The transition in the middle behaves similarly, for the thread with instance ID 2. However, it can only be executed if the thread with instance ID 1 is already active. The transition at the bottom indicates that our chosen thread limit β was not sufficient. This transition can be fired if both thread instances are active (there are tokens in $\mathsf{inUse}_1^\vartheta$ and $\mathsf{inUse}_2^\vartheta$).

To the right of Fig. 4, we see the transitions generated for a `join` command. For each thread instance (regardless of the template), there is a transition that corresponds to `join e`. This transition can be fired if the thread instance is active (i.e., there is a token in the corresponding inUse place). After firing the transition, the thread instance is no longer active (i.e., there is a token in the corresponding $\mathsf{notInUse}$ place). The transition label ensures that the thread ID has the value of `e` (we omit the keyword `assume` for brevity).

Example 3.4. The Petri net in Fig. 2 represents $PN_2(\mathcal{P})$, where $\mathcal{P}$ is the program shown in Fig. 1. For readability, we forgo annotating the places with the corresponding program control locations. For example, the place named $\ell_{0,1}$ represents the control location $\langle$`c:=c+i ; assert c<=2*i ; c:=c-i, w`$, 1\rangle$.

3.3   Properties of the Petrified Program

We show that the petrified program $PN_\beta(\mathcal{P})$ for a given program $\mathcal{P}$ satisfies certain properties, which allow it to be used in the verification of $\mathcal{P}$. Detailed proofs of these results can be found in the extended version of the paper [17].

Recall that the Petri program verification algorithm [9] only supports 1-safe Petri programs. We thus show that petrification ensures 1-safety. To this end, observe that all reachable markings of a petrified program $PN_\beta(\mathcal{P})$ satisfy the following conditions, for all $\vartheta \in$ **Templates** and $k \in \{1, \ldots, \beta\}, k' \in \{\perp, 1, \ldots, \beta\}$:

- The sum of the tokens in $\mathsf{inUse}_k^\vartheta$, $\mathsf{notInUse}_k^\vartheta$ and $\mathsf{insuff}^\vartheta$ is exactly 1.
- The place $\mathsf{inUse}_k^\vartheta$ has a token iff there exists some $X \in$ **Command** $\cup \{\Omega, \not\downarrow\}$ such that $\langle X, \vartheta, k\rangle$ has a token.
- The sum of the tokens in all places of the form $\langle X, \vartheta, k'\rangle$ (with some $X \in$ **Command** $\cup \{\Omega, \not\downarrow\}$) is at most 1.

We call a marking that satisfies these conditions *coherent*.

Lemma 3.5. *All reachable markings of $PN_\beta(\mathcal{P})$ are coherent.*

Proof. It is easy to see that the initial marking is coherent. Furthermore, observe that the successor of a coherent marking is again coherent: Each transition according to Eqs. (1) to (4) preserves coherence. $\square$

Proposition 3.6. (1-Safety). *The Petri program $PN_\beta(\mathcal{P})$ is 1-safe.*

Proof. Any coherent marking assigns at most one token to a place. Since all reachable markings are coherent, it follows that the Petri program is 1-safe. $\square$

Thus we can verify the petrified program using an existing algorithm [9]. It remains to give a specification against which the petrified program shall be verified. In fact, we define *two* specifications for the petrified program $PN_\beta(\mathcal{P})$: one that indicates the absence of assert violations, and another that indicates sufficiency of the thread limit β.

Definition 3.7. (Specifications). The *safety specification* $\mathcal{S}_{\text{safe}}$ and the *bound specification* $\mathcal{S}_{\text{bound}}$ of the petrified program $PN_\beta(\mathcal{P})$ are given by the sets of places

$$\mathcal{S}_{\text{safe}} = \{ \langle \notin, \vartheta, k \rangle \mid k \in \{\bot, 1, \ldots, \beta\} \wedge \vartheta \in \textbf{Templates} \}$$
$$\mathcal{S}_{\text{bound}} = \{ \, \text{insuff}^\vartheta \mid \vartheta \in \textbf{Templates} \}$$

It remains to be shown that the specifications of the petrified program actually correspond to the behaviour of the program $\mathcal{P}$. In order to show this, we first create a link between executions of the program $\mathcal{P}$ and firing sequences of the petrified program $PN_\beta(\mathcal{P})$. In particular, we map a given marking m and a state σ to a corresponding global configuration: We create local configurations for all program control locations in m, and extract the thread ID as well as local and global states from σ. Formally, $\text{conf}(m, \sigma) := (M, \sigma|_{\textbf{Globals}})$ where

$$M = \wr \langle C, \vartheta, t, \lfloor \sigma \rfloor_{\vartheta, k} \rangle \mid \langle C, \vartheta, k \rangle \in m \wedge (t = k = \bot \vee t = \sigma(\text{id}_k^\vartheta)) \wr$$

with $\lfloor \sigma \rfloor_{\vartheta, k}(x) = \sigma(x_k^\vartheta)$ for all local variables $x \in \textbf{Var} \setminus \textbf{Globals}$. The following lemmata allow us to associate an execution with firing sequences that are executable according to Petri program semantics:

Lemma 3.8. *Let* $m_0 \triangleright_{\tilde{st}_1} \ldots \triangleright_{\tilde{st}_n} m_n$ *be a firing sequence of* $PN_\beta(\mathcal{P})$ *where none of the markings* m_i *contains a place* insuff^ϑ *for any* ϑ. *Let* $\sigma_0, \ldots, \sigma_n$ *be states with* $(\sigma_{i-1}, \sigma_i) \in [\![\tilde{st}_i]\!]$ *for all* $i \in \{1, \ldots, n\}$. *There exist simple statements* $st_1, \ldots, st_n$ *such that* $\text{conf}(m_0, \sigma_0) \xrightarrow{st_1} \ldots \xrightarrow{st_n} \text{conf}(m_n, \sigma_n)$ *is an execution.*

Proof. (sketch). We proceed by induction over n. For the base case, observe that the global configuration $\text{conf}(m_{init}, \sigma_0)$ is initial. For the induction step, it remains to justify the last step in the execution, using the semantic relation. We proceed by case distinction over the rule (among Eqs. (1), (2) and (4)) that is responsible for the last transition in the firing sequence (transitions according to Eq. (3) are ruled out by our assumption on the firing sequence). Using the semantic rules corresponding to each case, in combination with the (FRAME) rule, we show that the above semantic transition indeed exists. □

Lemma 3.9. *Let* $(M_0, g_0) \xrightarrow{st_1} \ldots \xrightarrow{st_n} (M_n, g_n)$ *be an execution whose thread width is at most* β. *Then there exists a firing sequence* $m_0 \triangleright_{\tilde{st}_1} \ldots \triangleright_{\tilde{st}_n} m_n$ *of* $PN_\beta(\mathcal{P})$, *and a sequence* $\sigma_0, \ldots, \sigma_n$ *of states over* $\textbf{Var}_{\text{inst}}$ *such that*

- *$\text{conf}(m_i, \sigma_i) = (M_i, g_i)$ for all $i \in \{0, \ldots, n\}$;*
- *and $(\sigma_{i-1}, \sigma_i) \in [\![\tilde{st}_i]\!]$ for all $i \in \{1, \ldots, n\}$.*

Proof. (sketch). We first prove inductively that we can assign instance IDs to the local configurations in each step of the execution in a consistent manner. Given such an "augmented" execution, one can extract the markings $m_0, \ldots, m_n$ and the states $\sigma_0, \ldots, \sigma_n$ in a straightforward way. □

Theorem 3.10. (Thread Width Detection). *The Petri program $PN_\beta(\mathcal{P})$ satisfies the bound specification S_{bound} iff the thread width for $\mathcal{P}$ is at most β.*

Proof. (sketch). Given an execution with a thread width greater than β, we apply Lemma 3.9 to the longest prefix of the execution such that the thread width of the prefix is at most β. The firing sequence and the sequence of states given by the lema can be extended to reach a place $\mathsf{insuff}^\vartheta$. For the reverse implication, we proceed analogously by applying Lemma 3.8 to the longest prefix of a given firing sequence that does not put a token into a place $\mathsf{insuff}^\vartheta$. The resulting execution can be extended to an execution with thread width greater than β. □

Theorem 3.11. (Soundness). *If the Petri programs $PN_\beta(\mathcal{P})$ satisfies both its safety and its bound specifications, then the CONC program $\mathcal{P}$ is correct.*

Proof. From the fact that $PN_\beta(\mathcal{P})$ satisfies the bound specification, we conclude by Theorem 3.10 that the thread width for $\mathcal{P}$ is at most β. Contrapositively, we prove that if $\mathcal{P}$ is incorrect, then $PN_\beta(\mathcal{P})$ does not satisfy its safety specification. Suppose that $(M_0, g_0) \xrightarrow{st_1} \ldots \xrightarrow{st_n} (M_n, g_n)$ is an erroneous execution. Lemma 3.9 gives us a corresponding firing sequence $m_0 \rhd_{\tilde{st}_1} \ldots \rhd_{\tilde{st}_n} m_k$ and a sequence of states $\sigma_0, \ldots, \sigma_k$. Since M_n contains some local configuration $\langle \sharp, \vartheta, t, s \rangle$, by definition of conf we must have a place $\langle \sharp, \vartheta, k \rangle \in m_n$. Thus the firing sequence and the sequence of states form a counterexample to the safety specification S_{safe}. □

Theorem 3.12. (Completeness). *If the CONC program $\mathcal{P}$ is correct, then the corresponding Petri program $PN_\beta(\mathcal{P})$ satisfies its safety specification.*

Proof. Contrapositively, let us suppose that $PN_\beta(\mathcal{P})$ does not satisfy its safety specification. Then there exists a firing sequence $m_0 \rhd_{\tilde{st}_1} \ldots \rhd_{\tilde{st}_n} m_n$ and states $\sigma_0, \ldots, \sigma_n$ such that $(\sigma_{i-1}, \sigma_i) \in [\![\tilde{st}_i]\!]$ for all i, with some $\langle \sharp, \vartheta, k \rangle \in m_n$. Wlog. we can assume that the firing sequence does not run into a place $\mathsf{insuff}^{\vartheta'}$. By Lemma 3.8, we know that $\mathrm{conf}(m_0, \sigma_k) \xrightarrow{st_1} \ldots \xrightarrow{st_n} \mathrm{conf}(m_n, \sigma_n)$ is an execution, for some $st_1, \ldots, st_n$. By definition of conf, it follows that the execution is erroneous, i.e., $\langle \sharp, \vartheta, t, s \rangle \in \mathrm{conf}(m_n, \sigma_n)$. Thus $\mathcal{P}$ is incorrect. □

4   Verifying Programs Through Repeated Petrification

The previous section shows that one can verify a CONC program $\mathcal{P}$ by picking a suitable thread limit β, and proving that the petrified program $PN_\beta(\mathcal{P})$ satisfies both its safety and its bound specification. This gives rise to several possible verification algorithms, illustrated in Fig. 5. Each algorithm proves correctness

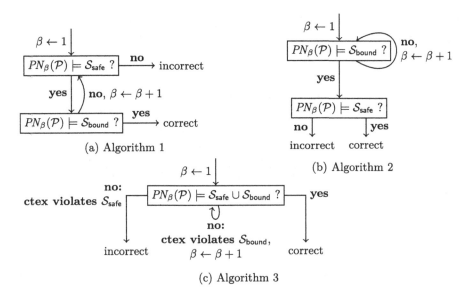

(a) Algorithm 1

(b) Algorithm 2

(c) Algorithm 3

Fig. 5. Three iterative algorithms that reduce the verification problem for a Conc program $\mathcal{P}$ to (several instances of) the verification problem for Petri programs.

of a given program $\mathcal{P}$ by iteratively determining a suitable thread limit β. In each iteration, the algorithms invoke a Petri program verification algorithm [9] to determine if the petrified program $PN_\beta(\mathcal{P})$ satisfies some given specification.

Algorithm 1 first checks safety, and only if the petrified program satisfies the safety specification, the algorithm checks the bound specification. If the bound specification is violated, the thread limit β is increased. The safety specification must then be checked again, as it may be violated for the increased β.

By contrast, Algorithm 2 first determines the thread width β of the program, by repeatedly checking if the petrified program satisfies the bound specification and, if appropriate, incrementing the thread limit. Only after the thread width has been established, the petrified program is checked against the safety specification.

Finally, Algorithm 3 combines the check of both specifications, by checking if the petrified program satisfies their union. If this is not the case, the counterexample returned by the verification is examined to determine if the program is incorrect, or if the thread limit should be increased.

Theorem 4.1. (Correctness). *For any of the algorithms in Fig. 5, if the algorithm terminates for a given program $\mathcal{P}$, then*

- *if the output is "correct", the program $\mathcal{P}$ is correct;*
- *and if the output is "incorrect", the program $\mathcal{P}$ is incorrect.*

Proof. The result is straightforward by application of Theorems 3.10 to 3.12. For the third algorithm, we also note that a Petri program satisfies the union of two specifications iff it satisfies both specifications individually. □

Theorem 4.2. (Relative Termination). *Given a program $\mathcal{P}$ with a finite thread width β, if each invocation of the Petri program verification algorithm [9] terminates, then all the verification algorithms in Fig. 5 terminate after at most β iterations.*

The first algorithm may terminate earlier if $\mathcal{P}$ is incorrect, and in fact, it even terminates if $\mathcal{P}$ is incorrect but has infinite thread width: If an erroneous execution exists, this execution has some thread width β. Thus, in iteration β at the latest, the petrified program $PN_\beta(\mathcal{P})$ does not satisfy the safety specification S_{safe}, and Algorithm 1 terminates. Algorithm 2 never terminates for programs with infinite thread width. For Algorithm 3, termination is not guaranteed and depends on the counterexample selection of the underlying Petri program verification algorithm.

5   Application: Verification of C Programs

We implemented petrification, as well as the three verification algorithms of Fig. 5, in the program analysis framework ULTIMATE [29]. Our implementation consumes C programs that use the POSIX threads (pthreads) API [1] for dynamic thread management. We presume an execution model that satisfies sequential consistency.

5.1   Translation from Pthreads to CONC Programs

The POSIX threads extension includes many features, including mutexes and condition variables. Our implementation supports a subset of these features through a symbolic encoding in CONC using assume statements. Here, we focus on the features relevant to dynamic thread management, specifically:

- There are unique thread IDs of type `pthread_t`.
  In CONC, to ensure unique thread IDs, we introduce a global integer variable `freshId` that is incremented after every `fork`.
- The function `pthread_create(id, attr, f, arg)` creates a new thread. It takes a pointer `id` to `pthread_t` (the unique thread ID will be stored at this address), some attributes `attr`, the function `f` which serves as thread template, and an argument `arg` (a pointer) passed to the function `f`.
  In CONC, a call to `pthread_create(id, attr, f, arg)` is translated to three statements. First, we write the current value `freshId` to the location of the pointer `id`. Second, we execute the statement `fork freshId f(arg)`. Finally, we increment `freshId`.
  Our implementation allows thread templates (created from C functions) to take parameters, and extends the syntax and semantics of CONC to allow passing such parameters in `fork` statements. We do not support non-standard attributes, so `attr` must be NULL.

```
 1  int c, i;
 2
 3
 4  void *w(void *x) {
 5    c += i;
 6    assert(c <= 2 * i);
 7    c -= i;
 8  }
 9
10
11  int main() {
12    pthread_t ids[10000];
13    while (i < 10000) {
14      pthread_create(&ids[i], NULL, w, NULL);
15      if (i > 0) {
16        pthread_join(ids[i-1], NULL);
17      }
18      i++;
19    }
20    return 0;
21  }
```

Fig. 6. C program using the pthreads API.

```
 1  c := 0;
 2  i := 0;
 3  while (i < 10000) {
 4    ids[i] := freshId;
 5    fork freshId w();
 6    freshId := freshId + 1;
 7    if (i > 0) {
 8      join ids[i-1];
 9    }
10    i := i + 1;
11  }
```

(a) The main thread

```
 1  c := c + i;
 2  assert c <= 2 * i;
 3  c := c - i;
```

(b) The worker thread w

Fig. 7. The representation of the program from Fig. 6 in CONC.

– The function pthread_join(id, ret) waits for the thread with the given ID to terminate. It takes the thread ID id (of type pthread_t) and a pointer ret to store the return value of the function to be joined (if ret is NULL, the return value is discarded).

In CONC, a call to pthread_join(id, ret) is translated to the CONC statement join id. To cover the case that ret is not NULL, our implementation supports statements of the form join id assigns x, which store the thread's return value in the variable x.

Figure 6 shows a version of our example program from Fig. 1, written in C using the pthreads API. The main thread creates 10 000 instances of the thread w using pthread_create, and stores the unique thread IDs in the array ids. In each iteration, the thread created in the the previous iteration is joined with pthread_join(ids[i-1], NULL). Figure 7 shows the translation of the program to CONC as described above. Petrification of this CONC program with thread limit $\beta = 2$ yields a Petri program similar to Fig. 2. ULTIMATE can then prove that this Petri program satisfies the bound and safety specification. Thus the C program in Fig. 6 is correct and has thread width 2.

5.2 Practical Performance of the Approach

The ULTIMATE framework contains three verification tools that verify Petri programs: AUTOMIZER [9,16], TAIPAN [10,14] and GEMCUTTER [12,20]. ULTIMATE can encode different specifications via assert statements, such as unreachability of an error function, or absence of certain undefined behaviours.

The feasibility of our presented approach is demonstrated by the success of the ULTIMATE tools in the *ConcurrencySafety* category of the *International*

Table 1. Comparison of SV-COMP'23 [3] results.

| | AUTOMIZER | | | CPACHECKER | | | GOBLINT | | |
	#	time (h)	mem (GB)	#	time (h)	mem (GB)	#	time (h)	mem (GB)
total (2 865)	1 516	35.3	1 590	973	16.1	1 210	847	0.4	29
safe	1 227	30.6	1 300	712	10.8	900	847	0.4	29
unsafe	289	4.8	290	261	5.3	310	0	0.0	0

Table 2. Comparison of the algorithms from Fig. 5.

| | Algorithm 1 | | | Algorithm 2 | | | Algorithm 3 | | |
	#	time (h)	mem (GB)	#	time (h)	mem (GB)	#	time (h)	mem (GB)
total (2 865)	1 580	26.1	2 208	1 571	26.6	2 227	1 573	26.1	2 176
safe	1 224	22.2	1 700	1 230	22.9	1 760	1 225	22.3	1 690
unsafe	356	4.0	508	341	3.7	467	348	3.8	486

Competition on Software Verification (SV-COMP'23) [3]. In this category, verification tools had to check 2865 verification tasks, each consisting of a concurrent C program and one of four different specifications: unreachability of a call to a distinguished error function, absence of data races, absence of invalid pointer dereferences and other memory safety issues, and absence of signed integer overflow. The ULTIMATE tools participated (using Algorithm 3 as shown in Fig. 5c) and occupied the 2nd (AUTOMIZER), 3rd (GEMCUTTER) and 4th (TAIPAN) place in the *ConcurrencySafety* category, with 1st place taken by the bounded model checker DEAGLE [15]. Thus the ULTIMATE tools placed ahead of all other tools that soundly verify concurrent programs. Table 1 shows an extract of the competition results, comparing AUTOMIZER against the next best sound verification tools in the *ConcurrencySafety* category, CPACHECKER [5] and GOBLINT [27].

We also evaluated ULTIMATE's implementation of all three verification algorithms from Fig. 5 on the SV-COMP'23 benchmark set, with AUTOMIZER [9] as a backend. This evaluation was performed using the BENCHEXEC benchmarking tool [6] on an AMD Ryzen Threadripper 3970X 32-Core processor, with a timeout of 15 min and a memory limit of 16 GB. Table 2 shows the results: Algorithm 1 succeeds on the largest number of unsafe benchmarks and overall. For unsafe benchmarks it succeeds on a strict superset of the verification tasks for which Algorithms 2 and 3 found a bug. However, Algorithm 2 and Algorithm 3 are both able to verify a few more safe benchmarks. The results are similar, likely because most programs in the benchmark set either have thread width 1 (but multiple thread templates), or they are safe and have infinite thread width.

6   Conclusion

We address the verification of programs with dynamic thread management, i.e., programs that fork and join threads at runtime. Our main contributions are:

- We present a reduction from the verification of programs with dynamic thread management to the verification of programs with a fixed number of threads. Our approach determines the maximum number of threads that are active at the same time (the program's *thread width*) and verifies that the program indeed satisfies this bound.
- We formalize our approach using Petri programs, an existing model for concurrent programs with a fixed number of threads, and a simple programming language that supports dynamic thread management. Our approach is sound, and it is relatively complete for programs with a finite thread width.
- We implemented our approach as a verification tool for C programs. Our implementation verifies programs that use the POSIX threads (pthreads) API, and competes with the best verifiers for concurrent programs at the International Competition on Software Verification (SV-COMP'23) [3].

References

1. Base Specifications POSIX.1-2017. Standard, The Open Group, San Francisco, CA, January 2018. https://pubs.opengroup.org/onlinepubs/9699919799/
2. Barrett, C., Fontaine, P., Tinelli, C.: The SMT-LIB standard: version 2.6. Technical Report, Department of Computer Science, The University of Iowa (2017), www.SMT-LIB.org
3. Beyer, D.: Competition on software verification and witness validation: SV-COMP 2023. In: Sankaranarayanan, S., Sharygina, N. (eds.) Tools and Algorithms for the Construction and Analysis of Systems. TACAS 2023. LNCS, vol. 13994, pp 495–522. Springer, Cham (2023). https://doi.org/10.1007/978-3-031-30820-8_29
4. Beyer, D., Friedberger, K.: A light-weight approach for verifying multi-threaded programs with CPAchecker. In: MEMICS. EPTCS, vol. 233, pp. 61–71 (2016). https://doi.org/10.4204/EPTCS.233.6
5. Beyer, D., Keremoglu, M.E.: CPACHECKER: a tool for configurable software verification. In: Gopalakrishnan, G., Qadeer, S. (eds.) CAV 2011. LNCS, vol. 6806, pp. 184–190. Springer, Heidelberg (2011). https://doi.org/10.1007/978-3-642-22110-1_16
6. Beyer, D., Löwe, S., Wendler, P.: Reliable benchmarking: requirements and solutions. Int. J. Softw. Tools Technol. Transf. **21**(1), 1–29 (2019). https://doi.org/10.1007/s10009-017-0469-y
7. Clarke, E.M., Grumberg, O.: Avoiding the state explosion problem in temporal logic model checking. In: PODC, pp. 294–303. ACM (1987). https://doi.org/10.1145/41840.41865
8. Clarke, E., Talupur, M., Touili, T., Veith, H.: Verification by network decomposition. In: Gardner, P., Yoshida, N. (eds.) CONCUR 2004. LNCS, vol. 3170, pp. 276–291. Springer, Heidelberg (2004). https://doi.org/10.1007/978-3-540-28644-8_18

9. Dietsch, D., Heizmann, M., Klumpp, D., Naouar, M., Podelski, A., Schätzle, C.: Verification of concurrent programs using Petri net unfoldings. In: Henglein, F., Shoham, S., Vizel, Y. (eds.) VMCAI 2021. LNCS, vol. 12597, pp. 174–195. Springer, Cham (2021). https://doi.org/10.1007/978-3-030-67067-2_9

10. Dietsch, D., Heizmann, M., Nutz, A., Schätzle, C., Schüssele, F.: ULTIMATE TAIPAN with symbolic interpretation and fluid abstractions. In: TACAS 2020. LNCS, vol. 12079, pp. 418–422. Springer, Cham (2020). https://doi.org/10.1007/978-3-030-45237-7_32

11. Emerson, E.A., Kahlon, V.: Reducing model checking of the many to the few. In: McAllester, D. (ed.) CADE 2000. LNCS (LNAI), vol. 1831, pp. 236–254. Springer, Heidelberg (2000). https://doi.org/10.1007/10721959_19

12. Farzan, A., Klumpp, D., Podelski, A.: Sound sequentialization for concurrent program verification. In: PLDI, pp. 506–521. ACM (2022). https://doi.org/10.1145/3519939.3523727

13. Gavrilenko, N., Ponce-de-León, H., Furbach, F., Heljanko, K., Meyer, R.: BMC for weak memory models: relation analysis for compact SMT encodings. In: Dillig, I., Tasiran, S. (eds.) CAV 2019. LNCS, vol. 11561, pp. 355–365. Springer, Cham (2019). https://doi.org/10.1007/978-3-030-25540-4_19

14. Greitschus, M., Dietsch, D., Podelski, A.: Loop invariants from counterexamples. In: Ranzato, F. (ed.) SAS 2017. LNCS, vol. 10422, pp. 128–147. Springer, Cham (2017). https://doi.org/10.1007/978-3-319-66706-5_7

15. He, F., Sun, Z., Fan, H.: Satisfiability modulo ordering consistency theory for multi-threaded program verification. In: PLDI, pp. 1264–1279. ACM (2021). https://doi.org/10.1145/3453483.3454108

16. Heizmann, M., et al.: ULTIMATE AUTOMIZER and the CommuHash Normal Form. In: Sankaranarayanan, S., Sharygina, N. (eds.) Tools and Algorithms for the Construction and Analysis of Systems. TACAS 2023, LNCS, vol. 13994, pp. 577–581. Springer, Cham (2023). https://doi.org/10.1007/978-3-031-30820-8_39

17. Heizmann, M., Klumpp, D., Nitzke, L., Schüssele, F.: Petrification: Software model checking for programs with dynamic thread management (extended version). CoRR abs/2311.01302 (2023). https://doi.org/10.48550/arXiv.2311.01302

18. Hoenicke, J., Majumdar, R., Podelski, A.: Thread modularity at many levels: a pearl in compositional verification. In: POPL, pp. 473–485. ACM (2017). https://doi.org/10.1145/3009837.3009893

19. Inverso, O., Tomasco, E., Fischer, B., La Torre, S., Parlato, G.: Bounded model checking of multi-threaded C programs via lazy sequentialization. In: Biere, A., Bloem, R. (eds.) CAV 2014. LNCS, vol. 8559, pp. 585–602. Springer, Cham (2014). https://doi.org/10.1007/978-3-319-08867-9_39

20. Klumpp, D., et al.: ULTIMATE GEMCUTTER and the axes of generalization. In: TACAS 2022. LNCS, vol. 13244, pp. 479–483. Springer, Cham (2022). https://doi.org/10.1007/978-3-030-99527-0_35

21. Ponce-de-León, H., Furbach, F., Heljanko, K., Meyer, R.: DARTAGNAN: bounded model checking for weak memory models (Competition contribution). In: TACAS 2020. LNCS, vol. 12079, pp. 378–382. Springer, Cham (2020). https://doi.org/10.1007/978-3-030-45237-7_24

22. Microsoft: documentation of the `System.Threading.Thread` class (2023). https://learn.microsoft.com/en-us/dotnet/api/system.threading.thread, Accessed 01 Feb 2023

23. Nguyen, T.L., Fischer, B., La Torre, S., Parlato, G.: Unbounded lazy-CSeq: a lazy sequentialization tool for C programs with unbounded context switches. In:

Baier, C., Tinelli, C. (eds.) TACAS 2015. LNCS, vol. 9035, pp. 461–463. Springer, Heidelberg (2015). https://doi.org/10.1007/978-3-662-46681-0_45

24. Oracle: Documentation of the `java.lang.Thread` class (2022). https://docs.oracle.com/en/java/javase/19/docs/api/java.base/java/lang/Thread.html, Accessed 01 Feb 2023

25. Owicki, S.S., Gries, D.: Verifying properties of parallel programs: an axiomatic approach. Commun. ACM **19**(5), 279–285 (1976). https://doi.org/10.1145/360051.360224

26. Qadeer, S., Wu, D.: KISS: keep it simple and sequential. In: PLDI, pp. 14–24. ACM (2004). https://doi.org/10.1145/996841.996845

27. Saan, S., et al.: GOBLINT: thread-modular abstract interpretation using side-effecting constraints. In: TACAS 2021. LNCS, vol. 12652, pp. 438–442. Springer, Cham (2021). https://doi.org/10.1007/978-3-030-72013-1_28

28. Schwarz, M., Saan, S., Seidl, H., Erhard, J., Vojdani, V.: Clustered relational thread-modular abstract interpretation with local traces. In: Wies, T. (eds.) Programming Languages and Systems. ESOP 2023. LNCS, vol. 13990, pp. 28–58. Springer, Cham (2023). https://doi.org/10.1007/978-3-031-30044-8_2

29. The ULTIMATE team: ULTIMATE program analysis framework (2023). https://ultimate-pa.org/, Accessed 24 Aug 2023

30. Yang, Q., Li, M.: A cut-off approach for bounded verification of parameterized systems. In: ICSE (1), pp. 345–354. ACM (2010). https://doi.org/10.1145/1806799.1806851

A Fully Verified Persistency Library

Stefan Bodenmüller[1]([✉])[iD], John Derrick[2][iD], Brijesh Dongol[3][iD],
Gerhard Schellhorn[1][iD], and Heike Wehrheim[4][iD]

[1] University of Augsburg, Augsburg, Germany
stefan.bodenmueller@informatik.uni-augsburg.de
stefan.bodenmueller@informatik.uni-augsburg.de
[2] University of Sheffield, Sheffield, UK
[3] University of Surrey, Guildford, UK
[4] University of Oldenburg, Oldenburg, Germany

Abstract. Non-volatile memory (NVM) technologies offer DRAM-like
speeds with the added benefit of failure resilience. However, developing
concurrent programs for NVM can be challenging since programmers
must consider both inter-thread synchronisation and durability aspects
at the same time. To alleviate this, libraries such as FliT have been
developed to manage *transformations to durability*, allowing a linearizable concurrent object to be converted into a durably linearizable one by
replacing the reads/writes to memory by calls to corresponding operations of the FliT library. However, a formal proof of correctness for FliT
is missing, and standard proof techniques for durable linearizability are
challenging to apply, since FliT itself is not durably linearizable. In this
paper, we study the problem of proving correctness of transformations
to durability. First, we develop an abstract *persistency library* (called
PLib) that operationally characterises transformations to durability. We
prove soundness of PLib via a forward simulation coupled with a *prophecy
variable* used as an oracle about future behaviour. Second, we show correctness of the library FliT by proving that FliT *refines* PLib under the
realistic PTSO memory model, i.e., the persistent version of TSO memory model implemented by Intel architectures. The proof of refinement
between FliT and PLib has been mechanised within the theorem prover
KIV. Taken together, these proofs guarantee that FliT is also sound wrt
transformations to durability.

Keywords: Durable Linearizability · Px86-TSO · Persistency
Libraries · FliT · Verification

1 Introduction

Byte-addressable *non-volatile (aka persistent) memory (NVM)*, e.g., Memory-
Semantic SSD [25] and XL-FLASH [7], offer higher performance than flash

Dongol is supported by EPSRC grants EP/X037142/1, EP/X015149/1, EP/V038
915/1, and EP/R025134/2. Wehrheim is supported by DFG project WE 2290/14-1.

R. Dimitrova et al. (Eds.): VMCAI 2024, LNCS 14500, pp. 26–47, 2024.
https://doi.org/10.1007/978-3-031-50521-8_2

memory while providing resilience to system-wide crashes. Unlike DRAM, the contents of NVM survive failures (e.g., a power outage). However, developing NVM programs is challenging since they must provide properties such as failure atomicity and recoverability, introducing a high level of additional programmer overhead. This is particularly acute for concurrent programs under weak memory, which have thread-safety and concurrency synchronisation requirements in addition to persistency for failure atomicity.

In recent years, there have been several proposals [5,13,26,30] that aim to help programmers with the transition to NVM, i.e., correct durability of concurrent programs developed for systems without NVM. Many of these approaches provide mechanisms that support *transformations to durability*, whereby program operations such as reads and writes to memory are replaced by calls to library operations that manage durability. In particular, given a (concurrent) library that is correct in a setting *without* crashes, the transformation mechanism can be used to ensure correctness *with* crashes. We note that the notion of "correctness" may also change as part of the transformation to durability.

Proposals such as TL4x and PMDK [1,26] support durable transactions that provide failure atomicity (but not necessarily thread safety [26]). Atlas [5] ensures thread-safe durability provided that a programmer can guarantee that the original program is race-free. Mirror [13] transforms linearizable [14] objects (viz., data structures like stacks or queues) into durably linearizable [15] objects, provided that the object in question is non-blocking. FliT [30] is a generic proposal that ensures correct durability by tracking read-write dependencies on behalf of a programmer using persistent-write-back (pwb) instructions. Neither FliT nor Mirror have a fully formalised correctness proof of such transformations.

In this paper, we take a more abstract approach to the question of transformations to durability for linearizable concurrent objects, which may implement concurrent data structures such as stacks or queues. As our first contribution, we develop a *persistency library* (called PLib) that abstractly models the essential requirements for durability. Such requirements capture the key *dependencies* between read and write accesses to shared variables and their persistence order. We prove that PLib guarantees sound transformation to durability. This proof proceeds by showing PLib refines a *canonical durable automaton* [9] that is derived from the sequential specification of the object. Our proof employs forward simulation using information about the future alike *prophecy variables* [8].

Our second contribution is a proof of correctness of the FliT library [30]. This proof leverages correctness of PLib by showing FliT in turn to refine PLib. This part of our overall soundness proof has been mechanised in the theorem prover KIV [28]. For this concrete implementation level, we model FliT over a real processor, i.e., to be subject to the effects of the (weak) memory model of the processor. Here, we take the memory model of persistent PTSO [16], the persistent version of the TSO memory model of Intel's x86 processor [23]. As a result, both refinements together show that FliT guarantees correct transformations to durability on PTSO.

```
1  put(x, arg){                    5  get(x){
2      write(x.val, arg);          6      r := read(x.flag)
3      write(x.flag, true);        7      if r {
4  }                               8          s := read(x.val);
                                   9          return s; }
                                  10      return ⊥; }
```

Fig. 1. Operations put and get (initially x.val = 0 and x.flag = *false*)

Overview. This paper is structured as follows. Section 2 motivates our work and Sect. 3 presents the formal background and key definitions. We present PLib and its correctness proof in Sect. 4, present the FliT library implementation (under the PTSO memory model) in Sect. 5 and its correctness proof in Sect. 6.

Auxiliary Material. The KIV mechanisation corresponding to the proofs in Sect. 6 may be found at [4].

2   Motivation

We start by illustrating the challenges involved in concurrent programming on NVM. Consider the object O implementing put and get operations in Fig. 1 that use a variable x.flag to determine whether a value for a given location x.val has been set. Operation get(x) only reads from x.val after x.flag has been set, otherwise it returns the value ⊥[1]. Without NVM, these operations are clearly *linearizable* [14] wrt an abstract specification that performs each of the operations atomically since the fine-grained implementation only sets x.flag to true after writing to x.val. Thus any (concurrent) history of put and get operations can be reordered (while preserving the real-time order of operations) into a valid sequential history.

Let us now consider the behavior of Fig. 1 under NVM. In NVM, writes are first cached in *volatile memory*. Written values only reach *persistent memory* on flush actions. We then talk about *persisted writes*. Read actions return values from volatile memory and – if this does not contain an entry for a location – then read from persistent memory. On a crash, only the contents of persistent memory is preserved; thus a crash action resets the volatile memory to persistent memory. It is straightforward to see that executing put and get in such a memory model results in histories that are not *durably linearizable* [15], where durable linearizability holds if a history after removing crashes is linearizable. In particular, it is possible to generate the history in Fig. 2a, where the get operation before the crash can read from unpersisted writes of the put operation, and the crash can occur before these writes are persisted (so that the get after the crash returns ⊥). This history is not linearizable after removing the crash, thus is not durably linearizable. Similarly, since writes do not persist in the order of

---

[1] Note that the put operation would in principle need to reset flag to *false* before writing a new value. For simplicity we elide this here.

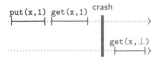
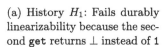
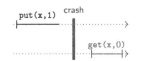
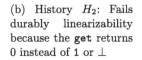
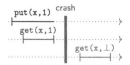

(a) History H_1: Fails durably linearizability because the second **get** returns $\bot$ instead of 1

(b) History H_2: Fails durably linearizability because the **get** returns 0 instead of 1 or $\bot$

(c) History H_3: Fails durably linearizability because the second **get** returns $\bot$ instead of 1

Fig. 2. Three histories (dotted lines denote operations of different threads, x-axis is time)

their occurrence, histories such as H_2 (Fig. 2b) would also be possible. Here, the **get** operation performs a read and sees the flag to be *true* (i.e., x.flag is persisted), but still returns the old (initial) value. A naive approach to addressing this issue is to modify the program so that each write is flushed to persistent memory immediately after it occurs. However, this approach is not only inefficient, it is also incorrect. For instance, under concurrency, it would be possible to generate the history in Fig. 2c since the second write to x.flag in **put** may have occurred, but not yet persisted, yet be read by the first **get**.

The key to achieving durable linearizability generically is to ensure that a client (i.e., a thread) executing an object's operation (e.g., **get**) that reads from another client operation (e.g., **put**) should *not* complete unless it can be assured that all writes it has read from have persisted. With this approach, histories such as H_3 (Fig. 2c) *cannot* occur even when the **put(x,1)** executes concurrently with **get(x)**. In particular, the **get** itself would then ensure that any unpersisted writes that it has read from have been persisted *before* returning. Thus it would be impossible for the second **get(x)** to return $\bot$. Second, to prevent writes from being persisted out-of-order, we must ensure prior writes executed by a thread have persisted before starting a new write by that thread. This prevents histories such as H_2 (Fig. 2b).

The ideas above have been implemented by libraries such as FliT [30] by providing read/write methods that efficiently track read-write dependencies between client operations. A programmer can gain durable linearizability guarantees by calling reads/writes of such a library in place of standard reads and writes, together with a call to a complete method before returning from an operation. The method complete ensures that any dependent unpersisted writes are persisted after executing complete. One of our main contributions in this paper is an abstract persistency library PLib (presented as an input/output automaton in Fig. 5) that we prove to abstractly capture this correctness principle. With this correctness proof for PLib at hand, we can then prove correctness of FliT by showing FliT to *refine* PLib.

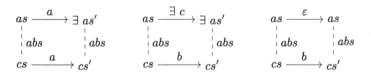

Fig. 3. Possible cases for a forward simulation abs, where $a \in external(C) = external(A)$, $b \in internal(C)$, $c \in internal(A)$, as, as' are abstract and cs, cs' concrete states. ε indicates no action of A. Existentially quantified states and actions must be found such that the diagram commutes.

3   Using IOAs to Verify (Durable) Linearizability

This section provides some standard definitions and known results, providing background for the rest of the paper.

Input/Output Automata (IOA). We use input/output automata (IOA) [20] for all of our sequential and concurrent models. We let Loc be the set of shared locations, Tid the set of thread identifiers and Val the values of locations.

Definition 1 (Input/Output   Automaton   (IOA)). An *Input/Output Automaton (IOA)* is a labeled transition system A with a set of *states* $states(A)$, a set of *actions* $acts(A)$, a set of *start states* $start(A) \subseteq states(A)$, and a *transition relation* $trans(A) \subseteq states(A) \times acts(A) \times states(A)$ (so that the actions label the transitions).

The set $acts(A)$ is partitioned into internal actions $internal(A)$ that represent events of the system that are hidden from the environment, and external actions $external(A)$ (typically atomic steps or invocations/responses to calling an operation) representing the IOA's interactions with its environment. We specify IOAs (like the one in Fig. 4) by giving the states in terms of state variables and their initial values. For every action $a \in acts(A)$ we give a precondition Pre on the state s that enables a step $(s, a, s') \in trans(A)$, and specify the result state s' by assignments given under Eff.

We next define the correctness conditions relevant for our approach. Linearizability is the standard correctness condition for concurrent objects [14]. Here, we provide a definition of linearizability in terms of refinement.

Refinement. An *execution* of an IOA A is a sequence $\sigma = s_0 a_0 s_1 a_1 \ldots s_n a_n s_{n+1}$ of alternating states and actions, such that $s_0 \in start(A)$ and for all states s_i, $(s_i, a_i, s_{i+1}) \in trans(A)$. We write $first(\sigma) = s_0$ for the initial and $last(\sigma)$ for the last state (if it exists) of an execution σ.

A *trace* of A (an element of $traces(A)$) is any sequence of (external) actions obtained by projecting the external actions of any execution of A. For IOAs C and A with $external(C) = external(A)$, we say that C is a *refinement* of A, denoted $C \leq A$, iff $traces(C) \subseteq traces(A)$. Refinement can be proven by establishing forward or backward simulations between IOAs (see e.g., [18]). Given that the methodology is well-known, we elide the formal definition, and provide an overview of the proof obligations for forward simulation diagrammatically in Fig. 3.

Canonical (Durable) Automata. We provide an operational definition of linearizability and durable linearizability using the notion of canonical (durable) automata, both of which are defined in terms of a sequential object specifying the expected behavior.

Definition 2 (Sequential Object). A *sequential object* is a 4-tuple (Σ, S, I, ρ) where Σ is an *alphabet* of operations, S is a set of states and $I \subseteq S$ are the initial states, and $\rho : (S \times \Sigma \times Val^*) \rightarrow 2^{S \times Val^*}$ is a *transition relation* generating a set of next states and output values (if any) for a given state, operation and input values.

The *canonical automaton* [19], $\mathrm{CAN}(\mathbb{S})$, and *canonical durable automaton* [9], $\mathrm{DAUT}(\mathbb{S})$, for a sequential object $\mathbb{S}$ are shown in Fig. 4. $\mathrm{CAN}(\mathbb{S})$ splits every operation $op \in \Sigma$ of $\mathbb{S}$ into separate external *invocation* and *response* actions and an internal *do*-action that occurs between the invocation and response. $\mathrm{DAUT}(\mathbb{S})$ additionally includes a crash action. The traces of a canonical (durable) automaton are well-formed *histories* [14,15], comprising invocations and responses of each operation plus the crash actions in the case of the durable automaton. Unlike the sequential object, the histories of a canonical automaton may be concurrent. However, since the effect of operations op of the sequential object (the do-action) occurs between the invocation and response, each history of the canonical (durable) automaton is (durably) linearizable wrt the given sequential specification, $\mathbb{S}$. Thus, in general, an automaton is (durably) linearizable wrt $\mathbb{S}$ iff it refines the canonical (durable) automaton for $\mathbb{S}$. Note that the durable automaton reflects the *persistent* state of the object, thus, only the program counters of the threads are affected by the crash.

Definition 3 (Linearizability and Durable Linearizability). Let C be an IOA. We say that C is *linearizable* wrt $\mathbb{S}$ if $C \leq \mathrm{CAN}(\mathbb{S})$, and that C is *durably linearizable* wrt $\mathbb{S}$ if $C \leq \mathrm{DAUT}(\mathbb{S})$.

Our main result is to show that a *linearizable object* can be transformed into a *durably linearizable object*. Formally, this requires (c.f. [2]) that we transform an IOA whose histories contain invocations and responses of operations to histories that contain invocations, responses and crashes. The addition of crash is straightforward if the crash has no effect on the original IOA except on the program counters of the threads. Given an IOA A, we let A_c be the IOA A augmented with the crash action from Fig. 4. Note that $\mathrm{DAUT}(\mathbb{S}) = \mathrm{CAN}(\mathbb{S})_c$. We have the following proposition.

Proposition 1. *If $A \leq \mathrm{CAN}(\mathbb{S})$ then $A_c \leq \mathrm{DAUT}(\mathbb{S})$.*

Composing Client and Library IOAs. In the following, we will have several instances of a client calling a library, written *Client*[*Lib*]. These communicate via external actions, which may be *invocation* and *response* actions or atomic calls. The instances of *Client*[*Lib*] that we use are shown in Fig. 8. In [2], *Client*[*Lib*] is defined formally as a product of two IOAs and the following theorem is shown, which establishes refinement for client-library compositions.

State variables:

$pc : Tid \rightarrow \{notStarted, crashed, idle\} \cup$ (initially: $\forall \, \tau \in Tid.\ pc(\tau) = notStarted$)
$\qquad \{\mathsf{inv}(op), \mathsf{res}(op) \mid op \in \Sigma\}$
$s : S$ (initially: $s \in I$)
$in, out : Tid \rightarrow Val^*$

Actions:

$\mathsf{inv}_\tau(op, v)$ $\mathsf{do}_\tau(op)$ $\mathsf{res}_\tau(op, v)$
Pre: $pc(\tau) \in \{notStarted, idle\}$ Pre: $pc(\tau) = \mathsf{inv}(op)$ Pre: $pc(\tau) = \mathsf{res}(op)$
Eff: $pc(\tau) := \mathsf{inv}(op)$ $(s', o') \in \rho(s, op, in(\tau))$ $v = out(\tau)$
     $in(\tau) := v$ Eff: $pc(\tau) := \mathsf{res}(op)$ Eff: $pc(\tau) := idle$
                               $(s, out(\tau)) := (s', o')$

crash
Eff: $pc := \lambda \, \tau : Tid.\ \textbf{if}\ pc(\tau) \neq notStarted\ \textbf{then}\ crashed\ \textbf{else}\ pc(\tau)$

Fig. 4. The canonical durable automaton $\mathrm{DAut}(\mathbb{S})$ for a sequential object $\mathbb{S} = (\Sigma, S, I, \rho)$; automaton $\mathrm{Can}(\mathbb{S})$ is derived by removing the highlighted program counter values and transitions

Theorem 1. (Refinement in context). *Let A, B, C be IOAs with external $(C) = external(A) \subseteq acts(B)$. If $C \leq A$, then $B[C] \leq B[A]$.*

4   Abstract Persistency Library and Its Correctness

In this section, we present our first contribution: the abstract persistency library PLib, which guarantees the transformation to durability for concurrent linearizable objects. An object O has to call the libWrite and libRead actions of PLib to access shared variables and additionally call complete before returning from operations. The key requirement on PLib to ensuring durability is that it tracks certain dependencies (see [30]) and persists writes in this order. Namely **1)** writes of the same thread must persist in the order in which they occur, and **2)** when a thread, say τ, executes complete, all of τ's writes as well as the writes τ has read from must have been persisted.

Persistency Library. Figure 5 gives the IOA of PLib. The state of PLib assumes a volatile memory *vmem*, a persistent memory *pmem* and a *log* of all *read/write events* together with their persistency status and a program counter. Note that the program counter tracks the control flow of the *library* and is unrelated to the program counter of the client thread. When a crash occurs, all currently running threads are stopped and cannot continue. To comply with durable linearizability [15], a thread that crashes is never restarted. Note that like the canonical (durable) automata, PLib specifies a concurrent system since each thread can independently execute its actions.

$$pdr(x, \varepsilon) = \varepsilon$$
$$pdr(x, \langle W_\tau(x, v), flag \rangle \cdot log) = \langle W_\tau(x, v), flag \rangle \cdot log$$
$$pdr(x, \langle R_\tau(x, v), flag \rangle \cdot log) = \langle R_\tau(x, v), true \rangle \cdot pdr(x, log)$$
$$pdr(x, \langle W_\tau(x', v), flag \rangle \cdot log) = \langle W_\tau(x', v), flag \rangle \cdot pdr(x, log) \quad \text{for} \quad x \neq x'$$
$$pdr(x, \langle R_\tau(x', v), flag \rangle \cdot log) = \langle R_\tau(x', v), flag \rangle \cdot pdr(x, log) \quad \text{for} \quad x \neq x'$$

$$lwp(x, log) = false \leftrightarrow \exists\ log_1, log_2.\ log = log_1 \cdot \langle W_\tau(x, v), false \rangle \cdot log_2 \wedge log_{2|Wr,x} = \varepsilon$$

State variables:

$vmem, pmem : Loc \rightarrow Val$	(initially: $\forall\ x \in Loc.\ vmem(x) = pmem(x) = 0$)
$log : ((Rd \cup Wr) \times \mathbb{B})^*$	(initially: $log = \varepsilon$)
$pc : Tid \rightarrow \{notStarted, idle, crashed\}$	(initially: $\forall\ \tau \in Tid.\ pc(\tau) = notStarted$)

Actions:

libWrite$_\tau(x, v)$
Pre: $pc(\tau) \in \{notStarted, idle\}$
$\quad log_{|\tau,false} = \varepsilon$
Eff: $vmem(x) := v$
$\quad log := log \cdot \langle W_\tau(x, v), false \rangle$
$\quad pc(\tau) := idle$

libRead$_\tau(x, v)$
Pre: $pc(\tau) \in \{notStarted, idle\}$
$\quad v = vmem(x)$
Eff: $log := log \cdot \langle R_\tau(x, v), lwp(x, log) \rangle$
$\quad pc(\tau) := idle$

complete$_\tau$
Pre: $pc(\tau) \in \{notStarted, idle\}$
$\quad log_{|\tau,false} = \varepsilon$
Eff: $pc(\tau) := idle$

persist(x)
Pre: $\exists\ log_1, log_2.$
$\quad log = log_1 \cdot \langle W_\tau(x, v), false \rangle \cdot log_2$
$\quad \wedge\ log_{1|Wr,x,false} = \varepsilon$
Eff: $pmem(x) := v$
$\quad log := log_1 \cdot \langle W_\tau(x, v), true \rangle \cdot pdr(x, log_2)$

crash
Eff: $pc := \lambda\ \tau \in Tid.\ \textbf{if}\ pc(\tau) \neq notStarted\ \textbf{then}\ crashed\ \textbf{else}\ pc(\tau)$
$\quad vmem := pmem$
$\quad log := \varepsilon$

Fig. 5. PLib specification (where $log_{|Wr}$ restricts log to all Wr entries, $log_{|\tau,false}$ to all entries of τ with flag *false* and so on)

Let Rd and Wr be the set of all read and write events. Each read event has the form $R_\tau(x, v)$, where τ is a thread identifier, $x \in Loc$ is a location and $v \in Val$ is a value. Similarly, a write event has the form $W_\tau(x, v)$. The *log* pairs write events with a boolean flag that determines whether the operation has been persisted. We call a write *persisted* if its effect is written to persistent memory and it has flag *true*, and a read *persisted* if the write that it reads from is persisted.

A libWrite and complete executed by thread τ may only proceed if it has neither unpersisted writes nor unpersisted reads in *log*, i.e. when $log_{|\tau,false} = \varepsilon$. We believe the condition is as weak as possible to allow a proof of durable linearizability for arbitrary data structures: A write is disallowed only if there

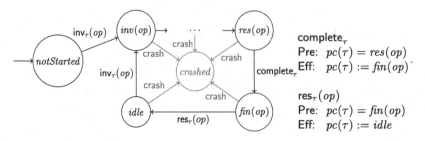

Fig. 6. States of a thread τ when calling operations of object O_{cc} ($\cdots$ are program counter values within the body of some $op \in \Sigma$)

Fig. 7. complete and res actions of O_{cc}

is a previous read or write by the same thread that has yet to be persisted. Returning from an operation must have persisted its values.

Execution of the write updates the value of the written location in *vmem*, then appends a new write event to *log* with flag *false*. Execution of read reads the value from volatile memory. It adds the read to the log together with a flag that is computed by checking whether the last write to this location (if there is any; otherwise the flag is set to true) has already been persisted using predicate *lwp*. Persisting a write also persists dependent reads (i.e., those which have read this value) using function *pdr*. The recursive definition of *pdr* ensures that the reads affected are those in the log before the next write to the same location.

Correctness of PLib. Next, we prove Theorem 2 below, i.e., that PLib guarantees a transformation to durability for linearizable objects. First, we explain the construction of the automaton $O_{cc}[\mathsf{PLib}]$. Let O_{cc} be the object O_c (as described in Proposition 1) further modified so that each operation calls a complete action before returning (see Fig. 6). The complete and res(op) actions of O_{cc} are depicted in Fig. 7. Let $O_{cc}[\mathsf{PLib}]$ be O_{cc} but with internal actions read, write, complete and crash replaced with calls to the libRead, libWrite, complete and crash actions of PLib, as depicted on the left of Fig. 8.

Theorem 2. *Suppose* $\mathbb{S}$ *is a sequential specification and* O *an object. Then we have that* $O \leq \mathrm{CAN}(\mathbb{S}) \Rightarrow O_{cc}[\mathsf{PLib}] \leq \mathrm{DAUT}(\mathbb{S})$.

We show that $O_{cc}[\mathsf{PLib}]$ is a refinement of $\mathrm{DAUT}(\mathbb{S})$ by proving three separate refinements:

1. $O_c \leq \mathrm{DAUT}(\mathbb{S})$. This follows from Proposition 1 and the assumption $O \leq \mathrm{CAN}(\mathbb{S})$.
2. $O_{cc} \leq O_c$. This change again only affects program counter values, thus it is trivial to show.
3. $O_{cc}[\mathsf{PLib}] \leq O_{cc}$. This is the challenging part, which we discuss in more detail below.

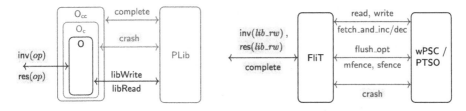

Fig. 8. Interface between client and library IOAs in our formal models; highlighted actions denote external events and $lib\_rw \in \{\text{libRead}, \text{libWrite}\}$

We start the proof of $O_{cc}[\text{PLib}] \leq O_{cc}$ by first clarifying the state and operations of O_{cc}. As we need to show correctness of PLib for *all* objects, we cannot assume any specific knowledge about the actions of O_{cc}. We only assume that O_{cc} uses the following state variables.

– A program counter *opc* for each thread τ, which in particular may take values *notStarted*, *idle*, *inv(op)*, *fin(op)*, *res(op)* and *crashed* as well as values for each atomic statement within the implementations of operations of the object *op*. The values are changed according to Fig. 6.
– The shared memory, *mem* : *Loc* → *Val*, that is written atomically by write operations and read atomically by read operations.
– Local registers of type *Reg* for each thread and a mapping *regs* : *Reg* → *Val* that describes the local state that the object uses in its concurrent implementation of the sequential specification $\mathbb{S}$.

The state of $O_{cc}[\text{PLib}]$ is similar, except that $O_{cc}[\text{PLib}]$ calls libWrite/libRead of PLib instead of reading and writing to shared memory using the write and read operations. Thus, in $O_{cc}[\text{PLib}]$ the shared memory *mem* is omitted since memory is only indirectly accessed via PLib (see Fig. 5).

Next, we describe the overall idea of the proof. Proving refinement in our setting amounts to proving trace inclusion, i.e., $traces(O_{cc}[\text{PLib}]) \subseteq traces(O_{cc})$. Concretely, from an execution of $O_{cc}[\text{PLib}]$ we will construct an execution of O_{cc} with the same sequence of external actions. This is guided by the following ideas:

Step 1. Crash events divide executions (and hence traces) of $O_{cc}[\text{PLib}]$ into so-called *eras* (of non-crash actions). We will reason about each era of $O_{cc}[\text{PLib}]$ in isolation and construct corresponding eras of O_{cc}. At the end, we join the constructed eras of O_{cc} to form a consistent execution of O_{cc}.

Step 2. Refinement is usually proven by showing a forward or backward simulation that uses an *abstraction relation* to relate states of a concrete automaton ($O_{cc}[\text{PLib}]$) to the states of an abstract automaton (O_{cc}), see Fig. 3 for conditions in a forward simulation. Here, we will employ a forward simulation together with a *prophecy variable containing information about the future* to define the abstraction relation. Our prophecy variable refers to the log from the end of the era, which we use to determine whether libRead/libWrite actions at the concrete level should be matched to corresponding read/write actions at the abstract level of O_{cc}, or to ε.

Step 3. For every execution σ of $O_{cc}[\mathsf{PLib}]$ (more specifically, every era) we use forward simulation to show the existence of an execution ρ (viz. era) of O_{cc} such that $\rho_{|\tau} \preceq \sigma_{|\tau}$ holds for all threads τ. Basically, the construction of ρ keeps the steps of τ which are "persisted" in σ and preserves the external actions of the trace.

Step 1 (Decompose into eras). Let σ be an execution of $O_{cc}[\mathsf{PLib}]$. Moreover, suppose $\sigma = \sigma_1 \,\mathsf{crash}\, \sigma_2 \,\mathsf{crash}\, \ldots \,\mathsf{crash}\, \sigma_k$ such that each σ_i contains no crashes. We refer to $\sigma_1, \sigma_2, \ldots, \sigma_k$ as the *eras* of σ.

Let σ_i be an era of a concrete execution of $O_{cc}[\mathsf{PLib}]$. We construct a corresponding abstract era ρ_i of O_{cc} as follows. In σ_i, execution starts with $vmem = pmem$ since this is true initially and each **crash** resets $vmem$ to $pmem$ (see Fig. 5). The abstraction relation we use (see below) ensures that in corresponding states, mem of O_{cc} is the same as $pmem$. The **crash** actions for $O_{cc}[\mathsf{PLib}]$ (and O_{cc}) leaves $pmem$ (and mem) unchanged. The abstraction relation below further ensures that the same set of threads are "active" in corresponding concrete and abstract eras and hence the effect of **crash** on abstract and concrete program counters is the same. Thus $\rho_1 \,\mathsf{crash}\, \rho_2 \ldots \mathsf{crash}\, \rho_n$ is an execution of O_{cc}.

Step 2 (Constructing the abstraction relation). We augment the states of $O_{cc}[\mathsf{PLib}]$ with information from the end of each era. More specifically, we consider the value $\log := s_n.log$ of the log variable at the end of the era $\sigma = s_0 a_0 s_1 a_1 \ldots s_n$, and use it as a prophecy variable. (Note that although this discussion is about era σ_i, we drop the index i for notational convenience). Using the (standard) construction of [18, Proposition 5.13], $\log$ is added to every state s_j resulting in an era $\sigma' = s_0' a_0 s_1' a_1 \ldots s_n'$.

We now work towards an abstraction relation that relates σ to a corresponding era $\rho = q_0 a_0' q_1 a_1' \ldots q_n$ of O_{cc} by relating each state s_j' to q_j. We show that this relation gives rise to a forward simulation below. Note that, as described in Fig. 3, the forward simulation ensures that each concrete step is matched by some abstract step, thus σ and ρ have the same length. Namely, in the case of stuttering steps, the concrete action is matched by an abstract ε step.

By construction, in the abstract execution, O_{cc} can only perform **reads** and **writes** from the *persistent part* of $\log$. Formally, the persistent part $pp(\log)$ of $\log$ is defined recursively as a list of read and write events. We use $l(j) \neq \langle \mathsf{RW}_\tau(\_,\_), false \rangle$ to mean $l(j) \neq \langle \mathsf{W}_\tau(\_,\_), false \rangle \wedge l(j) \neq \langle \mathsf{R}_\tau(\_,\_), false \rangle$.

$$pp(\varepsilon) = \varepsilon$$
$$pp(l \cdot \langle e_\tau, false \rangle) = pp(l)$$
$$pp(l \cdot \langle e_\tau, true \rangle) = \mathbf{if}\ \forall\, j.\ l(j) \neq \langle \mathsf{RW}_\tau(\_,\_), false \rangle\ \mathbf{then}\ pp(l) \cdot e_\tau\ \mathbf{else}\ pp(l)$$

Thus $pp(\log)$ contains all writes in era σ that are persisted by the end of σ. Note that the precondition libWrite for PLib (Fig. 5) ensures that there cannot be an earlier write with flag *false* followed by a write from the same thread with flag *true*. Persisted reads are dropped from the persistent part iff they are preceded by other unpersisted reads or writes of the same thread.

The reads and writes executed in era σ' when we reach state s_i' are recorded in $s_i'.log$ (see Fig. 5). Operations that are persisted before the end of the era σ are those in the persistent part of $\mathsf{plog}(i) := \log[0..\#(s_i'.log) - 1]$, i.e., in $pp(\mathsf{plog}(i))$. Thus, $\mathsf{plog}(i)$ contains the same events as $s_i'.log$ in the same order. However, while $s_i'.log$ indicates read/writes that have been persisted thus far, $\mathsf{plog}(i)$ records the reads/writes that are guaranteed to be persisted before the era σ ends.

We can use $\mathsf{plog}(i)$ to relate the persistent memory $s_i'.pmem$ to the abstract memory $q_i.mem$. In q_i we want to have exactly the memory state that corresponds to executing the events of $pp(\mathsf{plog}(i))$, the relation between persistent memory of s_i' and the abstract memory $q_i.mem$ is thus

$$q_i.mem = s_0'.pmem \oplus pp(\mathsf{plog}(i)) \tag{1}$$

where $pmem \oplus evs$ is inductively defined as

$$pmem \oplus \varepsilon = pmem$$
$$pmem \oplus ((\mathsf{R\_}(\_, \_), \_) \cdot evs) = pmem \oplus evs$$
$$pmem \oplus ((\mathsf{W\_}(x, v), \_) \cdot evs) = pmem[x \mapsto v] \oplus evs$$

So the abstract memory $q_i.mem$ is obtained by taking the persistent memory $pmem$ from s_0' and performing the updates of all writes of the persisted part of the log that have been executed up to state s_i' in order. Each individual thread τ has executed reads and writes that are persisted at the end of the era only, if and only if $\mathsf{plog}(i)|_\tau$ and $s_i.log|_\tau$ are equal, which is equivalent to $\mathsf{plog}(i)|_{\tau, false} = \varepsilon$.

The abstraction relation keeps registers and program counter values of a thread τ identical in the concrete and abstract levels until the thread reaches its first unpersisted read or write, i.e., until $\mathsf{plog}(i)|_{\tau, false}$ becomes nonempty. Therefore, the abstraction relation states

$$\mathsf{plog}(i)|_{\tau, false} = \varepsilon \implies s_i'.regs = q_i.regs \wedge s_i'.pc = q_i.pc \tag{2}$$

The full abstraction relation is the conjunction of (1) and (2).

Step 3 (Proving forward simulation). For the forward simulation, we have to map the steps $s_i' a_i s_{i+1}'$ of the augmented era σ' of $\mathsf{O_{cc}}[\mathsf{PLib}]$ to steps of the corresponding era in $\mathsf{O_{cc}}$ (in the sense of Fig. 3). Specifically,

- if $a_i = \mathsf{libWrite}(x, v)$ and $\mathsf{plog}(i+1)|_{\tau, false} = \varepsilon$ we map to $q_i \, \mathsf{write}(x, v) \, q_{i+1}$ in the abstract era ρ,
- if $a_i = \mathsf{libRead}(x, v)$ and $\mathsf{plog}(i+1)|_{\tau, false} = \varepsilon$ we map to $q_i \, \mathsf{read}(x, v) \, q_{i+1}$ in the abstract era ρ, and
- in all other cases, we map to $q_i \, \varepsilon \, q_{i+1}$ in ρ

Note that $\mathsf{plog}(i)|_{\tau, false} = \varepsilon$ iff $pp(\mathsf{plog}(i))|_\tau = \mathsf{plog}(i)|_\tau$.

Intuitively, a thread continues executing actions until it reaches a read or write that is not persisted at the end of the era (and hence will be lost when a crash occurs). The libReads and libWrites that are persisted in the concrete era are mapped to reads and writes. Both sets of corresponding operations see the same

```
libWrite(x, arg){                    libRead(x){
  sfence;                              val = read(x);
  fetch_and_inc(flit_ctr(x));          if (flit_ctr(x) > 0)
  write(x, arg);                          flush_opt(x);
  flush_opt(x);                        return val;
  sfence;                            }
  fetch_and_dec(flit_ctr(x));
}                                    complete(){ mfence; }
```

Fig. 9. Core algorithm of the FliT library

memory state at the abstract and concrete levels due to (1) of the abstraction relation. Steps of the object that are preserved can be mapped one-to-one since they see the same register values by (2).

This mapping of steps preserves invocations and responses (i.e., gives us the same sequence of external actions): A thread executing a return (the argument for an invocation is similar) must have all reads and writes persisted in $s_i.log_{|\tau}$, since it has just called the complete operation. Therefore, all events are persisted in $plog(i)_{|\tau}$ as well, so the step is kept in the abstract era. All earlier actions of the thread are kept as well, simply because $plog(j)$ for $j \leq i$ is a prefix of $plog(i)$.

5   The Library FliT and Memory Models

For our second contribution, we prove soundness (of a transformation to durability) for a concrete persistency library (FliT [30]) executing on a realistic memory model (PTSO [16,24]). We first explain FliT and PTSO, then prove soundness in Sect. 6.

Library **FliT.** The pseudo code of the core algorithm of FliT is given in Fig. 9. For our proofs, it is translated into an IOA as described in [10] with a step for every instruction. FliT comprises operations libRead and libWrite that perform reads from and writes to shared memory (i.e., locations x in Loc)[2]. Like $O_{cc}[PLib]$, the concurrent object O_{cc} calls libRead and libWrite when accessing memory. In addition, FliT implements an operation complete that the object O has to call before completing one of its operations $op \in \Sigma$.

The key idea behind FliT is that it manages persistent memory operations on behalf of a user. It guarantees that crucial dependencies among reads and writes are maintained and thereby achieves durable linearizability. The shared variable flit_ctr(x) for location x is used (for efficiency) to determine whether a shared write is in progress, i.e., another thread is currently writing to location

---

[2] In [30], the operations are called shared_load and shared_store, respectively. Since we aim to prove a durable linearizability theorem that assumes all reads and writes are durable, we omit the additional pflag of the original algorithm and assume that each FliT operation is called with the persistent flag. We also do not make use of private stores and loads (this is future work).

x so that the current value at x may not be persisted yet. libWrite performs a fetch_and_inc (i.e., a *read-modify-write*, RMW, operation) on flit_ctr(x) before performing the write itself, and a corresponding fetch_and_dec (another RMW) on flit_ctr(x) after the write. Both libRead and libWrite issue flush_opt instructions[3]. Finally, libWrite issues an sfence instruction at its start and end, and complete issues an mfence.[4] Basically, the flush_opt operation serves as a marker signifying a prior access of a thread τ to location x. RMW and mfence/sfence operations must wait until the marked access is persisted, i.e., reaches persistent memory.

Memory Models. FliT achieves its guarantees by employing a number of hardware instructions with a semantics defined by the *memory model* of the processor. Here, we study the usage of FliT on the memory model TSO of the Intel x86 processor [21], viz. its version PTSO for NVM as formalized in [16]. There are alternative definitions of the persistent version of x86 (e.g., [6,23]); we have taken PTSO here because it matches the intention of developers about program behaviour well and it has a clear connection to the persistent version of sequential consistency (which we will employ in our proof). Figure 10 provides an automaton specification of PTSO based on the operational semantics given in [16], and Fig. 8 (on the right) shows the events FliT uses for communication with the memory model. For simplicity, we do not give a specification of flush instructions here, as they are not used by FliT. For the same reason, we omit the generic RMW transitions and directly define transitions for fetch_and_inc and fetch_and_dec.

PTSO contains non-volatile memory $m : Loc \rightarrow Val$ and shared persistence buffers $P(x)$ storing a sequence of markers $W_\tau(v)$ and $FO(\tau)$ for every $x \in Loc$, representing pending write and flush_opt instructions by thread τ that have not yet been persisted. These markers in P are visible to all threads, i.e., threads can access the values of pending writes in P. Note that, compared to the original model in [16], $W_\tau(v)$ markers are augmented with the identifier τ of the thread that issued the write, which is relevant for verification only.

In addition to P, PTSO contains a private store buffer $B(\tau)$ for each thread τ that contains similar markers $W(x, v)$ and $FO(x)$ for write and flush_opt instructions, as well as markers SF issued by sfence instructions. When a write, flush_opt, or sfence is executed by a thread τ, the corresponding markers are first put into its store buffer $B(\tau)$ (actions write$_\tau$, flush_opt$_\tau$, and sfence$_\tau$ in Fig. 10). *Propagate* steps then move entries from store buffers to persistence buffers (under some conditions), which make the entries visible to other threads (actions prop_w, prop_fo, and prop_sf). Finally, *persist* steps move entries from persistence buffers to the main memory m (actions persist_w and persist_fo). As

---

[3] In [30], they employ an instruction pwb (persistent-write-back) for this. Here, we directly give the architecture-specific instruction.

[4] In [30], both operations use a generic PFENCE instruction for which no precise semantics is given. We instantiate these instructions with the appropriate concrete fence instructions based on the semantics of the memory models given by [16,24].

State variables:

$m : Loc \rightarrow Val$ (initially: $m = \lambda\, x.\ 0$)

$P : Loc \rightarrow \{\mathsf{FO}(\tau), \mathsf{W}_\tau(v) \mid \tau \in Tid, v \in Val\}^*$ (initially: $P = \lambda\, x.\ \varepsilon$)

$B : Tid \rightarrow \{\mathsf{FO}(x), \mathsf{W}(x,v), \mathsf{SF} \mid \tau \in Tid, x \in Loc, v \in Val\}^*$ (initially: $B = \lambda\, \tau.\ \varepsilon$)

Actions:

mfence_τ
Pre: $B(\tau) = \varepsilon$
      $\forall\, y.\ \mathsf{FO}(\tau) \notin P(y)$

sfence_τ
Eff: $B(\tau) := B(\tau) \cdot \mathsf{SF}$

$\mathsf{persist\_fo}(x)$
Pre: $P(x) = \mathsf{FO}(\tau) \cdot p$
Eff: $P(x) := p$

$\mathsf{prop\_fo}$
Pre: $B(\tau) = b_1 \cdot \mathsf{FO}(x) \cdot b_2$
      $\mathsf{W}(x, \_), \mathsf{FO}(x), \mathsf{SF} \notin b_1$
Eff: $B(\tau) := b_1 \cdot b_2$
      $P(x) := P(x) \cdot \mathsf{FO}(\tau)$

$\mathsf{flush\_opt}_\tau(x)$
Eff: $B(\tau) := B(\tau) \cdot \mathsf{FO}(x)$

$\mathsf{persist\_w}(x)$
Pre: $P(x) = \mathsf{W}_\tau(v) \cdot p$
Eff: $P(x) := p$
      $m(x) := v$

$\mathsf{prop\_w}$
Pre: $B(\tau) = \mathsf{W}(x,v) \cdot b$
Eff: $B(\tau) := b$
      $P(x) := P(x) \cdot \mathsf{W}_\tau(v)$

$\mathsf{prop\_sf}$
Pre: $B(\tau) = \mathsf{SF} \cdot b$
      $\forall\, y.\ \mathsf{FO}(\tau) \notin P(y)$
Eff: $B(\tau) := b$

$\mathsf{read}_\tau(x, v)$
Pre: $v = (m \oplus P \oplus B(\tau))(x)$

$\mathsf{write}_\tau(x, v)$
Eff: $B(\tau) := B(\tau) \cdot \mathsf{W}(x,v)$

$\mathsf{fetch\_and\_inc/dec}_\tau(x, v)$
Pre: $B(\tau) = \varepsilon$
      $\forall\, y.\ \mathsf{FO}(\tau) \notin P(y)$
      $v = (m \oplus P \oplus B(\tau))(x)$
Eff: $P(x) := P(x) \cdot \mathsf{W}_\tau(v \pm 1)$

crash
Eff: $P := \lambda\, x.\ \varepsilon$
      $B := \lambda\, \tau.\ \varepsilon$

Fig. 10. Automaton of the PTSO memory model (which is equivalent to Px86 [16])

given by the crash action, all pending markers in B and P are lost during a crash (the buffers are reset to ε), while only m is kept.

When thread τ intends to read location x (action read_τ), it first of all consults its store buffer, then the persistence buffer and finally persistent memory. This is modelled by the following function producing the combined memory $m \oplus P \oplus B(\tau)$ visible to thread τ.

$$m \oplus P \oplus B(\tau) = \lambda\, x. \begin{cases} v & \text{if } B(\tau) = b_1 \cdot \mathsf{W}(x,v) \cdot b_2 \wedge \mathsf{W}(x, \_) \notin b_2 \\ v & \text{else if } \mathsf{W}(x, \_) \notin B(\tau) \wedge P(x) = p_1 \cdot \mathsf{W}_\_(v) \cdot p_2 \\ & \qquad \wedge \mathsf{W}_\_(v) \notin p_2 \\ m(x) & \text{otherwise} \end{cases}$$

The read-modify-write actions $\mathsf{fetch\_and\_inc}_\tau(x)$ and $\mathsf{fetch\_and\_dec}_\tau(x)$ block until the store buffer is empty, and immediately propagate the updated value to the shared persistence buffer. Additionally, they require that all FO-markers of τ have left the persistence buffers, so that they have a similar effect to the strong mfence_τ action.

We prove soundness of FliT via a chain of refinements. This chain includes an intermediate memory model called wPSC, as an abstraction of PTSO. This memory model is based on the often assumed *sequential consistency* model (SC [17]), resp. PSC for NVM as given in [16]. The automaton of wPSC is given in Fig. 11. wPSC abstracts from the store buffers B of PTSO so that all markers are directly put into the shared persistence buffers P. Consequently, wPSC does not have any *propagate* actions, and write_τ and $\mathsf{flush\_opt}_\tau$ steps directly add markers to $P(x)$. Actions sfence_τ and mfence_τ are identical in wPSC as they both require all FO-markers of τ to be persisted. The combined memory for reading is identical for all threads: $m \oplus P = m \oplus P \oplus \varepsilon$.

State variables:
$m : Loc \to Val$ (initially: $m = \lambda\, x.\ 0$)
$P : Loc \to \{FO(\tau), W_\tau(v) \mid \tau \in Tid, v \in Val\}^*$ (initially: $P = \lambda\, x.\ \varepsilon$)

Actions:

mfence$_\tau$/sfence$_\tau$
Pre: $\forall\, y.\ FO(\tau) \notin P(y)$

weak persist_fo(x)
Pre: $P(x) = p_1 \cdot FO(\tau) \cdot p_2$
      $p_{1|\tau} = \varepsilon \qquad p_{1|W} = \varepsilon$
Eff: $P(x) := p_1 \cdot p_2$
crash
Eff: $P := \lambda\, x.\ \varepsilon$

flush_opt$_\tau$(x)
Eff: $P(x) := P(x) \cdot FO(\tau)$

weak persist_w(x)
Pre: $P(x) = p_1 \cdot W_\tau(v) \cdot p_2$
      $p_{1|\tau} = \varepsilon$
      $p_{1|W} = \varepsilon$
Eff: $P(x) := p_1 \cdot p_2$
      $m(x) := v$

read$_\tau$ (x, v)
Pre: $v = (m \oplus P)(x)$

write$_\tau$ (x, v)
Eff: $P(x) := P(x) \cdot W_\tau(v)$

fetch_and_inc/dec$_\tau$ (x, v)
Pre: $\forall\, y.\ FO(\tau) \notin P(y)$
      $v = (m \oplus P)(x)$
Eff: $P(x) := P(x) \cdot W_\tau(v \pm 1)$

Fig. 11. Memory models wPSC and PSC, where PSC is obtained by removing blue parts . (Color figure online)

The motivation for using this intermediate layer is to split the proof into two parts dealing with separate concerns. Based on wPSC, the first part of the proof (see Sect. 6.1) shows that the use of flush_opt and sfence resp. mfence in FliT is sufficient to guarantee a persist order that coincides with the persist constraints of our specification PLib. The second part of the proof (see Sect. 6.2) then shows that, when introducing per-thread store buffers in PTSO, propagations may only lead to uncritical reorderings in the persistence buffers. However, this is only the case in the context of FliT (or a similar persistency library).

Figure 11 also depicts the differences to the original PSC model of [16]. Basically, wPSC differs only in that the *persist* actions have slightly weaker preconditions than PSC: persisting an FO- or W-marker of location x is possible even if there are some other markers before it in $P(x)$ (as long as they are just FOs of other threads). This adjustment is necessary due to the possible propagation reorderings of PTSO, see also Sect. 6.2.

6   Proving Correctness of FliT

Our correctness proof of FliT proceeds via proving FliT to *refine* PLib, i.e., to satisfy the requirements that PLib imposes on persistence orderings. More specifically, we aim to prove $FliT[PTSO] \leq PLib^{IR}$. Therein, $PLib^{IR}$ (where IR = Invoke/Respond) is a slight modification of PLib which replaces libRead and libWrite with invocations to and responses from the library (i.e., inv(libRead), inv(libWrite), res(libRead), res(libWrite)) in addition to do-steps (i.e., do(libRead), do(libWrite)). This split is necessary because FliT's reads and writes do not occur atomically. The proof of $FliT[PTSO] \leq PLib^{IR}$ proceeds in two steps:

1. $FliT[wPSC] \leq PLib^{IR}$, where wPSC is given in Fig. 11.
2. $FliT[PTSO] \leq FliT[wPSC]$, where PTSO is given in Fig. 10.

Combining this result with the simple to prove refinement $PLib^{IR} \leq PLib$ and Theorems 2 and 1, we get correct transformation to durability for FliT on PTSO, i.e., $O \leq C_{AN}(\mathbb{S}) \Rightarrow O_{cc}[FliT[PTSO]] \leq D_{AUT}(\mathbb{S})$.

The proofs of the two refinements are non-trivial; each of them took around three weeks to verify. They have been mechanised in the theorem prover KIV [28] and are described below. The proofs follow the strategy detailed in [10,27], which splits invariants and the abstraction relation into a global part for the shared state and a thread-local part to get thread-local proof obligations.

6.1   Proof of FliT[wPSC] $\leq$ PLib^{IR}

We show the two refinements via forward simulations. This means that we need to define an abstraction relation, *abs*, and then have to map steps (occurrences of actions) of the concrete automaton to steps of the abstract one as depicted in Fig. 3.

The first refinement FliT[wPSC] $\leq$ PLib^{IR} maps all relevant steps of FliT[wPSC] (in particular, the inv, do, and res steps of read and write as well as the persist_w step) steps to the corresponding steps of PLib^{IR}. The only exception is persist_fo which is mapped to an empty step. The core of the verification then is to ensure that the precondition of do(write) of FliT[wPSC] is more restrictive than the one of PLib^{IR}.

Essentially, the modifications of the flit counter (which are RMW instructions of wPSC) are responsible for this: the `fetch_and_inc` instruction before writing ensures that the values read before the write get persisted, the `fetch_and_dec` at the end of writing ensures that the write just done gets persisted, so each thread never has more than one unpersisted write (the formal proof has corresponding assertions). However, formally the preconditions of RMW instructions (fetch_and_inc/dec$_\tau$ in Fig. 11) only guarantee that all FO-markers of the thread have left the persistence buffer $P(x)$, so some additional reasoning is needed to prove this. The crucial assertion for FliT[wPSC] is that a thread τ that has just added an FO-marker to $P(x)$ before the `fetch_and_dec` instruction will keep this marker longer in $P(x)$ than the write event W it has added to $P(x)$ in the write instruction right before it. Also, the FO-marker will always be after the write event, so the write must be persisted first. To ensure this, all global (persist) steps and all steps by other threads than τ must preserve this. The thread relies on the fact that all these steps can only remove an initial piece from each $P(x)$ (including some writes) and then some of the FO-markers before the next write (but not FO-markers after that write). They can also add new events to the end of $P(x)$, but those are not events of thread τ.

The abstraction relation between FliT[wPSC] and PLib^{IR} keeps both the volatile and persistent memories equal (where the volatile memory of FliT[wPSC] is $m \oplus P$). Not yet persisted events for any location x (the elements of $log_{|x,false}$) must have a corresponding entry in $P(x)$. For writes, the same write events must appear in the same order in $P(x)$. For reads, an FO-marker must appear later in $P(x)$ than their position in log, except if the thread executing the read is directly after the read instruction, before it has a chance to add an FO-marker.

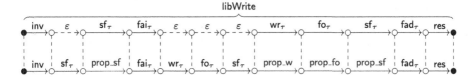

Fig. 12. Forward simulation (cf. Fig. 3) for FliT[PTSO] ≤ FliT[wPSC] for the libWrite operation. Internal steps of PTSO are depicted blue. wr_τ, fo_τ, sf_τ, fad_τ, fai_τ stand for $write_\tau$, $flush\_opt_\tau$, $sfence_\tau$, $fetch\_and\_dec_\tau$, $fetch\_and\_inc_\tau$, respectively. (Color figure online)

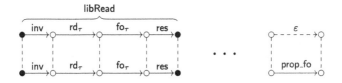

Fig. 13. Forward simulation (cf. Fig. 3) for FliT[PTSO] ≤ FliT[wPSC] for the libRead operation. Internal steps of PTSO are depicted blue. rd_τ, fo_τ stand for $read_\tau$, $flush\_opt_\tau$, respectively. (Color figure online)

6.2 Proof of FliT[PTSO] ≤ FliT[wPSC]

The main idea for the proof of FliT[PTSO] ≤ FliT[wPSC] is to keep the two volatile memories represented by the two persistence buffers P (denoted P_{wPSC} and P_{PTSO} in the following) approximately equal. In particular, both should contain the same sequence of write events, ensuring that reading on both levels yields the same result values. The refinement therefore has to delay writes of FliT[PTSO], which add the event to the store buffer B, until they are propagated to P_{PTSO} by the system via a prop_w step. To get a forward simulation, the flush_opt and sfence instructions called during libWrite (see Fig. 9) are delayed, too. They take effect only when they are propagated to P_{PTSO}. Formally, in the forward simulation, they all refine an empty step ε of FliT[wPSC], while the propagation steps refine the instructions of FliT[wPSC] (see Fig. 12). The delay is possible since propagation is ensured to happen in the same order, and since the fetch_and_dec$_\tau$ step at the end of writing can proceed only when the events have all been propagated and the store buffer B is empty (so the responses from writing at the end can be matched again). Informally, the "linearization point" of libWrite of FliT[PTSO] is not the write instruction which adds to B, but the propagation of the write to volatile memory, and this is the order in which writes happen in FliT[wPSC] as well as in PLib^{IR}.

For reading the situation is different. There, we cannot match adding an FO-marker (with flush_opt$_\tau$) on FliT[wPSC] to the corresponding prop_fo step on FliT[PTSO] since propagation may happen long after the read operation has finished. Therefore, it is necessary to match the flush_opt$_\tau$ steps one-to-one, and instead to match prop_fo for FO-markers issued by reading to an empty step ε as shown in Fig. 13. The two instances of prop_fo for an FO(τ) can be distinguished by checking whether thread τ is currently in its libWrite. However, the abstraction relation must now deal with the reordering of FO-markers caused by this mapping, which results in the somewhat tricky abstraction relation that essentially allows FO-markers of reads that are in P_{wPSC} to be somewhere later in P_{PTSO} or still in B.

The persist steps (persist_fo and persist_w, which are omitted in Fig. 12) of FliT[wPSC] and FliT[PTSO] are mapped one-to-one since W-events must have the same order in P_{PTSO} as in P_{wPSC}. However, the weakening of the original PSC model to wPSC as shown in Fig. 11 is essential for this refinement: PTSO always persists the leading marker in P_{PTSO} for some location x, and thus, wPSC must be able to persist the corresponding event in P_{wPSC}. But due to the potential reordering of FO-markers (see above), this event does not have to be the leading event in P_{wPSC}. The weakened preconditions of the wPSC persist steps take this into account while not being too liberal (in particular, the order of writes must be maintained).

7   Related Work

D'Osualdo, Raad and Vafeiadis [11] pursue the idea of reusing proofs of linearizability for showing concrete data structures to be durable linearizable. They formulate a so-called *Pathway Theorem* which presents a method for proving durable linearizability. This method is modular in the sense of separating the proof of linearizability from issues concerning volatile and persistent memory. They base their work on a declaratively defined memory model. A similar approach (but for durable opacity) has been investigated by Bila et al. [2] who aim at reusing proofs of opacity [3]. Durable opacity is the NVM analogue of the correctness criterion *opacity* employed for Transactional Memory [29] (not concurrent objects). Bila et al. assume sequential consistency.

Verification of FliT differs from these earlier works [2,10,11] since FliT is a generic instrumentation technique. It transforms *any* linearizable object into a durably linearizable object. Thus, the proof methods in [2,10,11] are orthogonal to ours since they provide proof techniques for concurrent objects that are already durably linearizable *without* using FliT. Moreover, FliT alone is not durably linearizable.

Besides FliT, there are a number of other works investigating techniques for making linearizable concurrent objects durably linearizable, most notably NVTraverse [12] and Mirror [13]. Both provide transformations to durability for lock-free concurrent objects whereby NVTraverse is limited to node-based tree data structures. Mirror is trimmed towards using a volatile replica of the data structure in RAM for speeding up reading (though it can be used solely with NVM as well). Its approach for ensuring durable linearizability is stricter than the one of FliT: the algorithm enforces modifications to be persisted before their linearization points. To achieve this, reading accesses only the volatile replica, which is only updated after the corresponding write has passed the persistence buffer. Moreover, writing is performed via a strong CAS implementation, which prevents multiple pending writes to the same location in the persistence buffers. This behaviour also matches the PLib specification presented in this paper (yielding a *log* that always contains only persisted R markers and at most one unpersisted W marker per location). Thus, a formal proof of Mirror's correctness should be possible by showing a refinement analogous to the one for FliT shown in Sect. 6, which we intend to do as future work.

Israelevitz and Scott [15] also propose a technique for converting linearizable objects into (buffered) durably linearizable objects by inserting pfence and pwb instructions. They prove correctness of their transformations. Montage [31] is a library for use by programmers when intending to achieve buffered durable linearizability [15]. To this end, programmers need to identify so-called payloads and follow several additional rules. Neither of these approaches is accompanied by a fully formal proof of its soundness.

Regarding other memory models, recent works [6] have defined persistency models for Arm (PArm) building on a prior operational semantics based on promises [22]. To prove correctness of FliT under this model, one would need to introduce other types of fence and flush instructions (that are compatible with PArm), then prove that this implementation refines PLib. However, NVM implementations and semantics for Arm are less developed than for TSO; the PArm semantics by Cho et al. [6] has not been validated against real hardware. Moreover, the current implementation of FliT is for Px86 only[5]. We therefore see a full verification of FliT[PArm] against PLib to be future work. Such work would be able to reuse the correctness of PLib described in Sect. 4. The focus instead would be on developing appropriate abstractions, invariants, and refinement relations as described in Sect. 5 for FliT[PTSO].

8   Conclusion

In this paper, we have studied approaches for transformations to durability of linearizable objects. We have proposed an abstract persistency library and have proven that it guarantees such a transformation. As a second step, we have shown that one of the existing concrete libraries (FliT) also correctly implements such a transformation, in particular when running on the weak memory model PTSO. As future work, we intend to consider a full proof for other such proposals, specifically for Mirror [13].

Acknowledgements. We thank Michal Friedman for helping clarify the behaviour of FliT.

Data Availability. The KIV specifications and mechanised proofs supporting this work's results are publicly available at figshare with the identifier https://doi.org/10.6084/m9.figshare.24132495.v1.

---

[5] See https://github.com/cmuparlay/flit.

References

1. Assa, G., Correia, A., Ramalhete, P., Schiavoni, V., Felber, P.: Tl4x: buffered durable transactions on disk as fast as in memory. In: Dehnavi, M.M., Kulkarni, M., Krishnamoorthy, S. (eds.) PPoPP, pp. 245–259. ACM (2023). https://doi.org/10.1145/3572848.3577495
2. Bila, E., Derrick, J., Doherty, S., Dongol, B., Schellhorn, G., Wehrheim, H.: Modularising verification of durable opacity. Log. Methods Comput. Sci. **18**(3) (2022). https://doi.org/10.46298/lmcs-18(3:7)2022
3. Bila, E., Doherty, S., Dongol, B., Derrick, J., Schellhorn, G., Wehrheim, H.: Defining and verifying durable opacity: correctness for persistent software transactional memory. In: Gotsman, A., Sokolova, A. (eds.) FORTE 2020. LNCS, vol. 12136, pp. 39–58. Springer, Cham (2020). https://doi.org/10.1007/978-3-030-50086-3_3
4. Bodenmüller, S., Derrick, J., Dongol, B., Schellhorn, G., Wehrheim, H.: Verification of FLIT refinements - KIV proofs (2023). https://kiv.isse.de/projects/FLIT.html. Executable artifact. https://doi.org/10.6084/m9.figshare.24132495.v1
5. Chakrabarti, D.R., Boehm, H.-J., Bhandari, K.: Atlas: leveraging locks for non-volatile memory consistency. ACM SIGPLAN Not. **49**(10), 433–452 (2014)
6. Cho, K., Lee, S.-H., Raad, A., Kang, J.: Revamping hardware persistency models: view-based and axiomatic persistency models for Intel-x86 and Armv8. In: PLDI, pp. 16–31 (2021)
7. Choe, J.: Review and things to know: flash memory summit 2022. TechInsights, August 2022. https://www.techinsights.com/blog/review-and-things-know-flash-memory-summit-2022
8. de Roever, W.P., Engelhardt, K.: Data Refinement: Model-Oriented Proof Theories and Their Comparison. Cambridge Tracts in Theoretical Computer Science, vol. 46. Cambridge University Press (1998)
9. Derrick, J., Doherty, S., Dongol, B., Schellhorn, G., Wehrheim, H.: Verifying correctness of persistent concurrent data structures. In: ter Beek, M.H., McIver, A., Oliveira, J.N. (eds.) FM 2019. LNCS, vol. 11800, pp. 179–195. Springer, Cham (2019). https://doi.org/10.1007/978-3-030-30942-8_12
10. Derrick, J., Doherty, S., Dongol, B., Schellhorn, G., Wehrheim, H.: Verifying correctness of persistent concurrent data structures: a sound and complete method. Formal Aspects Comput. **33**(4–5), 547–573 (2021). https://link.springer.com/article/10.1007/s00165-021-00541-8
11. D'Osualdo, E., Raad, A., Vafeiadis, V.: The path to durable linearizability. Proc. ACM Program. Lang. **7**(POPL), 748–774 (2023). https://doi.org/10.1145/3571219
12. Friedman, M., Ben-David, N., Wei, Y., Blelloch, G.E., Petrank, E.: NVTraverse: in NVRAM data structures, the destination is more important than the journey. In: Donaldson, A.F., Torlak, E. (eds.) PLDI, pp. 377–392. ACM (2020). https://doi.org/10.1145/3385412.3386031
13. Friedman, M., Petrank, E., Ramalhete, P.: Mirror: making lock-free data structures persistent. In: Freund, S.N., Yahav, E. (eds.) PLDI, pp. 1218–1232. ACM (2021). https://doi.org/10.1145/3453483.3454105
14. Herlihy, M., Wing, J.M.: Linearizability: a correctness condition for concurrent objects. ACM TOPLAS **12**(3), 463–492 (1990)
15. Izraelevitz, J., Mendes, H., Scott, M.L.: Linearizability of persistent memory objects under a full-system-crash failure model. In: Gavoille, C., Ilcinkas, D. (eds.) DISC 2016. LNCS, vol. 9888, pp. 313–327. Springer, Heidelberg (2016). https://doi.org/10.1007/978-3-662-53426-7_23

16. Khyzha, A., Lahav, O.: Taming x86-TSO persistency. Proc. ACM Program. Lang. **5**(POPL), 1–29 (2021). https://doi.org/10.1145/3434328
17. Lamport, L.: How to make a multiprocessor computer that correctly executes multiprocess programs. IEEE Trans. Comput. **28**(9), 690–691 (1979)
18. Lynch, N., Vaandrager, F.: Forward and backward simulations. Inf. Comput. **121**(2), 214–233 (1995)
19. Lynch, N.A.: Distributed Algorithms. Morgan Kaufmann (1996)
20. Lynch, N.A., Tuttle, M.R.: Hierarchical correctness proofs for distributed algorithms. In: PODC, pp. 137–151. ACM, New York, NY, USA (1987)
21. Owens, S., Sarkar, S., Sewell, P.: A better x86 memory model: x86-TSO. In: Berghofer, S., Nipkow, T., Urban, C., Wenzel, M. (eds.) TPHOLs 2009. LNCS, vol. 5674, pp. 391–407. Springer, Heidelberg (2009). https://doi.org/10.1007/978-3-642-03359-9_27
22. Pulte, C., Pichon-Pharabod, J., Kang, J., Lee, S.-H., Hur, C.-K.: Promising-ARM/RISC-V: a simpler and faster operational concurrency model. In: McKinley, K.S., Fisher, K. (eds.) PLDI, pp. 1–15. ACM (2019). https://doi.org/10.1145/3314221.3314624
23. Raad, A., Wickerson, J., Neiger, G., Vafeiadis, V.: Persistency semantics of the Intel-x86 architecture. Proc. ACM Program. Lang. **4**(POPL), 11:1–11:31 (2020). https://doi.org/10.1145/3371079
24. Raad, A., Wickerson, J., Vafeiadis, V.: Weak persistency semantics from the ground up: formalising the persistency semantics of ARMv8 and transactional models. PACMPL **3**(OOPSLA), 135:1–135:27 (2019)
25. Samsung Electronics: Samsung electronics unveils far-reaching, next-generation memory solutions at flash memory summit 2022, August 2022. https://news.samsung.com/global/samsung-electronics-unveils-far-reaching-next-generation-memory-solutions-at-flash-memory-summit-2022
26. Scargall, S.: Programming Persistent Memory. Apress, Berkeley, CA (2020). https://doi.org/10.1007/978-1-4842-4932-1
27. Schellhorn, G., Bodenmüller, S., Reif, W.: Thread-local, step-local proof obligations for refinement of state-based concurrent systems. In: Glässer, U., Creissac Campos, J., Méry, D., Palanque, P. (eds.) ABZ 2023: Rigorous State-Based Methods. LNCS, vol. 14010, pp. 408–436. Springer, Cham (2023). https://doi.org/10.1007/978-3-031-33163-3_6
28. Schellhorn, G., Bodenmüller, S., Bitterlich, M., Reif, W.: Software & system verification with KIV. In: Ahrendt, W., Beckert, B., Bubel, R., Johnsen, E.B. (eds.) The Logic of Software. A Tasting Menu of Formal Methods. LNCS, vol. 13360, pp. 408–436. Springer, Cham (2022). https://doi.org/10.1007/978-3-031-08166-8_20
29. Shavit, N., Touitou, D.: Software transactional memory. Distrib. Comput. **10**(2), 99–116 (1997)
30. Wei, Y., Ben-David, N., Friedman, M., Blelloch, G.E., Petrank, E.: FliT: a library for simple and efficient persistent algorithms. In: Lee, J., Agrawal, K., Spear, M.F. (eds.) PPoPP, pp. 309–321. ACM (2022). https://doi.org/10.1145/3503221.3508436
31. Wen, H., Cai, W., Du, M., Jenkins, L., Valpey, B., Scott, M.L.: A fast, general system for buffered persistent data structures. In: Sun, X.-H., Shende, S., Kalé, L.V., Chen, Y. (eds.) ICPP, pp. 73:1–73:11. ACM (2021). https://doi.org/10.1145/3472456.3472458

A Navigation Logic for Recursive Programs with Dynamic Thread Creation

Roman Lakenbrink[(✉)], Markus Müller-Olm, Christoph Ohrem,
and Jens Gutsfeld

Department of Computer Science, University of Münster, Münster, Germany
{roman.lakenbrink,markus.mueller-olm,
christoph.ohrem,jens.gutsfeld}@uni-muenster.de

Abstract. Dynamic Pushdown Networks (DPNs) are a model for multithreaded programs with recursion and dynamic creation of threads. In
this paper, we propose a temporal logic called NTL for reasoning about
the call- and return- as well as thread creation behaviour of DPNs. Using
tree automata techniques, we investigate the model checking problem for
the novel logic and show that its complexity is not higher than that of
LTL model checking against pushdown systems despite a more expressive logic and a more powerful system model. The same holds true for
the satisfiability problem when compared to the satisfiability problem
for a related logic for reasoning about the call- and return-behaviour of
pushdown systems. Overall, this novel logic offers a promising approach
for the verification of recursive programs with dynamic thread creation.

Keywords: Concurrency · Dynamic Pushdown Networks · Navigation
Logic · Model Checking · Satisfiability · Tree Automata

1 Introduction

Model Checking is an established technique for the verification of hardware and
software systems. Conceptually, it consists of checking whether a property given
in a specification logic holds for a model of a system. While logics such as LTL
or CTL and finite Kripke models were considered early on [11,22], later more
expressive logics as well as infinite state systems have been studied. The use of
pushdown systems, for instance, allows for a more precise analysis of recursive
software systems due to the presence of a call stack of a program while still
retaining a decidable model checking problem against LTL specifications [7]. In
the context of pushdown systems, an example of a logic more expressive than
LTL is the logic CaRet [4] which extends LTL by operators for non-regular
properties of the call and return behaviour of pushdown systems. This extension
does not lead to increased complexity for the model checking problem against
pushdown systems compared to LTL.

However, there are even more powerful system models than pushdown systems for which model checking of variants of LTL is decidable. In this paper, we
consider Dynamic Pushdown Networks (DPNs) [8], a model for software systems

© The Author(s), under exclusive license to Springer Nature Switzerland AG 2024
R. Dimitrova et al. (Eds.): VMCAI 2024, LNCS 14500, pp. 48–70, 2024.
https://doi.org/10.1007/978-3-031-50521-8_3

that cannot model only recursion, but also multithreading with dynamic thread creation. So far, the model checking problem for DPNs has only been considered for single indexed LTL, a variant of LTL for multithreaded systems in which an LTL formula is assigned to each thread [26]. Here, we consider the model checking problem for a more expressive logic. More specifically, we propose a fixpoint calculus with CaRet-like operators for the verification of DPNs via model checking. The logic allows to specify non-regular properties concerning the call and return behaviour of the different execution threads of a DPN. Unlike CaRet, it can additionally specify properties concerning the thread-spawn behaviour of programs. For example, consider a scenario where a program has a method for bookkeeping information about spawned threads and it is required that new threads be only spawned from this method in order to keep the bookkeeping consistent. The property $\mathcal{G}^r(\bigcirc^c\psi \rightarrow \mathcal{F}^-pr)$ specified in our new logic expresses that in all positions of all threads (expressed through the modality $\mathcal{G}^r$), new threads fulfilling the property ψ are only spawned (expressed through $\bigcirc^c\psi$) when the procedure pr is in the call stack (expressed through $\mathcal{F}^-pr$). This formalises the requirement. Properties regarding such relationships between parent and child threads cannot be expressed in the variant of LTL from [26] or other specification logics for DPNs we are aware of. Our logic thus constitutes the first specification logic able to reason about the thread spawning behaviour of DPNs.

Contributions and Structure of the Paper. After introducing some notation and results (Sect. 2), we present a semantics for DPNs based on graphs (Sect. 3). As our first main contribution, we then introduce a novel specification logic called Navigation Temporal Logic (NTL) with the ability to reason about the call/return and thread creation behaviour of DPNs (Sect. 4). We discuss some example properties and applications of our logic in Sect. 5. Towards algorithmic verification, we then switch from a semantics based on graphs to a semantics based on trees (Sect. 6). As our second main contribution, we then investigate the model checking and satisfiability problems for the new logic (Sect. 7). In particular, we show that the model checking problem is decidable in time exponential in the size of the specification and polynomial in the size of the system model, i.e. the same as for LTL model checking against pushdown systems, and that the satisfiability problem is solvable in time exponential in the size of the specification, i.e. the same as for the satisfiability problem for $VP\text{-}\mu\text{-}TL$ [9], a temporal logic subsuming CaRet and subsumed by our logic. For both problems, we establish matching lower bounds. Section 8 concludes the paper. Due to lack of space, some technical proofs can be found in an extended version made available via arxiv [18].

Related Work. There are several specification logics related to the logic we present in this paper. The temporal logic LTL was considered for model checking finite state systems [22] as well as pushdown systems [7]. For pushdown models, CaRet was developed with different successor types that allow the inspection of the call and return behaviour of the system [4]. Also, variants of CaRet have been studied in the literature [2,3,10,16]. As mentioned, CaRet is one inspiration for the logic presented in this paper and we adopt and complement its successor types in our logic. Other inspirations are the linear time μ-calculus from [28] and the logic *VP-μ-TL* from [9]. In these logics, fixpoint operators can be used to express arbitrary ω-regular (resp. ω-visibly pushdown) properties on paths, which makes them more expressive than LTL and CaRet, respectively. From these logics, we take fixpoint operators. There is also a plethora of work on dynamic pushdown networks. The model was first introduced in [8]. Different methods for reachability analysis of DPNs have been proposed [8,20]. Additionally, different variants of the model were investigated. [24] and [19] consider variants of DPNs that communicate via locks. Another variant is the model of Dynamic Networks of concurrent pushdown systems from [6] in which threads can communicate via global variables. However, none of the above works on DPNs is concerned with model checking. The only approach to model checking DPNs we are aware of consists of checking different variants of DPNs against a variant of LTL called single indexed LTL [12,26]. Compared to NTL, this variant cannot specify properties concerning the call and return behaviour of a thread or the relationship between different threads. In Sect. 5, we show that single indexed LTL can be embedded into NTL.

2   Preliminaries

Without further ado, we introduce tools and notation used throughout the paper. This section can be skipped on first reading and be consulted for reference later.

Trees. An $\mathbb{N}_0$-*tree* T is a prefix-closed subset of $\mathbb{N}_0^*$, i.e. for all nodes $t \in \mathbb{N}_0^*$ and directions $d \in \mathbb{N}_0$, $t \cdot d \in T$ (where we use $\cdot$ for concatenation) implies $t \in T$. Moreover, we require that $t \cdot d \in T$ for some $d \in \mathbb{N}_0$ implies $t \cdot d' \in T$ for all $d' \le d$. We call an element $t \in T$ a *node* of T with special node ε, which we call the *root*. A node of the form $t \cdot d$ is called a *child* of t and t is called the *parent* of $t \cdot d$. Additionally, for sequences $w \in \mathbb{N}_0^*$, we call $t \cdot w$ a *descendant* of t and t an *ancestor* of $t \cdot w$. Nodes $t \in T$ that have no children are called *leaves*. A *path* in a tree T is a finite or infinite sequence $t_0 t_1 \ldots$ of nodes such that $t_0 = \varepsilon$ and for all $i \in \mathbb{N}_0$, t_{i+1} is a child of t_i. An $\mathbb{N}_0$-tree that is a subset of $\{0,1\}^*$ is also called a *binary tree*. In this case, we call a node of the form $t \cdot 0$ the *left child* of t and a node of the form $t \cdot 1$ the *right child* of t. Let Σ be a finite set of labels and $ar: \Sigma \to \{0,1,2\}$ be a function assigning an arity to each label. A (Σ, ar)-*labelled binary tree* is a pair (T,l) where T is a binary tree and $l: T \to \Sigma$ is a labelling function such that each node $t \in T$ has exactly $ar(l(t))$ children.

2-Way Alternating Tree Automata. For a finite set X, let $\mathcal{B}^+(X)$ be the set of positive boolean combinations over X, i.e. boolean formulae built with elements of X, conjunction and disjunction. For $Y \subseteq X$ and $\vartheta \in \mathcal{B}^+(X)$, we say that Y *satisfies* ϑ iff assigning the value *true* to the elements of Y and *false* to the elements of $X \setminus Y$ makes the formula ϑ true. Let $\mathrm{Dir} = \{0,1,\varepsilon,\uparrow\}$ be the set of moves in the tree with directions 0 for the left child, 1 for the right child, ε for standing still and $\uparrow$ for moving upwards. We define $u \cdot \varepsilon = u$ and $u \cdot d \cdot \uparrow = u$ for all $u \in \{0,1\}^*$ and $d \in \{0,1\}$. A *2-way alternating tree automaton*

(2ATA) [29] over (Σ, ar)-labelled binary trees is a tuple $\mathcal{A} = (Q, q_0, \rho, \Omega)$ where Q is a finite set of states, $q_0 \in Q$ is an initial state, $\rho: Q \times \Sigma \to \mathcal{B}^+(\text{Dir} \times Q)$ is a transition function and $\Omega: Q \to \{0, \ldots, k\}$ is a priority mapping. The size $|\mathcal{A}|$ of a 2-way alternating tree automaton is defined as the sum of the sizes of its constituents. We sometimes also refer to the size of individual constituents of an automaton. In particular, we refer to the number of states, i.e. $|Q|$, and the size of the acceptance condition, i.e. k. If the transition function of a 2-way alternating tree automaton uses only symbols from $\{0, 1\}$ instead of Dir and additionally maps all nodes t either to *true* or to disjunctions over conjunctions that consist of exactly one pair (d, q) for each $d < ar(l(t))$, it is called a *nondeterministic parity tree automaton* (NPTA).

For a (Σ, ar)-labelled binary tree $\mathcal{T} = (T, l)$, a node $t \in T$ and a state $q \in Q$, a (t, q)-*run* of $\mathcal{A}$ over $\mathcal{T}$ is a pair (T_r, r) such that T_r is an $\mathbb{N}_0$-tree and $r: T_r \to T \times Q$ assigns a pair of a node of T and a state of $\mathcal{A}$ to all nodes in T_r. Additionally, (T_r, r) has to satisfy the following conditions: (i) $r(\varepsilon) = (t, q)$ and (ii) for all nodes $y \in T_r$ with $r(y) = (x, s)$ and $\rho(s, l(x)) = \vartheta$, there is a set $Y \subseteq \text{Dir} \times Q$ satisfying ϑ and for all $(d, s') \in Y$, there is $n \in \mathbb{N}_0$ such that $y \cdot n \in T_r$ and $r(y \cdot n) = (x \cdot d, s')$. In particular, for all leaves $y \in T_r$ with $r(y) = (x, s)$, we thus require $\rho(s, l(x)) = true$. A (t, q)-run (T_r, r) is *accepting* iff on each infinite path in T_r the lowest priority occurring infinitely often is even. If $\mathcal{A}$ is a nondeterministic parity tree automaton, a minimal set Y satisfying $\rho(s, l(x))$ in the above definition moves to each child of the current node x in the tree T. For an (ε, q)-run, we can thus simply identify T with T_r and consider a map $r_{\mathcal{A}}: T \to Q$ as an (ε, q)-run over $\mathcal{A}$. The set of nodes $t \in T$ such that there is an accepting (t, q)-run of $\mathcal{A}$ over $\mathcal{T}$ is denoted by $\mathcal{L}_q^{\mathcal{T}}(\mathcal{A})$. We say that $\mathcal{A}$ *accepts* a tree $\mathcal{T}$ iff there is an accepting (ε, q_0)-run of $\mathcal{A}$ over $\mathcal{T}$. The set of trees accepted by $\mathcal{A}$ is denoted by $\mathcal{L}(\mathcal{A})$. We use the following theorems:

Proposition 1 ([29]). *For every 2ATA $\mathcal{A}$, there is an equivalent NPTA $\mathcal{A}'$. The number of states in $\mathcal{A}'$ is at most exponential in the number of states of $\mathcal{A}$ and the size of the acceptance condition of $\mathcal{A}'$ is linear in the size of the acceptance condition of $\mathcal{A}$.*

Proposition 2 ([13,17,25]). *The emptiness problem for NPTA can be solved in time polynomial in the number of states and exponential in the size of the acceptance condition.*

Proposition 3 *(i) For any two NPTA $\mathcal{A}_1$ and $\mathcal{A}_2$, there is a NPTA $\mathcal{A}$ with*
$$\mathcal{L}(\mathcal{A}) = \mathcal{L}(\mathcal{A}_1) \cap \mathcal{L}(\mathcal{A}_2).$$
(ii) If either acceptance condition is trivial, the size of $\mathcal{A}$ is in $\mathcal{O}(|\mathcal{A}_1| \cdot |\mathcal{A}_2|)$.

Proposition 3 (i) can be found e.g. in [23]. For (ii), a straightforward product construction can be used and yields an automaton of the size claimed.

Dynamic Pushdown Networks. Let AP be a finite set of atomic propositions, Γ be a finite set of stack symbols and $\perp \notin \Gamma$ be a special bottom of stack symbol. A *Dynamic Pushdown Network* (DPN) [8] is a tuple $\mathcal{M} = (S, s_0, \gamma_0, \Delta, L)$ where

S is a finite set of control locations, $s_0 \in S$ is an initial control location, $\gamma_0 \in \Gamma$ is an initial stack symbol and $L\colon S \times \Gamma \to 2^{AP}$ is a labelling function. The transition relation $\Delta = \Delta_I \mathbin{\dot{\cup}} \Delta_C \mathbin{\dot{\cup}} \Delta_R \mathbin{\dot{\cup}} \Delta_S$ is a finite set of *internal* rules (Δ_I), *calling* rules (Δ_C), *returning* rules (Δ_R) and *spawning rules* (Δ_S). Internal rules $s\gamma \to s'\gamma' \in \Delta_I \subseteq S\Gamma \times S\Gamma$, call rules $s\gamma \to s'\gamma'\gamma'' \in \Delta_C \subseteq S\Gamma \times S\Gamma^2$ or returning rules $s\gamma \to s' \in \Delta_R \subseteq S\Gamma \times S$ enable transitions of a single pushdown process in control location s with top of stack γ to the new control location s' with new top of stack γ', $\gamma'\gamma''$ and ε, respectively. A spawning rule $s\gamma \to s'\gamma' \triangleright s_n\gamma_n \in \Delta_S \subseteq S\Gamma \times S\Gamma \times S\Gamma$ is an internal rule with the additional side effect of spawning a new process in control location s_n and stack content γ_n. We formally develop a semantics for DPNs in Sect. 3.

3   Graph Semantics of Dynamic Pushdown Networks

The semantics of DPNs is often defined as an *interleaving semantics*. In such semantics, a configuration of a DPN is a collection of local configurations of the underlying pushdown systems representing the currently active threads. A step in this semantics consists of a step of one of the active threads, possibly adding a configuration of a new thread to the collection. This way, the semantics accurately reflects different interleavings of the steps of the threads issued by arbitrary schedulers, hence the name. For our intents, interleaving semantics has some drawbacks, however. First, an encoding of the intersection problem for contextfree languages is often possible in interleaving semantics, which leads to undecidability of the investigated verification problem. Second, we are mostly interested in temporal properties of individual threads, not necessarily temporal properties of interleavings. This is because the behaviour of a thread is in most cases independent of what types of steps other threads currently make in a specific interleaving. Third, we want to reason about the parent-child relationship of threads which is lost in most formalisations of interleaving semantics.

We thus instead adopt a semantics based on graphs. Intuitively, in an *execution graph*, each thread is modelled by a linear sequence of positions connected by *int-*, *call-* and *ret*-edges based on the types of transitions taken in the thread. In order to model the parent-child relationship between threads, a position where a spawn-transition is taken is connected to the first position of the spawned thread via a *spawn*-edge. This is analogous to the notion of *action trees* from [14,20]. Additionally, similar to *nested words* [5], calls and their matching returns are connected via *nesting edges*. We formalise these graphs in the next paragraph.

Execution Graphs. Let Moves $= \{int, call, ret, spawn\}$ be the set of moves for dynamic pushdown networks, V be a set of nodes, $l\colon V \to 2^{AP}$ be a labelling function, $\to^d \subseteq V^2$ be a transition relation for all $d \in$ Moves and $\curvearrowright \subseteq V^2$ be a nesting relation. For $x, y \in V$, we call y a (d)-*successor* of x and x a (d)-*predecessor* of y if $x \to^d y$ for some $d \in$ Moves. A tuple $G = (V, l, (\to^d)_{d \in \text{Moves}}, \curvearrowright)$ is called an *execution graph*, iff the following conditions hold:

1. Every node has exactly one predecessor with respect to $\bigcup\{\to^d \mid d \in \text{Moves}\}$ except for a special node v_0 without predecessor.

2. For all $x \in V$ we have $(v_0, x) \in (\bigcup\{\to^d \mid d \in \mathsf{Moves}\})^*$.
3. Every node either has (a) exactly one *int*-successor and at most one *spawn*-successor, (b) exactly one *call*-successor, (c) exactly one *ret*-successor or (d) no successors.
4. On every finite path starting in v_0 or a node with a *spawn*-predecessor and following only $\mathsf{Moves} \setminus \{spawn\}$-successors, the number of *call*-moves on that path is greater than or equal to the number of *ret*-moves on that path.
5. For all $x \in V$, let A_x be the set of nodes $y \neq x$ such that there is a path π from x to y following only $\mathsf{Moves} \setminus \{spawn\}$-successors where the number of *call*-moves on π is equal to the number of *ret*-moves on π. Then we have $x \curvearrowright y$ for a node $y \in V$ iff x has a *call*-successor and y is a node in A_x such that the witnessing path has minimal length.

The set of execution graphs is denoted by $\mathsf{ExGraphs}$.

An example of an execution graph can be found in Fig. 1. In this example, a main thread spawns two additional threads. Additionally, there are two nested procedure calls in the main thread and one procedure call in a spawned thread.

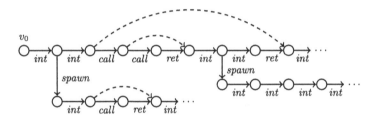

Fig. 1. Example of an execution graph. Labelled edges represent edges $\to^d$ for $d \in \mathsf{Moves}$ and dashed edges represent nesting edges $\curvearrowright$.

Graph Semantics. In most cases, we care only about graphs generated by a given DPN instead of arbitrary execution graphs. This is formalised in the *graph semantics of DPNs*. In the definition of this semantics, we make use of *configurations* of the processes of a given DPN $\mathcal{M} = (S, s_0, \gamma_0, \Delta, L)$. Formally, a configuration of a pushdown process of $\mathcal{M}$ is a pair $c = (s, u)$ where $s \in S$ is a control location and $u \in \Gamma^* \bot$ is a stack content ending in $\bot$. We define successor relations on configurations corresponding to the different types of transition rules of DPNs. For this purpose, let $c = (s, u)$, $c' = (s', u')$ and $c'' = (s'', u'')$ be configurations. We call c' an internal successor of c, denoted by $c \to_{int} c'$, if there is a transition $s\gamma \to s'\gamma' \in \Delta_I$ and $u = \gamma w$, $u' = \gamma' w$ for some stack content $w \in \Gamma^* \bot$. We call c' a call successor of c, denoted by $c \to_{call} c'$, if there is a transition $s\gamma \to s'\gamma'\gamma'' \in \Delta_C$ and $u = \gamma w$, $u' = \gamma'\gamma''w$ for some stack content $w \in \Gamma^* \bot$. We call c' a return successor of c, denoted by $c \to_{ret} c'$, if there is a transition $s\gamma \to s' \in \Delta_R$ and $u = \gamma w$, $u' = w$ for some stack content $w \in \Gamma^* \bot$. Finally, we call c' a successor of c with spawned process c'', denoted $c \to c' \triangleright c''$, if there is a transition $s\gamma \to s'\gamma' \triangleright s''\gamma'' \in \Delta_S$ and $u = \gamma w$, $u' = \gamma'w$ for some stack content $w \in \Gamma^* \bot$ as well as $u'' = \gamma'' \bot$. Using the notion of configurations and the successor relations just introduced, we now define the graph semantics. We say that an execution graph $(V, l, (\rightarrow^d)_{d \in \mathsf{Moves}}, \curvearrowright)$ *is generated* by $\mathcal{M}$ if there is an assignment $as \colon V \rightarrow S \times \Gamma^* \bot$ satisfying (i) $as(v_0) = (s_0, \gamma_0 \bot)$ and (ii) for all $x \in V$, we have $l(x) = L(s, \gamma)$ where $as(x) = (s, \gamma w)$ for some control location $s \in S$, stack symbol $\gamma \in \Gamma$ and stack content $w \in \Gamma^* \bot$ and

- if x has only one d-successor y with $d \in \{int, call, ret\}$, then $as(x) \rightarrow_d as(y)$,
- if x has an *int*-successor y and a *spawn*-successor z, then $as(x) \rightarrow as(y) \triangleright as(z)$ and
- if x has no successor, then $as(x)$ has no successor.

The set of execution graphs generated by $\mathcal{M}$ is denoted by $[\![\mathcal{M}]\!]$.

Successor Functions. On execution graphs $G = (V, l, (\rightarrow^d)_{d \in \mathsf{Moves}}, \curvearrowright)$, we define multiple successor functions $succ_g^G$, $succ_\uparrow^G$, $succ_a^G$, $succ_-^G$, $succ_p^G$ and $succ_c^G$ with signature $V \rightsquigarrow V$, i.e. partial functions from V to V. The first four of these successor functions come from logics like CaRet [4] and allow us to progress single threads and their call-return behaviour in different ways. The latter two functions are new and give means to reason about the thread spawning behaviour of DPNs. For $x \in V$, the functions are defined as follows:

- The *global successor* $succ_g^G(x)$ of x is the *int-*, *call-* or *ret*-successor of x, if it exists, and undefined otherwise.
- The *global predecessor* $succ_\uparrow^G(x)$ of x is the *int-*, *call-* or *ret*-predecessor of x, if it exists, and undefined otherwise.
- The *abstract successor* $succ_a^G(x)$ of x is the node y with $x \curvearrowright y$ or $x \rightarrow^{int} y$, if it exists, and undefined otherwise.
- The *caller* $succ_-^G(x)$ of x is the node y with a *call*-successor y' such that there is a path from y' to x following abstract successors, if it exists, and undefined otherwise.
- The *parent* $succ_p^G(x)$ of x is the node y with a *spawn*-successor z such that there is a path from z to x with only $\mathsf{Moves} \setminus \{spawn\}$-transitions, if it exists, and undefined otherwise.
- The *child* $succ_c^G(x)$ of x is the *spawn*-successor of x, if it exists, and undefined otherwise.

We illustrate these successor functions for parts of the execution graph from Fig. 1 in Fig. 2a and Fig. 2b. Abstract successors (seen in dashdotted red in Fig. 2a) follow the execution of a procedure on the same stack level and skip over executions of additional procedures via nesting edges. If a procedure is left in the next step, i.e. if the next step is a return, the abstract successor is undefined. Callers (seen in dotted blue in Fig. 2a) are defined if the stack level is at least one in a configuration and move to the latest previous call on a lower stack level. Parents (seen in dotted green in Fig. 2b) are defined in every branch of an execution graph representing a thread except for the thread starting in v_0 and move to the position in the graph where the current thread was spawned. Children (seen in dashdotted yellow in Fig. 2b) are defined only if the current thread currently executes a spawn transition and move to the initial position of the spawned thread.

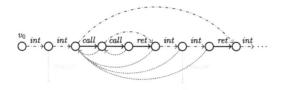 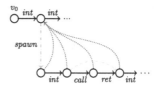

(a) Abstract successors (red, dashdotted) and callers (blue, dotted). Irrelevant edges are gray and some internal edges coinciding with abstract successors are omitted to improve readability.

(b) Parents (green, dotted) and children (yellow, dashdotted). Irrelevant edges are gray to improve readability.

Fig. 2. Successor types in parts of the execution graph from Fig. 1.

4   A Navigation Logic for Dynamic Pushdown Networks

Syntax. We now define the new logic Navigation Temporal Logic (NTL) for expressing properties of execution graphs. As mentioned in the introduction, we have three main inspirations. From the logics CaRet [4] and *VP-μ-TL* [9], we take different next operators inspecting the call and return behaviour of a thread. We complement these by additional next operators expressing parent and child relationships between different processes. From the linear time μ-calculus [28] and logics like *VP-μ-TL* [9], we take fixpoint operators for additional expressivity beyond LTL modalities. First, we define the syntax of NTL.

Definition 1 (Syntax of NTL). *The syntax of NTL formulae is defined by*

$$\varphi ::= \quad ap \mid \neg\varphi \mid \varphi_1 \vee \varphi_2 \mid X \mid \bigcirc^f \varphi \mid \mu X.\varphi$$

where $ap \in AP$ is an atomic proposition, X is a fixpoint variable and $f \in \{g, \uparrow, a, -, p, c\}$ is a successor type.

An NTL formula φ is called *closed*, if every fixpoint variable X is bound in φ, i.e. it only appears in a subformula of the form $\mu X.\psi$. A formula φ is called *well-formed*, if every fixpoint variable X occurring in φ (i) is bound by only one fixpoint formula which we then denote by $fp(X)$, (ii) appears only in the scope of an even number of negations inside $fp(X)$ and (iii) is in scope of at least one next operator inside $fp(X)$. We use $Sub(\varphi)$ for the set of subformulae of a formula φ. The size $|\varphi|$ of a formula φ is defined as the number of its distinct subformulae. We also need a notion of substitution: $\varphi[\varphi'/X]$ is the formula that is obtained from φ by replacing every occurrence of the fixpoint variable X with φ'.

Let us explain the intuition behind each construct. Atomic formulae ap express that $ap \in AP$ holds in the current node of the graph. Next operators $\bigcirc^f \varphi$ can be used to navigate and express that the corresponding successor exists in the current node and additionally satisfies φ. Negation and disjunction are interpreted as usual. Finally, we have fixpoint variables X and least fixpoint operators $\mu X.\varphi$ for more involved properties. Informally, $\mu X.\varphi$ is the least fixpoint of a function that *unrolls* the formula by replacing $\mu X.\varphi$ with $\varphi[\mu X.\varphi/X]$.

We use some common syntactic sugar such as $true \equiv ap \vee \neg ap$, $false \equiv \neg true$, $\varphi_1 \wedge \varphi_2 \equiv \neg(\neg\varphi_1 \vee \neg\varphi_2)$, $\varphi_1 \rightarrow \varphi_2 \equiv \neg\varphi_1 \vee \varphi_2$, $\varphi_1 \leftrightarrow \varphi_2 \equiv (\varphi_1 \rightarrow \varphi_2) \wedge (\varphi_2 \rightarrow \varphi_1)$ and $\nu X.\varphi \equiv \neg\mu X.\neg\varphi[\neg X/X]$. We also introduce a dual operator $\overline{\bigcirc}^f \varphi \equiv \neg \bigcirc^f \neg\varphi$ of $\bigcirc^f \varphi$ for each successor type $f \in \{g, \uparrow, a, -, p, c\}$ which is needed for a special form in the next paragraph. It is necessary to explicitly define these dual operators since $\neg \bigcirc^f \varphi$ is not equivalent to $\bigcirc^f \neg\varphi$ as the corresponding successors can be undefined for certain nodes in an execution graph. Unlike $\bigcirc^f \varphi$, $\overline{\bigcirc}^f \varphi$ is equivalent to $true$ for nodes that do not have an f-successor. We also introduce some variants of LTL modalities as derived operators. In particular we use $\varphi_1 \, \mathcal{U}^f \varphi_2 \equiv \mu X.(\varphi_2 \vee (\varphi_1 \wedge \bigcirc^f X))$, $\mathcal{F}^f \varphi \equiv true \, \mathcal{U}^f \varphi$ and $\mathcal{G}^f \varphi \equiv \neg\mathcal{F}^f \neg\varphi$ for $f \in \{g, \uparrow, a, -, p, c\}$. For $f = g$, we sometimes omit the superscript and write $\mathcal{F}\varphi$ etc. Intuitively, these modalities correspond to the usual LTL modalities evaluated on the path starting in the current position and taking f-successors. Additionally, we introduce modalities $\mathcal{F}^r \varphi \equiv \mu X.(\varphi \vee \bigcirc^g X \vee \bigcirc^c X)$ and $\mathcal{G}^r \varphi \equiv \neg\mathcal{F}^r \neg\varphi$ to express that φ holds in some position or all positions, respectively, reachable from the current position. Using these abbreviations, dual operators and the equivalence $\neg\neg\varphi \equiv \varphi$ we can transform every well-formed formula into an equivalent formula in which negation only appears in front of atomic propositions. We call this form *positive normal form* and assume formulae to be given in this form in the algorithms presented in this paper.

Semantics. We now formally define the semantics of NTL. It is defined with respect to an execution graph $G = (V, l, (\rightarrow^d)_{d \in \mathsf{Moves}}, \curvearrowright)$ and a fixpoint variable assignment $\mathcal{V}$ assigning sets of nodes of G to fixpoint variables. Intuitively, $\llbracket\varphi\rrbracket_{\mathcal{V}}^G$ is the set of nodes of G satisfying φ when each free fixpoint variable X is interpreted to hold at nodes $\mathcal{V}(X)$. In the following, for a fixpoint variable assignment $\mathcal{V}$, a fixpoint variable X and a set of nodes $M \subseteq V$, we write $\mathcal{V}[X \mapsto M]$ for the fixpoint variable assignment with $\mathcal{V}[X \mapsto M](X) = M$ and $\mathcal{V}[X \mapsto M](Y) = \mathcal{V}(Y)$ for all variables $Y \neq X$.

Definition 2 (Semantics of NTL). *Let* $G = (V, l, (\rightarrow^d)_{d \in \mathsf{Moves}}, \curvearrowright)$ *be an execution graph and* $\mathcal{V}$ *be a fixpoint variable assignment. The semantics of an NTL formula with respect to* G *and* $\mathcal{V}$ *is defined by*

$$
\begin{aligned}
\llbracket ap \rrbracket_{\mathcal{V}}^G &:= \{x \in V \mid ap \in l(x)\} \\
\llbracket \neg\varphi \rrbracket_{\mathcal{V}}^G &:= V \setminus \llbracket\varphi\rrbracket_{\mathcal{V}}^G \\
\llbracket \varphi_1 \vee \varphi_2 \rrbracket_{\mathcal{V}}^G &:= \llbracket\varphi_1\rrbracket_{\mathcal{V}}^G \cup \llbracket\varphi_2\rrbracket_{\mathcal{V}}^G \\
\llbracket X \rrbracket_{\mathcal{V}}^G &:= \mathcal{V}(X) \\
\llbracket \bigcirc^f \varphi \rrbracket_{\mathcal{V}}^G &:= \{x \in V \mid succ_f^G(x) \text{ is defined and } succ_f^G(x) \in \llbracket\varphi\rrbracket_{\mathcal{V}}^G\} \\
\llbracket \mu X.\varphi \rrbracket_{\mathcal{V}}^G &:= \bigcap\{M \subseteq V \mid \llbracket\varphi\rrbracket_{\mathcal{V}[X \mapsto M]}^G \subseteq M\}
\end{aligned}
$$

where $ap \in AP$ *is an atomic proposition,* X *is a fixpoint variable and* $f \in \{g, \uparrow, a, -, p, c\}$ *is a successor type.*

In this semantics definition, two remarks are in order. First, it is easy to see using Knaster-Tarski's fixpoint theorem [27] that for formulae φ in positive normal form, $\llbracket \mu X.\varphi \rrbracket_{\mathcal{V}}^G$ characterises the least fixpoint of the monotone function $\alpha_S \colon 2^V \rightarrow 2^V$ with $\alpha_S(M) = \llbracket \varphi \rrbracket_{\mathcal{V}[X \mapsto M]}^G$ for $S = (G, \mathcal{V}, X, \varphi)$. Second, for closed NTL formulae φ, the semantics does not depend on the fixpoint variable assignment. For such formulae, we introduce additional semantic notations. We write $\llbracket \varphi \rrbracket^G$ for $\llbracket \varphi \rrbracket_{\mathcal{V}}^G$ where $\mathcal{V}$ is an arbitrary fixpoint variable assignment and set $\llbracket \varphi \rrbracket := \{ G \in \mathsf{ExGraphs} \mid v_0 \in \llbracket \varphi \rrbracket^G \}$. For an execution graph G, we write $G \models \varphi$ for $G \in \llbracket \varphi \rrbracket$. Finally, for a DPN $\mathcal{M}$, we write $\mathcal{M} \models \varphi$, iff $G \models \varphi$ for all $G \in \llbracket \mathcal{M} \rrbracket$.

In this paper, we consider the following decision problems for NTL:

- *Model Checking:* Given a DPN $\mathcal{M}$ and a closed well-formed NTL formula φ, does $\mathcal{M} \models \varphi$ hold?
- *DPN Satisfiability:* Given a closed well-formed NTL formula φ, is there a DPN $\mathcal{M}$ such that $\mathcal{M} \models \varphi$?
- *Graph Satisfiability:* Given a closed well-formed NTL formula φ, is there an execution graph G such that $G \models \varphi$?

5   Example Properties

We motivate the introduction of our new logic with some examples.

Locking Policies. In programming languages like Java, mutual exclusion between different threads on certain procedures or code blocks is realised via *synchronized* procedures or blocks. Internally, this feature works by acquiring a lock upon entering a synchronized procedure or block that is released when leaving the synchronized part of the code [1]. Locks are thus acquired and released in a nested manner. In DPNs, this synchronization mechanism can be modelled by including symbols for locks in the stack alphabet that are pushed onto the stack when acquiring a lock and removed from the stack when releasing it. A call or return of a synchronized procedure is then modelled by taking two *call-* or *ret*-transitions of the DPN, respectively, one for pushing or popping the lock symbol and another one as usual. We also include the lock symbols as atomic propositions that are assigned to corresponding configuration heads. In this setup, the formula $\varphi_l := \mathcal{F}^- l$ expresses that the lock l is currently held using the caller modality $\mathcal{F}^-$. This form of modelling also works for reentrant locks, i.e. locks that can be acquired multiple times. When threads acquire multiple locks, problems with deadlocks can occur when different threads acquire locks in a different order. Assume, for example, that we have two locks where thread one acquires lock one first and then lock two and thread two acquires lock two first and then lock one. In this case, a deadlock can occur when the threads are scheduled such that thread one acquires lock one and thread two acquires lock two. A common policy to avoid deadlocks is to ensure that all threads acquire locks in the same order. The formula $\varphi_{ij} := \mathcal{F}^-(l_i \wedge \mathcal{G}^- \neg l_j)$ expresses that lock l_i is currently held and when it was acquired, lock l_j was not held. It can be used in the formula $\mathcal{G}^r(\varphi_{l_i} \wedge \varphi_{l_j}) \rightarrow \varphi_{ij}$ to express that l_i is always acquired before l_j, if both locks are held. The disjunction $\mathcal{G}^r(\varphi_{l_i} \wedge \varphi_{l_j}) \rightarrow \varphi_{ij} \vee \mathcal{G}^r(\varphi_{l_i} \wedge \varphi_{l_j}) \rightarrow \varphi_{ji}$ then expresses the existence of a global order for locks l_i and l_j and the existence of a global order for all locks can be expressed by a boolean combination of a quadratic number of variants of this formula. Another problem with locking arises when certain threads wait for a lock that is held by another thread for an infinite amount of time, e.g. if a synchronized method is never left. A policy addressing this problem is to ensure that all locks that are acquired are released at some point in the future. We can express this using the formula $\mathcal{G}^r \bigwedge_{l\in Locks} \varphi_l \to \mathcal{F}\neg\varphi_l$. Under these two policies, a necessary and sufficient condition for mutual exclusion of two program points labelled s_1 and s_2 is that a common lock is held at the two program points. This can also be expressed in a formula from our logic: $\bigvee_{l\in Locks} \mathcal{G}^r((s_1 \to \varphi_l) \wedge (s_2 \to \varphi_l))$.

Behaviour of Main and Worker Threads. We elaborate on a motivating example for single indexed LTL from [26] expressible in NTL. In this example, a main thread of a server processes requests from clients by starting a worker thread responding to the specific request. Then, the main thread should repeatedly accept new requests, expressed by the formula $\mathcal{G}^r(main \to \mathcal{GF}accept)$. Also, each worker thread should respond with a correct acknowledgement to each type of request, i.e. it should respond exactly with ack to req and exactly with ack' to req'. This is expressed by the formula $\mathcal{G}^r(worker \to (req \to (\mathcal{F}ack \wedge \mathcal{G}\neg ack') \wedge req' \to (\mathcal{F}ack' \wedge \mathcal{G}\neg ack)))$. Such requirements were already expressible in single indexed LTL. However, using the different types of successor operators in NTL, we can further expand on this scenario and express properties not expressible in single indexed LTL. For example, it is a reasonable requirement that worker threads are only spawned by the main thread and only if the main thread has accepted a request. This requirement can be expressed in the formula $\mathcal{G}^r(worker \to \bigcirc^p(main \wedge accept))$. Another desirable property in this scenario is a variant of the property from the introduction. In particular, we may want worker threads to only be spawned from a procedure pr which performs bookkeeping about the currently active worker threads. This is expressed by the formula $\mathcal{G}^r(\bigcirc^c worker \to \mathcal{F}^- pr)$.

Single Indexed LTL Model Checking. It is no surprise that the previous motivating example for single indexed LTL is expressible in NTL. Indeed, we show that the full approach of single indexed LTL DPN model checking from [26] can also be handled using our logic. We first sketch their setup. In [26], a DPN $\mathcal{M} = \{\mathcal{P}_1,\ldots,\mathcal{P}_n\}$ is defined as a set of pushdown systems $\mathcal{P}_i$ with the ability to spawn threads executing one of the pushdown systems of $\mathcal{M}$. A single indexed LTL formula is a conjunction $\varphi = \bigwedge_{i=1}^n \varphi_i$ of LTL formulae φ_i that are each assigned to a specific pushdown system $\mathcal{P}_i$. Then, $\mathcal{M} \models \varphi$ holds iff $\mathcal{M}$ has a *global run* such that for all i, every *local run* of $\mathcal{P}_i$ in the global run satisfies φ_i. In our setup, their global runs correspond to execution graphs and their local runs correspond to the paths in the execution graph starting in positions where new threads are spawned and following the global successors. Since in single indexed LTL model checking, the existence of a global run is checked, whereas in NTL model checking, it is checked that all execution graphs satisfy a property, we can check that $\mathcal{M} \not\models \varphi$ for a single indexed LTL formula $\varphi = \bigwedge_{i=1}^n \varphi_i$ using NTL model checking. This is done as follows. We model the partition of a DPN $\mathcal{M}$

from their setup into its pushdown systems $\mathcal{P}_i$ by labelling every control location of $\mathcal{P}_i$ with a fresh atomic proposition p_i in its translation $\bar{\mathcal{M}}$ in our setup. LTL formulae φ_i can trivially be translated to NTL by encoding until operators using least fixpoints. Then, the NTL formula $\bar{\varphi} = (p_1 \wedge \neg\varphi_1) \vee \mathcal{F}^r(\bigvee_{i=1}^n \bigcirc^c(p_i \wedge \neg\varphi_i))$ expresses that there is a local run of $\mathcal{P}_i$ for some i that does not satisfy φ_i. In this formula, the disjunct $(p_1 \wedge \neg\varphi_1)$ identifies a violation by the root process $\mathcal{P}_1$ and the disjunct $\mathcal{F}^r(\bigvee_{i=1}^n \bigcirc^c(p_i \wedge \neg\varphi_i))$ identifies violations by spawned processes. Accordingly, $\mathcal{M} \models \varphi$ (in the single indexed LTL setup) iff $\bar{\mathcal{M}} \not\models \bar{\varphi}$ (in our setup).

6 From Graph Semantics to Tree Semantics

In order to enable algorithmic verification with tree automata, we introduce an additional structure called *execution tree*. In a nutshell, these trees are obtained from execution graphs by keeping the same set of nodes and adjusting the edge relation a little. In particular, we discard *ret*-edges. In order to properly interpret left and right children in this adjusted structure, we add labels (l, d, p) where l represents the label of the current node, d represents the transition types from this node to its children and p represents the transition type from the parent to this node. This yields us a structure simpler than execution graphs that still contains the same information and can be analysed using tree automata.

Execution Trees. Let $G = (V, l, (\rightarrow^d)_{d \in \mathsf{Moves}}, \curvearrowright)$ be an execution graph. We inductively define a map $\delta_G \colon V \rightarrow \{0, 1\}^*$ assigning a tree node to each graph node $x \in V$ as follows.

- If $x = v_0$, we set $\delta_G(x) := \varepsilon$.
- If x has a *call-* or *int*-predecessor y, we set $\delta_G(x) := \delta_G(y) \cdot 0$. In this case, we also call $\delta_G(x)$ a *call-* or *int-child* of $\delta_G(y)$, respectively.
- If there is $y \in V$ such that y is a *spawn*-predecessor of x or $y \curvearrowright x$, we set $\delta_G(x) := \delta_G(y) \cdot 1$. If y is a *spawn*-predecessor of x, we also call $\delta_G(x)$ a *spawn-child* of $\delta_G(y)$ and if $y \curvearrowright x$, we also call $\delta_G(x)$ a *ret-child* of $\delta_G(y)$.

Additionally, for a subset $M \subseteq \mathsf{Moves}$ and nodes $x, y \in V$, we call $\delta_G(y)$ an M-descendant of $\delta_G(x)$ and $\delta_G(x)$ an M-ancestor of $\delta_G(y)$, if there is a path from $\delta_G(x)$ to $\delta_G(y)$ in the tree following only M-children.

Let $TL = 2^{AP} \times \{int, call, callRet, spawn, ret, end\} \times (\mathsf{Moves} \cup \{\bot\})$ be the set of labels for tree nodes and the arity function $ar \colon TL \rightarrow \{0, 1, 2\}$ be defined by $ar(l, ret, p) = ar(l, end, p) = 0$, $ar(l, int, p) = ar(l, call, p) = 1$ and $ar(l, callRet, p) = ar(l, spawn, p) = 2$. The *tree representation* $\mathcal{T}(G)$ of G is the (TL, ar)-labelled binary tree $(\mathrm{im}(\delta_G), r)$ where $\mathrm{im}(\delta_G) = \{\delta_G(x) \mid x \in V\}$ denotes the image of δ_G and for all $x \in V$ we have $r(\delta_G(x)) = (l(x), d(x), p(x))$ where (i) either $p(x) \neq \bot$ and x has a $p(x)$-predecessor or $p(x) = \bot$ and $x = v_0$ and (ii) one of the following conditions hold:

- x has only one $d(x)$-successor and $d(x) \in \{int, ret\}$.
- x has only one *int-* and one *spawn*-successor and $d(x) = spawn$.
- x has only one *call*-successor, there is no $y \in V$ with $x \curvearrowright y$, and $d(x) = call$.

- x has only one *call*-successor, there is $y \in V$ with $x \curvearrowright y$, and $d(x) = callRet$.
- x has no successors and $d(x) = end$.

A tree representation of an execution graph is also called an *execution tree*. An example of an execution tree can be found in Fig. 3.

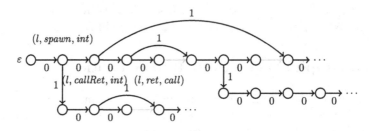

Fig. 3. Execution tree for the execution graph in Fig. 1. An edge from node t to node t' labelled d means that $t' = t \cdot d$. Labels are depicted for nodes 0, 00 and 0000. Gray edges exist in the execution graph but not in the execution tree.

Adapted Successor Functions. We adapt the successor functions previously defined on execution graphs to execution trees in order to allow us to check the satisfaction of formulae directly on execution trees. Specifically, we define multiple successor functions $succ_g^T$, $succ_\uparrow^T$, $succ_a^T$, $succ_-^T$, $succ_p^T$ and $succ_c^T$ with signature $T \rightsquigarrow T$ for execution trees $\mathcal{T} = (T, r)$. For $t \in T$ with $r(t) = (l, d, p)$, the successor functions are given as follows:

- The *abstract successor* $succ_a^T(t)$ of t is defined as the left child of t, if $d \in \{int, spawn\}$, the right child of t, if $d = callRet$, and undefined else.
- The *caller predecessor* $succ_-^T(t)$ of t is defined as the parent node of t, if $p = call$, the caller predecessor of its parent node, if $p \in \{int, ret\}$ and this is defined, and undefined else.
- The *global successor* $succ_g^T(t)$ of t is defined as the left child of t, if $d \in \{int, call, callRet, spawn\}$, $succ_a^T(succ_-^T(t))$, if $d = ret$, and undefined else.
- The *global predecessor* $succ_\uparrow^T(t)$ of t is defined as the parent node of t, if $p \in \{int, call\}$, the $\{int, ret\}$-descendant leaf of the left child of its parent node, if $p = ret$, and undefined else.
- The *parent predecessor* $succ_p^T(t)$ of t is defined as the parent node of t, if $p = spawn$, the parent predecessor of its parent node, if $p \in \{int, call, ret\}$ and this is defined, and undefined else.
- The *child successor* $succ_c^T(t)$ of t is defined as the right child of t, if $d = spawn$, and undefined else.

We show in the following lemma that these adapted successor functions behave exactly like their counterparts on execution graphs.

Lemma 1. *Let $G = (V, l, (\to^d)_{d \in \text{Moves}}, \curvearrowright)$ be an execution graph with $T(G) = T$. For all $f \in \{g, \uparrow, a, -, p, c\}$ we have $\delta_G \circ succ_f^G = succ_f^T \circ \delta_G$, i.e. for all nodes $x \in V$, $\delta_G(succ_f^G(x))$ is defined iff $succ_f^T(\delta_G(x))$ is defined and in this case $\delta_G(succ_f^G(x)) = succ_f^T(\delta_G(x))$.*

A detailed proof of this lemma can be found in the extended version [18].

7   Model Checking and Satisfiability

We now use execution trees to decide the model checking and satisfiability problems for NTL. For this, we construct three tree automata: one automaton for checking whether a tree is an execution tree, a second automaton for checking whether an execution graph (given by its tree representation) satisfies a given formula, and another automaton for checking whether a tree represents an execution graph generated by a given DPN.

An Automaton for Execution Trees. We first construct a nondeterministic parity tree automaton that checks whether a (TL, ar)-labelled binary tree is an execution tree. At each node labelled by (l, d, p), the automaton needs to ensure that the node is a p-child, if $p \neq \bot$, and that it is the root, if $p = \bot$. Moreover, if $d = callRet$, it has to check that its *call*-child does have an $\{int, ret\}$-descendant leaf. Finally, it has to ensure that for leaves t labelled by (l, d, p) we have $d = ret$ iff t is the $\{int, ret\}$-descendant leaf of a *call*-child of a node labelled by $(l', callRet, p')$ for some $l' \in 2^{AP}$ and $p' \in \text{Moves} \cup \{\bot\}$. Thus, we can define the automaton as $\mathcal{A}_{ET} = (Q, q_0, \rho, \Omega)$ with state set $Q = (\text{Moves} \cup \{\bot\}) \times \{0, 1\}$ and initial state $q_0 = (\bot, 0)$. Intuitively, in a state (p, c), p denotes the parent edge type and the bit c indicates whether the current node is an $\{int, ret\}$-descendant of a *call*-child of a node labelled by $(l', callRet, p')$ for some $l' \in 2^{AP}$ and $p' \in \text{Moves} \cup \{\bot\}$. The transition function ρ is defined by

$$\rho((p,c),(l,d,p')) := \begin{cases} (0,(int,c)) & \text{if } d = int \\ (0,(call,0)) & \text{if } d = call \text{ and } c = 0 \\ (0,(call,1)) \wedge (1,(ret,c)) & \text{if } d = callRet \\ (0,(int,c)) \wedge (1,(spawn,0)) & \text{if } d = spawn \\ true & \text{if } (d,c) \in \{(ret,1),(end,0)\} \end{cases}$$

for $p = p'$ and $\rho((p,c),(l,d,p')) := false$ in all other cases. The priority assignment is given by $\Omega(p,c) = c$ for all $(p,c) \in Q$.

A proof of the following theorem can be found in the extended version [18].

Theorem 1. *One can construct a NPTA $\mathcal{A}_{ET}$ over (TL, ar)-labelled binary trees with a constant size such that $\mathcal{L}(\mathcal{A}_{ET}) = \{T(G) \mid G \text{ is an execution graph}\}$.*

An Automaton for Formulae. For the next automaton, we define a 2-way alternating tree automaton evaluating φ on execution trees, intersect it with the automaton recognising execution trees and then transform this automaton into a nondeterministic parity tree automaton. In the following, let φ be a closed, well-formed NTL formula in positive normal form. We define the automaton for φ as $\tilde{A}_\varphi = (Q, q_0, \rho, \Omega)$ where Q, q_0, ρ and Ω are described in more detail in the following paragraphs.

The states of the automaton are given by

$$Q = Sub(\varphi) \cup Q_1 \cup Q_2 \text{ where}$$
$$Q_1 = \{\bigcirc^- \bigcirc^a \psi, \bigcirc^a \psi \mid \bigcirc^g \psi \in Sub(\varphi) \text{ or } \bigcirc^{\bar{g}} \psi \in Sub(\varphi)\} \text{ and}$$
$$Q_2 = \{call, leaf\} \times \{\psi \mid \bigcirc^\uparrow \psi \in Sub(\varphi) \text{ or } \bigcirc^{\bar{\uparrow}} \psi \in Sub(\varphi)\}$$

with initial state $q_0 = \varphi$. Since we use another automaton to check that the given tree indeed represents an execution graph, we care only about execution trees as inputs in this construction. Intuitively, being in a state $\psi \in Sub(\varphi) \cup Q_1$ at the position $\delta_G(x)$ in the input execution tree $T(G)$, the automaton checks whether the node x satisfies ψ, i.e. whether $x \in [\![\psi]\!]^G$. The states in Q_2 are used to handle the global predecessor next modality and its dual version. We use states of the form $(call, \psi)$ to denote that we should move to the $call$-child of the current node and switch to state $(leaf, \psi)$; states of the form $(leaf, \psi)$ denote that we should check ψ for the $\{int, ret\}$-descendant leaf of the current node.

The transition function ρ is defined as described next. Recall that $\tilde{A}_\varphi$ operates on execution trees which are labelled by triples (l, d, p) where $l \in 2^{AP}$ are the atomic propositions, $d \in \{int, call, callRet, spawn, ret, end\}$ specifies the successor types of the current node and $p \in \text{Moves} \cup \{\perp\}$ denotes the type of its predecessor. If the current state is an atomic proposition or a negation of an atomic proposition, we can check directly whether the tree node is labelled by this proposition and thus determine whether the formula holds:

$$\rho(ap, (l, d, p)) := \begin{cases} true \text{ if } ap \in l \\ false \text{ if } ap \notin l \end{cases} \qquad \rho(\neg ap, (l, d, p)) := \begin{cases} false \text{ if } ap \in l \\ true \text{ if } ap \notin l. \end{cases}$$

For a disjunction or conjunction of two formulae, we can use the power of alternation and set

$$\rho(\psi_1 \vee \psi_2, \sigma) := (\varepsilon, \psi_1) \vee (\varepsilon, \psi_2) \text{ and } \rho(\psi_1 \wedge \psi_2, \sigma) := (\varepsilon, \psi_1) \wedge (\varepsilon, \psi_2).$$

For a formula of the form $\bigcirc^f \psi$, we move to the corresponding successor of the given node and then switch to state ψ. In most cases, the according transitions can be defined straightforwardly using the characterisation from the successor functions on execution trees:

$$\rho(\bigcirc^g \psi, (l, d, p)) := \begin{cases} (0, \psi) \text{ if } d \in \{int, call, callRet, spawn\} \\ (\varepsilon, \bigcirc^- \bigcirc^a \psi) \text{ if } d = ret \\ false \text{ if } d = end \end{cases}$$

$\rho(\bigcirc^a \psi, (l, d, p))$
$$:= \begin{cases} (0, \psi) \text{ if } d \in \{int, spawn\} \\ (1, \psi) \text{ if } d = callRet \\ false \text{ if } d \in \{call, ret, end\} \end{cases}$$

$\rho(\bigcirc^- \psi, (l, d, p))$
$$:= \begin{cases} (\uparrow, \psi) \text{ if } p = call \\ (\uparrow, \bigcirc^- \psi) \text{ if } p \in \{int, ret\} \\ false \text{ if } p \in \{spawn, \bot\} \end{cases}$$

$\rho(\bigcirc^P \psi, (l, d, p))$
$$:= \begin{cases} (\uparrow, \psi) \text{ if } p = spawn \\ (\uparrow, \bigcirc^P \psi) \text{ if } p \in \{int, call, ret\} \\ false \text{ if } p = \bot \end{cases}$$

$\rho(\bigcirc^c \psi, (l, d, p))$
$$:= \begin{cases} (1, \psi) \text{ if } d = spawn \\ false \text{ if } d \neq spawn \end{cases}$$

In the above definition, we move to *false* when we see that the desired successor does not exist and the formula is not satisfied. The transition function for dual next operators is defined analogously but moves to *true* instead of *false* in case the successor does not exist.

For the global predecessor, we additionally use states of the form $(call, \psi)$ and $(leaf, \psi)$ for moving to the $\{int, ret\}$-descendant leaf of the *call*-child of the parent of a node in cases where the global predecessor is defined this way:

$\rho(\bigcirc^\uparrow \psi, (l, d, p))$
$$:= \begin{cases} (\uparrow, \psi) \text{ if } p \in \{int, call\} \\ (\uparrow, (call, \psi)) \text{ if } p = ret \\ false \text{ if } p \in \{spawn, \bot\}, \end{cases}$$

$\rho((leaf, \psi), (l, d, p))$
$$:= \begin{cases} (0, (leaf, \psi)) \text{ if } d \in \{int, spawn\} \\ (1, (leaf, \psi)) \text{ if } d = callRet \\ (\varepsilon, \psi) \text{ if } d \in \{ret, call, end\} \end{cases}$$

and $\rho((call, \psi), \sigma) := (0, (leaf, \psi))$.

Note that if we are in a state $(leaf, \psi)$ at position $\delta_G(x)$ in the tree, $d(x) \in \{call, end\}$ cannot hold if the tree represents an execution graph since in this case x lies on the path between nodes y and z following $\mathsf{Moves} \setminus \{spawn\}$-successors with $y \curvearrowright z$ and $x \neq z$.

Finally, fixpoint formulae lead to loops:

$$\rho(\lambda X.\psi, \sigma) := (\varepsilon, \psi) \text{ for } \lambda \in \{\mu, \nu\} \text{ and } \rho(X, \sigma) := (\varepsilon, fp(X)).$$

The acceptance condition specifies whether a fixpoint formula may be visited at most a finite number of times or an infinite number of visits is allowed. In this definition, higher priorities are assigned to fixpoint formulae binding variables which *depend* on other fixpoint variables. Formally, we say that a fixpoint variable X' depends on the variable X in φ, written $X \prec_\varphi X'$, if X is a free variable in $fp(X')$. We consider all maximal chains $X_1 \prec_\varphi \ldots \prec_\varphi X_n$ of fixpoint variables appearing in φ. If $fp(X_1)$ is a formula of the form $\mu X.\psi$, we set $\Omega(fp(X_1)) = 1$, otherwise we set $\Omega(fp(X_1)) = 0$. Then, we move through the chains and assign this priority to $fp(X_i)$ as long as the fixpoint type does not change. In that case, we increase the currently assigned priority by one and keep going. Then, we set $\Omega(q)$ to the highest priority assigned so far for all other states q.

We establish the following theorem.

Theorem 2. *Let φ be a closed, well-formed NTL formula, G be an execution graph and $\tilde{\mathcal{A}}_\varphi$ be the 2ATA defined above. Then $\tilde{\mathcal{A}}_\varphi$ accepts $\mathcal{T}(G)$ iff $G \in [\![\varphi]\!]$.*

Proof Sketch. The proof is by induction on the structure of φ. Therefore, we also have to deal with non-closed subformulae and consider valuations to decide whether a subformula is satisfied. In order to do this in a formal way, we consider automata with special states $X_1, \ldots, X_n$, called *holes* [21], that can be filled with sets of nodes $L_1, \ldots, L_n$ of a given tree. Intuitively, such an automaton can operate on a tree as before, but when a hole X_i is encountered during a run and we are at the tree node t, then we do not continue on the current path and say that it is accepting iff $t \in L_i$. By $\mathcal{L}_q^{\mathcal{T}}(\mathcal{A}[X_1 : L_1, \ldots, X_n : L_n])$ we denote the set of nodes $t \in T$ such that there is an accepting (t, q)-run over $\mathcal{A}$ where the states $X_1, \ldots, X_n$ are holes filled by $L_1, \ldots, L_n$.

For the inductive proof, we assume that the free variables of the current formula $\psi \in Sub(\varphi)$ are holes in the automaton and show that the language of this automaton corresponds to the semantics of ψ. Intuitively, we fill the holes in the automaton, i.e. the free variables of ψ, with the same sets of nodes as specified by a given valuation that we consider for the semantics of ψ. More formally, the holes are filled by sets of tree nodes that correspond to given sets of graph nodes in the valuation.

We consider the case for subformulae of the form $\psi \equiv \mu X.\psi'$ with free variables $X_1, \ldots, X_n$. Let $\mathcal{V}$ be a fixpoint variable assignment, $\mathcal{T}(G) = \mathcal{T} = (T, r)$ and R be a (t, ψ)-run over $\tilde{\mathcal{A}}_\varphi$ for a $t \in T$, where the states $X_1, \ldots, X_n$ are holes filled by $\delta_G(L_1), \ldots, \delta_G(L_n)$ with $L_i = \mathcal{V}(X_i)$. We observe that R can only visit states φ' of the form $\mu X.\psi''$ or $\nu X.\psi''$ if φ' is a subformula of ψ. Therefore, $\Omega(\psi)$ is the lowest priority occurring in the run so that the state ψ can only be visited finitely often if the run is accepting. This means we can characterize $\mathcal{L}_\psi^{\mathcal{T}}(\tilde{\mathcal{A}}_\varphi[X_1 : \delta_G(L_1), \ldots, X_n : \delta_G(L_n)])$ as the least fixpoint of the function $f : 2^T \to 2^T$ with $f(\delta_G(L)) := \mathcal{L}_{\psi'}^{\mathcal{T}}(\tilde{\mathcal{A}}_\varphi[X_1 : \delta_G(L_1), \ldots, X_n : \delta_G(L_n), X : \delta_G(L)])$. Thus, we can use the induction hypothesis and the fixpoint characterization of the semantics of ψ obtained by Knaster-Tarski's fixpoint theorem to get the desired result in this inductive step.

Since φ is closed, the induction establishes in particular that $\tilde{\mathcal{A}}_\varphi$ accepts $\mathcal{T}(G)$ iff $G \in [\![\varphi]\!]$. Details of this proof can be found in [18].   □

As mentioned, we do not use this automaton directly but instead intersect it with $\mathcal{A}_{ET}$ from Theorem 1 and then transform it into a nondeterministic parity tree automaton using Proposition 1. We obtain:

Corollary 1. *Let φ be a closed, well-formed NTL formula. Then we can construct an NPTA $\mathcal{A}_\varphi$ over (TL, ar)-labelled binary trees with a number of states exponential and an acceptance condition linear in $|\varphi|$ such that $\mathcal{L}(\mathcal{A}_\varphi) = \{\mathcal{T}(G) \mid G$ is an execution graph with $G \in [\![\varphi]\!]\}$.*

An Automaton for DPNs. We proceed with an automaton for a DPN $\mathcal{M} = (S, s_0, \gamma_0, \Delta, L)$. We define $\mathcal{A}_\mathcal{M}$ as an NPTA that checks whether an execution tree represents an execution graph generated by $\mathcal{M}$. We set $\mathcal{A}_{\mathcal{M}} := (Q, q_0, \rho, \Omega)$ where Q, q_0, ρ and Ω are described in more detail next.

The state set is given by $Q = S \times \Gamma \times ((S \times \Gamma) \cup \{\bot\})$ with initial state $q_0 = (s_0, \gamma_0, \bot)$. Being in a state $(s, \gamma, c) \in Q$ at the position $\delta_G(x)$ in the tree labelled by (l, d, p) means that there is a suitable assignment as assigning configurations to the graph nodes whose corresponding tree nodes have been visited so far where $as(x) = (s, \gamma w)$ for some stack content $w \in \Gamma^* \bot$. If $d = callRet$, we also have to know the configuration assigned to the global predecessor of the ret-child of the current node to check that we can extend as suitably for the children of the current node. We thus guess this configuration in this case and use $c \in S \times \Gamma$ to indicate that we must assign c to the $\{int, ret\}$-descendant leaf of the call successor of the current node in order to fulfill the requirements for the assignment as. Note that the $\{int, ret\}$-descendant leaf exists in this case, if the input tree is an execution tree. The transition function ρ then checks that (i) $l = L(as(x))$, (ii) if $c \in S \times \Gamma$, then the configuration c is assigned to the $\{int, ret\}$-descendant leaf of $\delta_G(x)$ and (iii) the assignment as can be properly extended to the children of $\delta_G(x)$. We set

$$\rho((s, \gamma, c), (l, int, p)) := \bigvee \{(0, (s', \gamma', c)) \mid s\gamma \to s'\gamma' \in \Delta_I\},$$

$$\rho((s, \gamma, \bot), (l, call, p)) := \bigvee \{(0, (s', \gamma', \bot)) \mid \exists \gamma'' \in \Gamma \text{ s.t. } s\gamma \to s'\gamma'\gamma'' \in \Delta_C\},$$

$$\rho((s, \gamma, c), (l, callRet, p)) := \bigvee \{(0, (s', \gamma', (s_r, \gamma_r))) \wedge (1, (s'', \gamma'', c)) \mid$$
$$s\gamma \to s'\gamma'\gamma'' \in \Delta_C \text{ and } s_r\gamma_r \to s'' \in \Delta_R\},$$

$$\rho((s, \gamma, c), (l, spawn, p)) := \bigvee \{(0, (s', \gamma', c)) \wedge (1, (s_n, \gamma_n, \bot)) \mid$$
$$s\gamma \to s'\gamma' \triangleright s_n\gamma_n \in \Delta_S\},$$

$$\rho((s, \gamma, (s, \gamma)), (l, ret, p)) := true \text{ and}$$

$$\rho((s, \gamma, \bot), (l, end, p)) := \begin{cases} true \text{ if there is no transition for } s\gamma \text{ in } \Delta \\ false \text{ else} \end{cases}$$

for $l = L(s, \gamma)$ and $\rho((s, \gamma, c), (l, d, p)) := false$ in all other cases. Since we are only concerned with execution trees as inputs, all conditions necessary to determine if the input tree is generated by $\mathcal{M}$ are already checked by the transition function of $\mathcal{A}_{\mathcal{M}}$. We thus set $\Omega(q) := 0$ for all $q \in Q$. We establish the following theorem. A detailed proof can be found in the extended version [18].

Theorem 3. *Let $\mathcal{M}$ be a DPN. We can construct an NPTA $\mathcal{A}_{\mathcal{M}}$ over (TL, ar)-labelled binary trees with a number of states quadratic in $|\mathcal{M}|$ and a trivial acceptance condition such that for all execution graphs G, $T(G) \in \mathcal{L}(\mathcal{A}_{\mathcal{M}})$ iff $G \in [\![\mathcal{M}]\!]$.*

Complexity of Model Checking and Satisfiability. These automata constructions can be used to obtain a decision procedure for the model checking and satisfiability problems. For the former, we obtain the following theorem:

Theorem 4. *The model checking problem for NTL is* EXPTIME*-complete. For fixed formulae, the problem is in* PTIME*.*

Proof. For the upper bound, we construct an automaton for the negation of the formula using Corollary 1 and intersect it with an automaton for the DPN from Theorem 3. Since the acceptance condition of the latter is trivial, the size of the resulting automaton is quadratic in the size of the DPN and exponential in the size of the formula by Proposition 3 (ii). It is tested for emptiness using Proposition 2 in time exponential in $|\varphi|$ and polynomial in $|\mathcal{M}|$ to answer the model checking problem.

The lower bound follows by a reduction from the *LTL* pushdown model checking problem which was shown to be EXPTIME-hard in [7]. The reduction is trivial since *LTL* is a sublogic of NTL for single threads and pushdown systems can be trivially embedded into DPNs with a single thread.   □

For satisfiability, we can show that the two problems defined in Sect. 4 are equivalent and thus only need to solve one of the problems by a direct procedure.

Theorem 5. *The graph and DPN satisfiability problems are equivalent.*

Proof. For the first direction, assume that a formula φ is satisfiable by a DPN $\mathcal{M}$. Then $G \models \varphi$ for all $G \in [\![\mathcal{M}]\!]$. Since $[\![\mathcal{M}]\!] \neq \emptyset$ (this indeed holds for all DPNs), we can thus choose an arbitrary graph $G \in [\![\mathcal{M}]\!]$ to show that φ is satisfiable by a graph.

For the other direction, assume that a formula φ is satisfiable by a graph G. By Corollary 1, we know that $\mathcal{T}(G) \in \mathcal{L}(\mathcal{A}_\varphi)$. Since $\mathcal{L}(\mathcal{A}_\varphi)$ is a nonempty ω-regular tree language, we know that $\mathcal{T} \in \mathcal{L}(\mathcal{A}_\varphi)$ for a regular tree $\mathcal{T} = (T, r)$, i.e. a tree with only finitely many non-isomorphic subtrees (see e.g. Cor 8.20. in [15]). Let $x_1, \ldots, x_n$ be the finitely many classes of nodes associated with the roots of the distinct subtrees of T such that x_1 is the class of ε and let (l_i, d_i, p_i) be the label of the nodes from class x_i. We construct a DPN $\mathcal{M} = (\{s\}, s, x_1, \Delta, L)$ with stack alphabet $\Gamma = \{x_1, \ldots, x_n\}$. The labeling L is defined such that $L(s, x_i) = l_i$. Transition rules are defined from the parent-child relationships between the different classes of nodes: (i) if $d_i = int$, then nodes of class x_i have exactly one child of class x_j and we include $sx_i \to sx_j \in \Delta$, (ii) if $d_i = spawn$, then nodes of class x_i have exactly one left child of class x_j and one right child of class x_k and we include $sx_i \to sx_j \triangleright sx_k \in \Delta$, (iii) if $d_i = callRet$, then nodes of class x_i have exactly one left child of class x_j and one right child of class x_k and we include $sx_i \to sx_jx_k \in \Delta$, (iv) if $d_i = call$, then nodes of class x_i have exactly one child of class x_j and we include $sx_i \to sx_jx_i \in \Delta$, (v) if $d_i = ret$, then nodes of class x_i have no children and we include $sx_i \to s \in \Delta$ and (vi) if $d_i = end$, then nodes of class x_i have no children and we do not include a transition.

It is easy to see that $[\![\mathcal{M}]\!]$ is a singleton set since $\mathcal{M}$ is deterministic. We choose H such that $[\![\mathcal{M}]\!] = \{H\}$ and show that $\mathcal{T} = \mathcal{T}(H)$ and thus $\mathcal{M} \models \varphi$. For this, let $\mathcal{T}(H) = (T_H, r_H)$.

We show by induction on the length of x that for all $x \in \{0, 1\}^*$, $x \in T$ iff $x \in T_H$ and in that case (a) $r(x) = r_H(x)$ and (b) if x belongs to class x_i, then the configuration in $\delta_G^{-1}(x)$ is $(s, x_i w)$ for some stack content w. In the base case, we know that $\varepsilon \in T$ and $\varepsilon \in T_H$. We know that $r(\varepsilon) = (l_1, d_1, p_1)$ since T is rooted in x_1 and $p_1 = \bot$ since T is an execution tree. Let $r_H(\varepsilon) = (l, d, p)$. Since $(s, x_1 \bot)$ is the starting configuration of $\mathcal{M}$, we know that it is also the configuration in $\delta_G^{-1}(\varepsilon)$ and that $l = l_1$. Additionally, we can show that $d = d_1$ by a case distinction on d_1. We only sketch the case $d_1 = int$, the other cases are similar. In this case, the only enabled transition in $(s, x_1 \bot)$ is $sx_1 \rightarrow sx_j$, an internal transition. Thus, $\delta_G^{-1}(\varepsilon)$ has exactly one int-successor in H which means that $d = int$. Finally, since $\mathcal{T}(H)$ is an execution tree, we have $p = \bot$.

In the inductive step, we consider $x \cdot d$ for $d \in \{0, 1\}$. From the induction hypothesis, we know that the claim holds for x. If $x \notin T$ and $x \notin T_H$, then also $x \cdot d \notin T$ and $x \cdot d \notin T_H$ since trees are prefix-closed. In the other case, let x_i be the class of x. We have $x \in T$ and $x \in T_H$ with $r(x) = r_H(x) = (l_i, d_i, p_i)$ and the configuration in $\delta_G^{-1}(x)$ is $(s, x_i w)$ for some stack content w. We distinguish cases based on d_i. We consider the most involved case where $d_i = callRet$. Since $\mathcal{T}$ is an execution tree, we know that $x \cdot d \in T$ for $d \in \{0, 1\}$. Let x_j be the class of $x \cdot 0$ and x_k be the class of $x \cdot 1$. We know that the only enabled transition in $(s, x_i w)$ is $sx_i \rightarrow sx_j x_k$. Since $d_i = callRet$ and since $\mathcal{T}(G)$ is an execution tree, we know that $\delta_G^{-1}(x \cdot 0)$ continues with the configuration after this call transition and $\delta_G^{-1}(x \cdot 1)$ continues with the configuration after the matching return transition (which exists in this case). Thus, the configuration in $\delta_G^{-1}(x \cdot 0)$ is $(s, x_j x_k w)$ and the configuration in $\delta_G^{-1}(x \cdot 1)$ is $(s, x_k w)$, establishing this part of the claim. We now establish that $r(x \cdot d) = r_H(x \cdot d)$. For the first and second component, this is established by the fact that the configuration in $\delta_G^{-1}(x \cdot d)$ determines both the label and the unique enabled transition. For the third component, this follows from the fact that both $\mathcal{T}$ and $\mathcal{T}(H)$ are execution trees and the fact that $r(x)$ and $r_H(x)$ match in the second component. $\qquad\square$

We obtain the following theorem for the two satisfiability problems:

Theorem 6. *The graph satisfiability problem and DPN satisfiability problem for NTL are* EXPTIME-*complete.*

Proof. Since the two problems are equivalent by Theorem 5, we need to only give an upper and lower bound for the graph satisfiability problem.

For the upper bound, we can construct an automaton for the formula using Corollary 1 and test it for emptiness using Proposition 2 in time exponential in $|\varphi|$ for an answer to the graph satisfiability problem.

The lower bound follows by a reduction from the *VP-μ-TL* satisfiability problem which was shown to be EXPTIME-hard in [9]. The reduction is straightforward since *VP-μ-TL* is a sublogic of NTL and we can easily extract a nested word satisfying a formula interpreted in *VP-μ-TL* from the execution graph satisfying the same formula interpreted in NTL. $\qquad\square$

8 Conclusion

We introduced a novel specification logic called NTL for reasoning about the call-return and thread creation behaviour of dynamic pushdown networks. We showed that a variety of interesting properties regarding the behaviour of multithreaded software is expressible in NTL. Further, the model checking and satisfiability problems were investigated. The complexity of these problems is not higher than that of the corresponding problems for related logics for pushdown systems despite a more powerful logic and system model. For future work, it would be interesting to consider more powerful variants of DPNs that allow communication and synchronization of different threads via locking or messages.

Acknowledgments. This work was partially funded by DFG project Model-Checking of Navigation Logics (MoNaLog) (MU 1508/3).

References

1. Oracle Java docs: Intrinsic locks and synchronization. https://docs.oracle.com/javase/tutorial/essential/concurrency/locksync.html
2. Alur, R., Arenas, M., Barceló, P., Etessami, K., Immerman, N., Libkin, L.: First-order and temporal logics for nested words. Log. Methods Comput. Sci. **4**(4) (2008). https://doi.org/10.2168/LMCS-4(4:11)2008
3. Alur, R., Chaudhuri, S., Madhusudan, P.: A fixpoint calculus for local and global program flows. In: Morrisett, J.G., Jones, S.L.P. (eds.) Proceedings of the 33rd ACM SIGPLAN-SIGACT Symposium on Principles of Programming Languages, POPL 2006, Charleston, South Carolina, USA, 11–13 January 2006, pp. 153–165. ACM (2006). https://doi.org/10.1145/1111037.1111051
4. Alur, R., Etessami, K., Madhusudan, P.: A temporal logic of nested calls and returns. In: Jensen, K., Podelski, A. (eds.) TACAS 2004. LNCS, vol. 2988, pp. 467–481. Springer, Heidelberg (2004). https://doi.org/10.1007/978-3-540-24730-2_35
5. Alur, R., Madhusudan, P.: Adding nesting structure to words. In: Ibarra, O.H., Dang, Z. (eds.) DLT 2006. LNCS, vol. 4036, pp. 1–13. Springer, Heidelberg (2006). https://doi.org/10.1007/11779148_1
6. Atig, M.F., Bouajjani, A., Qadeer, S.: Context-bounded analysis for concurrent programs with dynamic creation of threads. Log. Methods Comput. Sci. **7**(4) (2011). https://doi.org/10.2168/LMCS-7(4:4)2011
7. Bouajjani, A., Esparza, J., Maler, O.: Reachability analysis of pushdown automata: application to model-checking. In: Mazurkiewicz, A., Winkowski, J. (eds.) CONCUR 1997. LNCS, vol. 1243, pp. 135–150. Springer, Heidelberg (1997). https://doi.org/10.1007/3-540-63141-0_10
8. Bouajjani, A., Müller-Olm, M., Touili, T.: Regular symbolic analysis of dynamic networks of pushdown systems. In: Abadi, M., de Alfaro, L. (eds.) CONCUR 2005. LNCS, vol. 3653, pp. 473–487. Springer, Heidelberg (2005). https://doi.org/10.1007/11539452_36
9. Bozzelli, L.: Alternating automata and a temporal fixpoint calculus for visibly pushdown languages. In: Caires, L., Vasconcelos, V.T. (eds.) CONCUR 2007. LNCS, vol. 4703, pp. 476–491. Springer, Heidelberg (2007). https://doi.org/10.1007/978-3-540-74407-8_32
10. Bozzelli, L., Lanotte, R.: Hybrid and first-order complete extensions of CaRet. In: Brünnler, K., Metcalfe, G. (eds.) TABLEAUX 2011. LNCS (LNAI), vol. 6793, pp. 58–72. Springer, Heidelberg (2011). https://doi.org/10.1007/978-3-642-22119-4_7

11. Clarke, E.M., Emerson, E.A.: Design and synthesis of synchronization skeletons using branching time temporal logic. In: Kozen, D. (ed.) Logic of Programs 1981. LNCS, vol. 131, pp. 52–71. Springer, Heidelberg (1982). https://doi.org/10.1007/BFb0025774
12. Diaz, M., Touili, T.: Model checking dynamic pushdown networks with locks and priorities. In: Podelski, A., Taïani, F. (eds.) NETYS 2018. LNCS, vol. 11028, pp. 240–251. Springer, Cham (2019). https://doi.org/10.1007/978-3-030-05529-5_16
13. Emerson, E.A., Jutla, C.S.: The complexity of tree automata and logics of programs (extended abstract). In: 29th Annual Symposium on Foundations of Computer Science, White Plains, New York, USA, 24–26 October 1988, pp. 328–337. IEEE Computer Society (1988). https://doi.org/10.1109/SFCS.1988.21949
14. Gawlitza, T.M., Lammich, P., Müller-Olm, M., Seidl, H., Wenner, A.: Join-lock-sensitive forward reachability analysis for concurrent programs with dynamic process creation. In: Jhala, R., Schmidt, D. (eds.) VMCAI 2011. LNCS, vol. 6538, pp. 199–213. Springer, Heidelberg (2011). https://doi.org/10.1007/978-3-642-18275-4_15
15. Grädel, E., Thomas, W., Wilke, T. (eds.): Automata Logics, and Infinite Games. LNCS, vol. 2500. Springer, Heidelberg (2002). https://doi.org/10.1007/3-540-36387-4
16. Gutsfeld, J.O., Müller-Olm, M., Nordhoff, B.: A branching time variant of CaRet. In: Gallardo, M.M., Merino, P. (eds.) SPIN 2018. LNCS, vol. 10869, pp. 153–170. Springer, Cham (2018). https://doi.org/10.1007/978-3-319-94111-0_9
17. Kupferman, O., Vardi, M.Y.: Weak alternating automata and tree automata emptiness. In: Vitter, J.S. (ed.) Proceedings of the Thirtieth Annual ACM Symposium on the Theory of Computing, Dallas, Texas, USA, 23–26 May 1998, pp. 224–233. ACM (1998). https://doi.org/10.1145/276698.276748
18. Lakenbrink, R., Müller-Olm, M., Ohrem, C., Gutsfeld, J.: A navigation logic for recursive programs with dynamic thread creation (2023). https://arxiv.org/abs/2310.19579
19. Lammich, P., Müller-Olm, M., Seidl, H., Wenner, A.: Contextual locking for dynamic pushdown networks. In: Logozzo, F., Fähndrich, M. (eds.) SAS 2013. LNCS, vol. 7935, pp. 477–498. Springer, Heidelberg (2013). https://doi.org/10.1007/978-3-642-38856-9_25
20. Lammich, P., Müller-Olm, M., Wenner, A.: Predecessor sets of dynamic pushdown networks with tree-regular constraints. In: Bouajjani, A., Maler, O. (eds.) CAV 2009. LNCS, vol. 5643, pp. 525–539. Springer, Heidelberg (2009). https://doi.org/10.1007/978-3-642-02658-4_39
21. Lange, M.: Weak automata for the linear time μ-Calculus. In: Cousot, R. (ed.) VMCAI 2005. LNCS, vol. 3385, pp. 267–281. Springer, Heidelberg (2005). https://doi.org/10.1007/978-3-540-30579-8_18
22. Lichtenstein, O., Pnueli, A.: Checking that finite state concurrent programs satisfy their linear specification. In: Deusen, M.S.V., Galil, Z., Reid, B.K. (eds.) Conference Record of the Twelfth Annual ACM Symposium on Principles of Programming Languages, New Orleans, Louisiana, USA, January 1985, pp. 97–107. ACM Press (1985). https://doi.org/10.1145/318593.318622
23. Löding, C.: Automata on infinite trees. In: Pin, J. (ed.) Handbook of Automata Theory, pp. 265–302. European Mathematical Society Publishing House, Zürich (2021). https://doi.org/10.4171/Automata-1/8

24. Nordhoff, B., Müller-Olm, M., Lammich, P.: Iterable forward reachability analysis of Monitor-DPNs. In: Banerjee, A., Danvy, O., Doh, K., Hatcliff, J. (eds.) Semantics, Abstract Interpretation, and Reasoning about Programs: Essays Dedicated to David A. Schmidt on the Occasion of his Sixtieth Birthday, Manhattan, Kansas, USA, 19–20th September 2013. EPTCS, vol. 129, pp. 384–403 (2013). https://doi.org/10.4204/EPTCS.129.24
25. Pnueli, A., Rosner, R.: On the synthesis of an asynchronous reactive module. In: Ausiello, G., Dezani-Ciancaglini, M., Della Rocca, S.R. (eds.) ICALP 1989. LNCS, vol. 372, pp. 652–671. Springer, Heidelberg (1989). https://doi.org/10.1007/BFb0035790
26. Song, F., Touili, T.: Model checking dynamic pushdown networks. Formal Aspects Comput. **27**(2), 397–421 (2015). https://doi.org/10.1007/s00165-014-0330-y
27. Tarski, A.: A lattice-theoretical fixpoint theorem and its applications. Pacific J. Math. **5**(2), 285–309 (1955). https://projecteuclid.org:443/euclid.pjm/1103044538
28. Vardi, M.Y.: A temporal fixpoint calculus. In: POPL, pp. 250–259. ACM Press (1988). https://doi.org/10.1145/73560.735822
29. Vardi, M.Y.: Reasoning about the past with two-way automata. In: Larsen, K.G., Skyum, S., Winskel, G. (eds.) ICALP 1998. LNCS, vol. 1443, pp. 628–641. Springer, Heidelberg (1998). https://doi.org/10.1007/BFb0055090

Neural Networks

Taming Reachability Analysis of DNN-Controlled Systems via Abstraction-Based Training

Jiaxu Tian[1], Dapeng Zhi[1], Si Liu[2], Peixin Wang[3], Guy Katz[4], and Min Zhang[1(✉)] (iD)

[1] Shanghai Key Laboratory of Trustworthy Computing, East China Normal University, Shanghai, China
zhangmin@sei.ecnu.edu.cn
[2] ETH Zurich, Zurich, Switzerland
[3] University of Oxford, Oxford, UK
peixin.wang@cs.ox.ac.uk
[4] The Hebrew University of Jerusalem, Jerusalem, Israel

Abstract. The intrinsic complexity of deep neural networks (DNNs) makes it challenging to verify not only the networks themselves but also the hosting DNN-controlled systems. Reachability analysis of these systems faces the same challenge. Existing approaches rely on over-approximating DNNs using simpler polynomial models. However, they suffer from low efficiency and large overestimation, and are restricted to specific types of DNNs. This paper presents a novel abstraction-based approach to bypass the crux of over-approximating DNNs in reachability analysis. Specifically, we extend conventional DNNs by inserting an additional abstraction layer, which abstracts a real number to an interval for training. The inserted abstraction layer ensures that the values represented by an interval are indistinguishable to the network for both training and decision-making. Leveraging this, we devise the first black-box reachability analysis approach for DNN-controlled systems, where trained DNNs are only queried as black-box oracles for the actions on abstract states. Our approach is sound, tight, efficient, and agnostic to any DNN type and size. The experimental results on a wide range of benchmarks show that the DNNs trained by using our approach exhibit comparable performance, while the reachability analysis of the corresponding systems becomes more amenable with significant tightness and efficiency improvement over the state-of-the-art white-box approaches.

1 Introduction

Deep neural networks (DNNs) have demonstrated their remarkable capability of driving systems to perform specific tasks intelligently in open environments.

R. Dimitrova et al. (Eds.): VMCAI 2024, LNCS 14500, pp. 73–97, 2024.
https://doi.org/10.1007/978-3-031-50521-8_4

They determine optimal actions during interactions between the hosting systems and their surroundings. Formally verifying DNNs can provide safety guarantees [27,48,55], which is, however, difficult in practice due to their black-box nature and lack of interpretability [10,61]. Furthermore, their hosting systems aggravate the difficulty since determining system actions requires computations over nonlinear system dynamics [19,54].

Reachability analysis, one of the powerful formal methods, has been widely applied to the verification of continuous and hybrid systems [7,11,15]. Its successful applications include invariant checking [24,29], robust control [37,49], fault detection [50,53], set-based predication [6,46], etc. The essence of reachability analysis is to compute all reachable system states from given initial state(s), which can be used in various verification tasks such as model checking [9]. As an emerging approach to verifying DNN-controlled systems, reachability analysis has already been shown to be promisingly effective [21,23,32].

The Problem. Compared to continuous hybrid systems, it is significantly more challenging to compute reachable states for DNN-controlled systems due to the embedded complex and inexplicable DNNs. In addition to over-approximating nonlinear system dynamics [14,25,38], one also has to over-approximate the embedded DNNs for computing overestimated action sets [30,32] of such systems. Specifically, given a set S_i of continuous system states at time step i,[1] one first overestimates a set $\tilde{A}_i$ of actions that will be applied to S_i by over-approximating the neural network on S_i, and then overestimates a set $\tilde{S}_{i+1}$ of successors by applying $\tilde{A}_i$ to S_i using over-approximated system dynamics. We consider such dual over-approximations as *white-box* approaches since all the information of DNNs, such as architectures, activation functions, and weights, shall be known before defining appropriate over-approximated models [21,23,32]. Consequently, these approaches are restricted to certain types of DNNs. For instance, Verisig 2.0 [32] does not support neural networks with the ReLU activation functions; Sherlock [21] is only applicable to ReLU-based networks; ReachNN* [23] is not scalable against the network size and introduces more overestimation; Polar [30] also suffers from the efficiency problem when dealing with networks with differentiable activation functions (e.g., Tanh). Moreover, dual over-approximations introduce large overestimation accumulatively, which results in a considerable number of unreachable states in the overestimated sets.

Our Approach. We present a novel abstraction-based approach for bypassing the over-approximation of DNNs in computing the reachable states of DNN-controlled systems. Our approach introduces an *abstraction layer* into the neural network before training, which abstracts concrete system states into abstract ones. This abstraction ensures that concrete states that are abstracted into the same state share the same action determined by the trained DNN. Leveraging this property, we can therefore compute the actions of a set of concrete states by mapping them to the corresponding abstract states and by feeding the abstract states into the trained DNNs to query for the output action. As DNNs are used

---

[1] Continuous time is uniformly discretized into time steps.

as *black-box* oracles during the entire process, it suffices to know how system states are abstracted and to query the trained DNNs with the abstract states for the actions. Hence, the over-approximation of DNNs for computing actions is decently bypassed. Consequently, the overestimation due to the embedded network is avoided and no assumption is made on a network including its size, weight, architecture, and activation function.

The abstraction-based training also allows us to avoid state explosion during the computation of reachable states. This is because adjacent abstract states, e.g., the two intervals $[0, 1]$ and $[1, 2]$, can be efficiently aggregated, e.g., to $[0, 2]$, which substantially restrains the exponential growth in the number of computed reachable states. Additionally, we propose a parallel optimization via initial-set partitioning, which further accelerates the process of computing reachable states.

We have implemented our proposed approach into a tool called BBReach and extensively evaluated over a wide range of benchmarks. The experimental results show that DNNs trained by using our abstraction-based approach achieve competitive performance in terms of system cumulative reward. Our approach provides a black-box alternative to the reachability analysis of DNN-controlled systems, which bypasses the crux of DNN over-approximation and significantly improves the state-of-the-art white-box counterparts with respect to the tightness and efficiency in reachable state computation.

Contributions. Overall, we provide:

1. a novel abstraction-based training approach of DNNs, which mitigates the limitation of DNN over-approximation in the reachability analysis of DNN-controlled systems, without sacrificing the performance of trained DNNs (Sect. 4);
2. the first, sound black-box approach for the reachability analysis of trained DNN-controlled systems, which not only enhances computational tightness and efficiency, but also are compatible with various DNNs (Sect. 5); and
3. a prototype BBReach and an extensive assessment, which shows that BBReach improves existing white-box tools with respect to both the precision of results and the computational efficiency (Sect. 6).

2   Preliminaries

2.1   DNN-Controlled Systems

A DNN-controlled system is typically a cyber-physical system where a DNN is planted and trained as a decision-making controller. It can be modeled as a 6-tuple $\mathcal{D} = \langle S, S^0, A, \pi, f, \delta \rangle$, where S is the set of n-dimensional system states on n continuous variables, $S^0 \subseteq S$ is the set of initial states, A is the set of system actions, $\pi : S \rightarrow A$ is a policy function realized by the DNN in the system, $f : S \times A \rightarrow \dot{S}$ is a non-linear continuous environment dynamics represented by an ordinary differential equation (ODE) [26] that maps the current state and control input (i.e., action) into the derivative of states with respect to time t_c, and δ is the time step size.

In a DNN-controlled system, an agent reacts to the environment over time. The time is usually discretized by a time scale δ called the time step size, assuming that actions during each time scale δ are constants [44]. At each time step $i \in \mathbb{N}$, the agent first observes a state s_i from the environment and feeds the state into the network to compute a constant action a_i. The agent then transits to the successor state s_{i+1} by performing a_i on s_i according to some environment dynamics f. During the training phase of DNN-controlled systems, the agent also receives a reward r_i which is determined by a reward function $r_i = R(s_i, a, s_{i+1})$ from the environment after each state transition. Once the task is finished, e.g., the agent reaches the goal region at time step T, we obtain the sequence of traversed system states from an initial state, called a *trace*, and the cumulative reward $\sum_{i=0}^{T} r_i$ which quantitatively measures the system performance.

Example 1. (A DNN-Controlled System). Figure 1(a) shows a DNN-controlled system where a two-dimensional agent moves from the region $x_1 \in [0.7, 0.9]$, $x_2 \in [0.7, 0.9]$ to the goal region $x_1' \in [-0.3, 0.1]$, $x_2' \in [-0.35, 0.05]$, trying to avoid the red unsafe regions. The environment dynamics f is defined by the following ODEs:

$$\dot{x_1} = x_2 - x_1^3 \qquad \dot{x_2} = a \qquad (1)$$

The action $a = \pi(x_1, x_2)$ is computed by applying the DNN π to the values of x_1 and x_2. Based on a and f, the successor state s' can be computed for the agent to move. Figure 1(b) shows the traces (colored lines) of the system from some selected concrete initial states and an over-approximated set of reachable states (blue area) on the x_1 dimension from all the initial states.

The DNN planted in a system must be trained first so that it can determine optimal actions to complete a task. After making a decision, a loss is computed by a predefined loss function based on the reward that the agent receives for the decision. The parameters in the neural network are updated based on the loss by backpropagation [35].

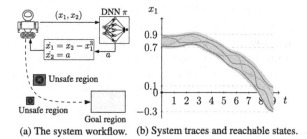

(a) The system workflow. (b) System traces and reachable states.

Fig. 1. The workflow of the DNN-controlled system in Example 1, the execution traces (colored lines) and an over-estimated set of reachable states (blue region) with respect to the dimension of x_1. (Color figure online)

The objective of the training phase is to maximize the cumulative reward. Once the training is completed, the network implements a state-action policy function that maps each system state to its optimal action. It drives the system to run and to interact with the environment.

2.2 Reachability Problem of DNN-Controlled Systems

Given a DNN-controlled system $\mathcal{D}$, whether or not a state is reachable is known as the *reachability problem*. The verification of safety properties can be reduced to the reachability problem. For instance, one can verify whether a system never moves to unsafe states or not, such as those in Example 1. Unfortunately, the problem is *undecidable* even for conventional cyber-physical systems that are controlled by explicit programmable rules, let alone uninterpretable neural networks. This is because such systems are more expressive than two-counter state machines whose reachability problem is proved to be undecidable [41].

When the set of initial states is a singleton, it is straightforward to compute the reachable state at any given time t_c. Let δ be a time scale during which system actions can be considered constant. Given an initial state s_0, the state at time $t_c = k\delta + t_c'$ for some integer $k \geq 0$ and $0 \leq t_c' \leq \delta$ is defined as follows:

$$\varphi_f(s_0, \pi, t_c) = s_k + \int_0^{t_c'} f(s, \pi(s_k))dx,$$

where $s_{i+1} = s_i + \int_0^\delta f(s, \pi(s_i))dx$ for all $i \in \{0, \dots, k-1\}$. Intuitively, we can compute the state s_{i+1} at the $(i+1)$-th time step based on the state s_i at its preceding time step i and the corresponding action $\pi(s_i)$. The state at t_c can be computed based on s_k, plus the offset caused by performing action $\pi(s_k)$ on state s_k with t_c' time scale.

Definition 1. (Reachable States of DNN-controlled Systems). Given a DNN-controlled system $\mathcal{D} = \langle S, S^0, A, \pi, f, \delta \rangle$, the sets of all the reachable states of the system at and during time t_c are denoted as $Reach_f^{t_c}(S_0)$ and $Reach_f^{[0,t_c]}(S_0)$, respectively. We have $Reach_f^{t_c}(S_0) = \{\varphi_f(s, \pi, t_c) | s \in S_0\}$ and $Reach_f^{[0,t_c]}(S_0) = \{\varphi_f(s, \pi, t) | s \in S_0, t \in [0, t_c]\}$.

Figure 2 depicts an example of the reachable states from S_0. For each time step i, we compute the set $Reach_f^{[0,\delta]}(S_i)$ of all the reachable states during the time period from i to $i+1$. The actions used for comput-

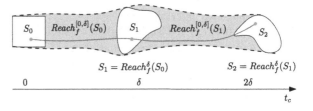

Fig. 2. Reachable states of DNN-controlled systems.

ing $Reach_f^{[0,\delta]}(S_i)$ are the constants determined by the DNN π on the states in S_i. In particular, we compute the set $S_{i+1} = Reach_f^\delta(S_i)$ of the reachable states at step $i+1$. Note that S_{i+1} is a subset of $Reach_f^{[0,\delta]}(S_i)$. We need to compute S_{i+1} independently from $Reach_f^{[0,\delta]}(S_i)$ because it is the basis of computing the reachable states in next step.

The procedure depicted in Fig. 2 indicates that the problem of computing $Reach_f^{[0,t_c]}(S_0)$ can be reduced to the problem of computing one-time-step reachable states, i.e., $Reach_f^{[0,\delta]}(S_0)$ and $Reach_f^{\delta}(S_0)$. However, the reduced problem is still intractable. This is because S_0 is usually an infinite set, meaning that it is impractical to enumerate each state in S_0, feed it into the DNN to compute the corresponding action, and then compute the state by Formula 2 for the set $Reach_f^{\delta}(S_0)$. Computing the states in $Reach_f^{[0,\delta]}(S_0)$ is even more challenging due to the continuous time in $[0, \delta]$.

3   Motivation

The combination of nonlinear dynamics and neural network controllers makes the calculation of $Reach_f^{[0,\delta]}(S_0)$ intractable. This is because the function φ_f (Formula 2) can not be expressed in a known closed form for most nonlinear dynamics f [13]. Additionally, a DNN π neither can be replaced by a known form equivalent function. A pragmatic solution is to compute tight over-approximation for φ_f and π. Most of the state-of-the-art approaches, such as Verisig [33], Polar [30], and ReachNN [31], adopt this strategy.

Without loss of generality, we show the process of over-approximating $Reach_f^{\delta}(S_0)$ in Example 1 using Polar. Given a set S_0 of states, Polar first over-approximates the neural network using a Taylor model (p, I_r) [39] on domain S_0 such that $\forall s \in S_0, \pi(s) \in p(s) + [-\epsilon, \epsilon]$, where p is a polynomial over the set of state variables $x_1, \ldots, x_n$ such as $p(x_1, x_2) =$

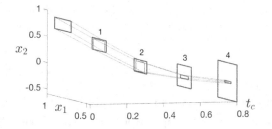

Fig. 3. An example of overestimation blowup of computed reachable states.

$0.5 + 0.1x_1 + 0.6x_1x_2 + 0.3x_1^2x_2$ and $I_r = [-\epsilon, \epsilon]$ is called the remainder interval. The range of $\pi(s)$ can be overestimated based on the Taylor model. Next, Polar over-approximates the solution of environment dynamics φ_f using another Taylor model over domains $s_0 \in S_0, \pi(s_0) \in p(s_0) + [-\epsilon, \epsilon], t_c \in [0, \delta]$ and obtains $x_1' \in p_1(x_1, x_2, t_c) + [-\epsilon_1, \epsilon_1], x_2' \in p_2(x_1, x_2, t_c) + [-\epsilon_2, \epsilon_2]$. Finally, Polar produces an overestimated set of S_1 at time δ based on x_1' and x_2'. A smaller range of I_r means less over-approximation error.

Suppose the initial region in Example 1 is $x_1 \in [0.7, 0.9], x_2 \in [0.7, 0.9]$. The overestimated reachable states can be calculated over 4 time steps according to the aforementioned method, which are depicted as red boxes ($\square$) in Fig. 3. For comparison, Fig. 3 also shows the reachable states by simulation with 1000 samples, which are shown as the small violet boxes ($\square$). We observe that the overestimation is amplified at the third and the fourth time step. At the third time step, the calculated remainder interval of the Taylor model for network is

$[-0.98, 0.98]$ while the one at the fourth time step is $[-4.47, 4.47]$. Correspondingly, the remainder intervals of the Taylor model for dynamics are $[-0.17, 0.17]$ and $[-0.46, 0.46]$ at the third and the fourth time step. The overestimation is accumulated and amplified step by step.

The above example shows that overestimation is mainly introduced by the over-approximation of the DNN. We further observe that if we could group the states in S_0 into several subsets such that all the states in the same subset have the same action according to π, we do not need to over-approximate π but, instead, replace $\pi(s)$ with its corresponding action. That is, if we know that all the states in a set S_0' share the same action, e.g., a, according to π, the problem of computing $Reach_f^\delta(S_0')$ can be simplified to solving the following problem:

$$\bigcup_{s_0' \in S_0'} \{s_0' + \int_0^\delta f(s_0', a)dx\}. \tag{2}$$

Naturally, we only need to over-approximate φ_f to solve the above problem. Therefore, we identify a condition of bypassing over-approximating π: S_0 can be divided into a finite number of subsets such that the states in the same subset have the same action according to π. This will be elaborated in our following abstraction-based approach.

4   Abstraction-Based Training

Given a DNN π and a set S_0 of system states, it is almost intractable to group the states in S_0 that have the same action according to π. It becomes even worse when actions are continuous, where each state in S_0 may have a different action from others. Instead of calculating these states *ex post facto*, we propose an *ex ante* approach by abstraction-based training, in which system states are first grouped by abstraction before training, and a trained DNN provably yields a unique action for the states in the same group.

4.1   Approach Overview

The process of grouping a set of system states and making them indistinguishable to neural networks is called *abstraction*. A group is considered as an abstract state. The indistinguishability of the states in the same group guarantees that a DNN computes a unique action for those states. This idea is inspired by the abstraction approaches in formal methods, by which system states are abstracted to reduce state space and improve verification scalability without losing the soundness of verification results [17]. The same idea is also studied in the AI communities. State abstraction has been proved useful for conventional Reinforcement Learning (RL) [2,4,51] and recently applied to Deep RL for training DNN controllers [34]. Studies show that one can train nearly optimal system policies via approximate state abstraction, while the trained policies are more concise and amenable for reasoning and verification than those trained on concrete states [2,34].

To implement state abstraction into deep learning, we extend ordinary DNN architectures by introducing an *abstraction layer* between the input layer and the first hidden layer. This layer is used to map concrete system states in a group to the same abstract representation, which is propagated throughout the hidden layers for training. We call a neural network

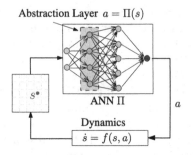

Fig. 4. Abstraction-based training.

that contains such an abstraction layer an *abstract neural network* (ANN). Note that an ANN is a special model of DNN. In what follows, we call the systems with ANN controllers *ANN-controlled systems* to differ from those controlled by conventional DNNs.

The training of ANNs is almost the same as for conventional DNNs. Figure 4 shows the training workflows with ANNs. We can simply replace DNNs with ANNs in existing training algorithms, such as Deep Q-Network (DQN) [42] and Deep Deterministic Policy Gradient (DDPG) [36], as the inserted abstraction layers in ANNs are invisible to these algorithms.

Therefore, an algorithm that supports training DNN-controlled systems can be seamlessly adapted to train ANN-controlled systems. When applying these algorithms to ANNs, the only difference is that we need to freeze the parameters on the edges between the input layer and the abstraction layer because they are determined and fixed according to the way in which system states are abstracted. Parameter freezing is a common operation in deep learning and is supported by most of the training platforms such as TensorFlow [1] and PyTorch [45]. After encoding the abstraction layer and freezing the parameters, a network can be trained just like conventional DNNs by these training algorithms.

4.2 Interval-Based State Abstraction

We propose a general approach for encoding interval-based abstractions into equivalent abstraction layers. Interval-based state abstraction is a very primitive, yet effective abstraction approach. In the domain of abstract interpretation [17], it is known as *interval abstract domain* and has been well studied for system [3] and program verification [28], as well as neural network approximation [59]. By interval-based abstraction, the domain of each dimension is evenly divided into several intervals. The Cartesian product of the intervals in all the dimensions constitutes a finite and discrete set, with each element representing an infinite set of concrete states.

Definition 2. (Interval-Based State Abstraction). Given an n-dimensional continuous state space S and an abstract state space S_ϕ obtained by discretizing S based on an abstraction granularity γ, for every concrete state

$s = (x_1, \ldots, x_n) \in S$ and abstract state $s_\phi = (l_1, u_1, \ldots, l_n, u_n) \in S_\phi$, the interval-based abstraction function $\phi : S \to S_\phi$ is defined as $\phi(s) = s_\phi$ if and only if for each dimension $1 \le i \le n : l_i \le x_i < u_i$.

Specifically, the abstract state space S_ϕ is obtained by dividing each dimension in the original n-dimensional state space S into a set of intervals, which means that each abstract state can be represented as a $2n$-

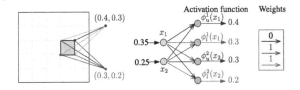

Fig. 5. An example of defining abstraction layers.

dimensional vector $(l_1, u_1, \ldots, l_n, u_n)$. We also call the $2n$-dimensional vector as interval box. In what follows, an interval box is used to represent a set of concrete states that fall into it. That is, for a $2n$-dimensional vector $(l_1, u_1, \ldots, l_n, u_n)$, we use it to represent the set of n-dimensional concrete states $\{(x_1, \ldots, x_n) \mid l_i \le x_i < u_i, \forall 1 \le i \le n\}$. In this work, we divide the state space uniformly for better scalability so that we do not need extra data structure to store the mapping between S and S_ϕ. More specifically, let L_i and U_i be the lower and upper bounds for the i-th dimension of S. We define the abstraction granularity as an n-dimensional vector $\gamma = (d_1, d_2, \ldots, d_n)$, and then evenly divide the i-th dimension into $(U_i - L_i)/d_i$ intervals.

An interval-based abstraction can be naturally encoded as an abstraction layer. The layer consists of $2n$ neurons, each of which represents an element in the $2n$-dimensional vector $(l_1, u_1, \ldots, l_n, u_n)$. Each neuron has an activation function in either of the following two forms:

$$\phi_l^i(x_i) = L_i + \lfloor \frac{(x_i - L_i)}{d_i} \rfloor d_i, \quad \phi_u^i(x_i) = L_i + \lfloor \frac{(x_i - L_i + d_i)}{d_i} \rfloor d_i$$

for converting the value x_i in a concrete state to its lower and upper bounds, respectively. The sign $\lfloor \cdot \rfloor$ is the floor function. The weights of the edges connecting the i-th neuron in the input layer to the $(2i - 1)$-th and $2i$-th neurons in the abstraction layer are assigned a value of 1, whereas the weights of all other edges are set with 0.

Example 2. Suppose that the ranges of both x_1 and x_2 in Example 1 are $[0, 0.5]$, and they are evenly partitioned into 5 intervals. The state space $[0, 0.5] \times [0, 0.5]$ is then uniformly partitioned into 25 interval boxes, as shown in Fig. 5. A concrete state such as $(0.35, 0.25)$ is mapped to an *interval box* represented by the corresponding lower bounds $(0.3, 0.2)$ of the first dimension and upper bounds $(0.4, 0.3)$ of the second dimension.

This abstraction can be realized by an abstraction layer, where there are four neurons and their activation functions are $\phi_u^1(x) = \phi_u^2(x) = \lfloor \frac{x+0.1}{0.1} \rfloor \times 0.1$ and $\phi_l^1(x) = \phi_l^2(x) = \lfloor \frac{x}{0.1} \rfloor \times 0.1$, respectively.

5   Abstraction-Based Reachability Analysis

5.1   Approach Overview

With the abstraction layer, we propose our abstraction-based black-box reachability analysis approach for ANN-controlled systems. Given an ANN-controlled system, a set S_0 of initial states and a maximal time step T, our task is to calculate a sequence of over-approximation sets consisting of interval boxes, denoted by $X_0, X_1, \ldots, X_T$, which are over-approximations of the actually reachable state sets $S_0, S_1, \ldots, S_T$ with $S_{t+1} = Reach_f^\delta(S_t)$, $0 \le t < T$. The overall process is presented in Algorithm 1. It is an iterative process of calculating an over-approximated set X_t of states that are reachable from a set X_{t-1} of states after time δ. After we determine the range of $\pi(s)$ over $s \in X_{t-1}$, the reachable states during the time slot $(t\delta, (t+1)\delta]$ can be over-approximated as a continuous system without a neural network. In what follows, we focus on the computation of the over-approximation sets $X_0, X_1, \ldots X_T$.

Algorithm 1: Overall process.

Input : Initial set S_0, ANN π, step size δ, dynamics f, abstraction function ϕ, maximal time step T

Output: Over-approximation sets $\bigcup_{t=1}^{T} X_t$

1 Compute I_0 satisfying $S_0 \subseteq I_0$, $X_0 \leftarrow [I_0]$

2 **foreach** t *in* $\{1, ..., T\}$ **do**

3 $\quad$ interval_arr $\leftarrow \{\}$

4 $\quad$ **foreach** I *in* X_{t-1} **do**

5 $\quad\quad$ $\mathbf{B}_I \leftarrow segment(I, \phi)$

6 $\quad\quad$ **foreach** $\mathcal{I}$ *in* $\mathbf{B}_I$ **do**

7 $\quad\quad\quad$ $a \leftarrow \pi(\hat{s})$ for some $\hat{s} \in \mathcal{I}$

8 $\quad\quad\quad$ $\mathcal{I}' \leftarrow post(\mathcal{I}, a, f)$

9 $\quad\quad\quad$ interval_arr $\leftarrow$ interval_arr $\cup \{\mathcal{I}'\}$

10 $\quad$ $X_t = aggregate(\text{interval\_arr})$

11 **return** $\bigcup_{t=1}^{T} X_t$

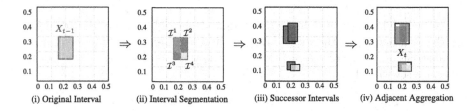

(i) Original Interval   (ii) Interval Segmentation   (iii) Successor Intervals   (iv) Adjacent Aggregation

Fig. 6. An example of over-approximating one-step reachable states.

Figure 6 depicts an example of one time-step iteration. Without loss of generality, we suppose that X_{t-1} is a singleton, e.g., $X_{t-1} = \{I\}$, where I is an interval box. We segment I into four smaller interval boxes (Fig. 6(ii) and Line 5 of Algorithm 1) based on the abstraction function ϕ that is used for training the network.

We then compute the action for the states in each segmented interval box by arbitrarily selecting a state $\hat{s}$ in the box and then feeding $\hat{s}$ into π to get the output (Line 7), e.g., a. Next, we compute a set $\mathcal{I}'$ of successor states of the states in $\mathcal{I}$ by over-approximating the environment dynamic f in Formula 2 (Fig. 6(iii) and Line 8 of Algorithm 1). Finally, we aggregate those adjacent successor interval boxes (Fig. 6(iv) and Line 10 of Algorithm 1) and obtain an over-estimated set X_t of reachable states at time step t.

5.2 Key Operations in Algorithm 1

We now describe in detail three key operations in Algorithm 1, namely *interval segmentation*, *post operation*, and *adjacent interval aggregation*. We fulfill the interval set propagation at each time step $t \in \mathbb{N}$ for the ANN-controlled systems based on these interval operations.

Interval Segmentation. Given an interval box I and an abstraction function ϕ, $segment(I, \phi)$ returns a set B_I of interval boxes which satisfy the following three *segmentation conditions*:

1. All the interval boxes constitute I;
2. Interval boxes do not overlap each other;
3. All the states in the same interval box have a unique action according to the trained ANN.

For conventional DNNs, one has to resort to brute-force interval splitting to find consistent regions that satisfy the above three conditions; this approach is only applicable to discrete action space [8]. We can easily partition I into such a set B_I, thanks to the specialized design of ANN. First, we determine the set of abstract states that intersect with I and denote the set by $\mathbf{S}_I = \{s_\phi \mid s_\phi \cap I \neq \emptyset\}$. We then calculate the intersection part between I and the abstract states in $\mathbf{S}_I$ individually. Each intersection part is a segmented interval box. In this way, we obtain a set of segmented interval boxes that satisfy the aforementioned three conditions and denote it by $\mathbf{B}_I = \{\mathcal{I} \mid \mathcal{I} = s_\phi \cap I \wedge s_\phi \in \mathbf{S}_I\}$. With the interval segmentation, through feeding an arbitrary state in the segmented interval box $\mathcal{I} \in B_I$ into ANN, we can obtain the corresponding unique action performed on $\mathcal{I}$. Since B_I is a finite set, the decisions of the network controller on I can be directly obtained without the layer-by-layer analysis process as in the white-box approaches [30,32]. This makes our reachability analysis approach a black-box one.

Recall the example in Fig. 6(i), where the black dotted lines denote the partition of the state space with abstraction granularity $\gamma = (0.1, 0.1)$. There exists an interval box $I = (0.15, 0.25, 0.15, 0.3)$ that intersects with four abstract states. The intersection of each abstract state with I is a segmented interval box. We have four interval boxes $B_I = \{\mathcal{I}^1, \mathcal{I}^2, \mathcal{I}^2, \mathcal{I}^4\}$. Apparently, the segmented interval boxes in B_I satisfy the three segmentation conditions.

Post Operation. Given an interval box $\mathcal{I}$, the action a applied to $\mathcal{I}$ and environment dynamics f, $post(\mathcal{I}, a, f)$ returns an interval box $\mathcal{I}'$, which is an over-approximation set of all the successor states by applying a to the states in $\mathcal{I}$ after δ.

We can solve $post(\mathcal{I}, a, f)$ as an ordinary continuous system without neural networks. Suppose that the environment dynamics is an ODE $\dot{s} = f(s, a)$. We use a Taylor model $p'(s, a, t_c) + I'_r$ to over-approximate the function $\varphi_f(s, a, t_c)$ over the domain $s \in \mathcal{I}, t_c \in [0, \delta]$. That is,

$$Reach_f^{[0,\delta]}(\mathcal{I}) = \bigcup_{s\in\mathcal{I}, t_c\in[0,\delta]} \{\varphi_f(s, a, t_c)\} \subseteq p'(s, a, t_c) + I'_r,$$

where I'_r is a remainder interval. The successor interval box $\mathcal{I}'$ can be calculated through evaluating the range of $p'(s, a, \delta) + I'_r$.

Let us consider an example for the segmented interval box $\mathcal{I}^1 = (0.15, 0.2, 0.2, 0.3)$ in Fig. 6(ii). The dynamics is defined as in Example 1. Suppose the action for the states in the interval box is $a = 0.5$ and the time scale $\delta = 0.1$. We can compute an over-approximated Taylor model for the solution of dynamics $f : \dot{x}_1 = x_2 - x_1^3, \dot{x}_2 = 0.5$ over $s \in \mathcal{I}^1, t_c \in [0, 0.1]$. The Taylor models for state variable x_1, x_2 are as follows:

$$x'_1 = 1.75 \times 10^{-1} + 1.91 \times 10^{-8}x_2 + 2.5 \times 10^{-2}x_1 + 0.245t_c$$
$$- 1.25 \times 10^{-10}x_1^2 + 5 \times 10^{-2}x_2t_c - 2.3 \times 10^{-3}x_1t_c + 0.239t_c^2$$
$$- 5 \times 10^{-10}x_1x_2t_c + \ldots + [-1.03 \times 10^{-4}, 8.94 \times 10^{-5}]$$
$$x'_2 = x_2 + 0.5t_c + [-0, 0]$$

Using these two expressions, we can over-approximate the set of reachable states at every moment during $[0, 0.1]$. In particular, we have $(0.172, 0.232, 0.25, 0.35)$ when $t_c = 0.1$.

Adjacent Interval Aggregation. Interval segmentation may lead to the exponential blowup in the number of intervals as the number of time steps increases. As exemplified in Fig. 6(iii), four successor intervals are obtained after applying corresponding actions and environment dynamics to the states in $\mathcal{I}^1, \ldots, \mathcal{I}^4$.

To cope with the explosion of successor intervals, we provide a dual operation of segmentation called *adjacent interval aggregation*, which aggregates multiple intervals together at the price of introducing a little overestimation. This operation is based on the interval hull

Algorithm 2: Adjacent interval aggregation.

Input : An interval array *IntArr*
Output: The aggregation results *Arr*
1 Initialize flag ← [False, False,...], *Arr* ← []
2 Construct the adjacency matrix M
3 **foreach** I_p *in IntArr* **do**
4   **if** *not flag[I_p]* **then**
5   Initialize queue ← [I_p]
6   flag[I_p] ← True
7   **while** *queue is not empty* **do**
8   I ← queue.pop()
9   I_{adjs} ← getAdjacent(I, M)
10   **foreach** *item in I_{adjs}* **do**
11   I_p ← aggInterval(I_p, item)
12   **if** *not flag[item]* **then**
13   queue.put(item)
14   flag[item] ← True
15   *Arr*.add(I_p)
16 **return** *Arr*

operation [43] except that we establish a criterion for determining which intervals can be aggregated into their interval hull. For instance, the green and brown intervals in Fig. 6(iii) can be aggregated, while the other small ones can be aggregated too. However, large overestimation would be introduced if the four interval boxes were aggregated to be one.

To balance the number of intervals and the overestimation introduced by aggregation, we define three cases for the adjacency relation between interval boxes, i.e., *inclusion*, *intersection*, and *separation*. Only the intervals in the three cases are aggregated. Given two interval boxes $A = (l_1, u_1, \ldots, l_n, u_n)$ and $B = (l'_1, u'_1, \ldots, l'_n, u'_n)$, as well as a preset distance threshold $h = (h_1, \ldots, h_n)$, the three cases are defined as follows:

1. **Inclusion:** An interval box is completely included in the other, i.e., $\forall i : (l_i \leq l'_i \wedge u_i \geq u'_i) \vee (l_i \geq l'_i \wedge u_i \leq u'_i)$.
2. **Intersection:** A and B have a partial overlap, i.e., $\exists! d : l'_d \leq l_d \leq u'_d \leq u_d \vee l_d \leq l'_d \leq u_d \leq u'_d$ and $\forall i, i \neq d : |l_i - l'_i| \leq h_i \wedge |u_i - u'_i| \leq h_i$.
3. **Separation:** A is isolated from B, i.e., $\exists! d : l_d - u'_d \leq h_d \vee l'_d - u_d \leq h_d$; and $\forall i, i \neq d : |l_i - l'_i| \leq h_i \wedge |u_i - u'_i| \leq h_i$.

To accelerate interval aggregation, we devise an efficient algorithm to aggregate three or more interval boxes each time if they constitute a sequence of adjacent intervals. Algorithm 2 shows the pseudo code. We first pre-construct an adjacency matrix (Line 2) to store the adjacent relations between the interval boxes in $IntArr$ firstly. Then, we implement this adjacent interval aggregation procedure using breadth-first search (Lines 5-14). Specifically, we consider each interval box in $IntArr$ as a node and each adjacent relation as an undirected edge. For each interval box I_p that is not traversed, all the interval boxes connected to I_p will be aggregated into their minimum bounding rectangle.

In Algorithm 2, the time complexity of building the adjacency matrix is $O(n^2)$. In the aggregation procedure, each interval box is traversed at most once, and the complexity of searching for the adjacent interval boxes for each interval box is $O(n)$. Therefore, Algorithm 2 is in $O(n^2)$.

Example 3. Let us revisit the system in Example 1 and suppose that $IntArr$ consists of 4 interval boxes, i.e., $\widehat{I}_1 = (0.08, 0.16, 0.3, 0.4)$, $\widehat{I}_2 = (0.17, 0.25, 0.32, 0.42)$, $\widehat{I}_3 = (0.19, 0.27, 0.07, 0.2)$, $\widehat{I}_4 = (0.2, 0.28, 0.1, 0.21)$, and the distance threshold is $h = (0.02, 0.02)$. According to the definition of adjacent relations, $\widehat{I}_1$ is adjacent to $\widehat{I}_2$ (Separation) and $\widehat{I}_3$ is adjacent to $\widehat{I}_4$ (Intersection). Hence, $\widehat{I}_1$ is aggregated with $\widehat{I}_2$, and $\widehat{I}_3$ is aggregated with $\widehat{I}_4$. Finally, we obtain $Arr = \{I_{1,2} = (0.08, 0.25, 0.3, 0.42), I_{3,4} = (0.19, 0.28, 0.07, 0.21)\}$.

5.3 The Soundness

We show a proof sketch for the soundness of Algorithm 1. The soundness means that any state that is reachable at time t_c from some initial state of an ANN-controlled system must be in the over-approximation set at t_c.

Theorem 1. (Soundness of Algorithm *1***).** *Given an ANN-controlled system with a set* S_0 *of initial states and an environment dynamic* f, *if a state* s' *is reached at time* $t_c = k\delta + t'_c, k \in \mathbb{N}, t'_c \in [0, \delta)$ *from some initial state* $s_0 \in S_0$, *then we must have* $s' = Reach_f^{t_c}(s_0) \in Reach_f^{t'_c}(X_k)$.

To prove Theorem 1, we first show the soundness of the *post* operation and interval aggregation. The soundness of the two operations is formulated by the following two lemmas, respectively.

Lemma 1. (Soundness of *post* **Operation).** *For each interval box* $\mathcal{I} \in B_I$, *there is* $s_{t+1} \in post(\mathcal{I}, \pi(s_t), f)$ *for all* $s_t \in \mathcal{I}$ *where* $s_{t+1} = \varphi_f(s_t, \pi(s_t), \delta)$.

Proof. After the segmentation process, we have $\forall s \in \mathcal{I} : \pi(s) = \pi(s_t) = a$ where a is a constant. With a constant action and the Lipschitz continuity of f, we can guarantee that there exists a unique solution of the ODE for a single initial state [40]. Then the solution of the ODE namely $\varphi_f(s, a, t_c)$ could be enclosed by a Taylor model [39] over $s(0) \in \mathcal{I}$ and $t_c \in [0, \delta]$. Thus, we could obtain the conservative result $s_{t+1} = \varphi_f(s_t, a, \delta) \in Reach_f^\delta(\mathcal{I}) \subset post(\mathcal{I}, a, f)$.   □

Lemma 2. (Soundness of Interval Aggregation). *Suppose* A *is the aggregated set of successor intervals for a set* X *of interval boxes. For all* $I \in X$, *there exists* $\widehat{I} \in A$ *such that* $I \subseteq \widehat{I}$.

Proof. In Algorithm 2, every interval box in X needs to be traversed. For each interval box $I \in X$, there exist two cases: (i) I is not involved in the adjacent interval aggregation process. In this case, I will be directly added to A, thus $\exists \widehat{I} = I : I \subseteq \widehat{I}$. (ii) I is aggregated into another interval box I'. Since the *aggregate* operation produces the minimum bounding rectangle which encloses all interval boxes involved, we have $\exists \widehat{I} = I' : I \subseteq \widehat{I}$. Consequently, we conclude that $\forall I \in X, \exists \widehat{I} : I \subseteq \widehat{I} \wedge \widehat{I} \in A$.   □

According to Algorithm 1, Theorem 1 can be proved by induction on the steps t_c based on Lemmas 1 and 2. The base case is straightforward when $t_c = 0$. In the induction case, we can prove that Theorem 1 holds on $[t\delta, (t+1)\delta]$ according to the two lemmas and the hypothesis that it holds on an arbitrary $t_c = t\delta$.

Proof. (Theorem 1). Starting from s_0, we can obtain the trajectory as $s_0, a_0, s_1, a_1, \dots$ in which $a_t = \pi(s_t)$ and $s_{t+1} = \varphi(s_t, a_t, \delta)$. Then by induction on the time step t, the induction schema is as follows:

Base Case: $t_c = 0$. Since $s_0 \in S_0 \wedge S_0 \subseteq X_0$, we have $s_0 \in X_0 = Reach_f^0(X_0)$.

Induction Step: $t_c = t\delta$. Assume $s' = s_t \in X_t = Reach_f^0(X_t)$ holds. Since X_t consists of a set of interval boxes, there exists an interval box $I_{X_t}^{n_1}$ satisfying $s_t \in I_{X_t}^{n_1} \wedge I_{X_t}^{n_1} \in X_t$. Then, let us consider the segmentation process for $I_{X_t}^{n_1}$, we divide $I_{X_t}^{n_1}$ into a set of interval boxes $\mathbf{B}_{I_{X_t}^{n_1}} = \{\mathcal{I}_{X_t}^1, \mathcal{I}_{X_t}^2, \dots, \mathcal{I}_{X_t}^{max}\}$ with $I_{X_t}^{n_1} = \bigcup_{n=1}^{max} \mathcal{I}_{X_t}^n$. Thus, there exists some $n_2 \in \mathbb{Z}^+$ such that $s_t \in \mathcal{I}_{X_t}^{n_2}$.

For $t_c \in [t\delta, (t+1)\delta)$, we have $s' = Reach_f^{t'_c}(s_t)$. Since $s_t \in \mathcal{I}_{X_t}^{n_2}$, we have $s' = Reach_f^{t'_c}(s_t) \in Reach_f^{t'_c}(\mathcal{I}_{X_t}^{n_2}) \subseteq Reach_f^{t'_c}(X_t)$.

For $t_c = (t+1)\delta$, we have $s' = s_{t+1}$, Based on Lemma 1, we have $s_{t+1} \in post(\mathcal{I}_{X_t}^{n_2}, \pi(s_t), f)$. After the adjacent interval aggregation process, X_{t+1} consists of the aggregation result. According to Lemma 2, we have $\exists \widehat{I} : post(\mathcal{I}_{X_t}^{n_2}, \pi(s_t), f) \subseteq \widehat{\mathcal{I}} \wedge \widehat{\mathcal{I}} \in X_{t+1}$. Therefore, we have $s_{t+1} \in \widehat{\mathcal{I}} \wedge \widehat{\mathcal{I}} \in X_{t+1}$ and we can conclude that $s' = s_{t+1} \in X_{t+1} = Reach_f^0(X_{t+1})$.

Theorem 1 is proved.   □

6 Implementation and Experiments

We conduct a comprehensive assessment of our approach and compare it with the state-of-the-art white-box tools. Our goal is to demonstrate the advances of the proposed abstraction-based training and black-box reachability analysis approaches. These include (i) comparable performance of trained systems and negligible time overhead in the training (Sect. 6.2), (ii) tighter over-approximated sets of reachable states, as well as higher scalability and efficiency (Sect. 6.3), and (iii) the effectiveness of the adjacent interval aggregation algorithm in reducing state explosion (Sect. 6.4). We also explore how our approach performs under different abstraction granularity levels (Sect. 6.4).

6.1 Implementation and Benchmarks

Implementation. We implement our approach in a tool called BBReach in Python. We use Ariadne [16] to solve the reachability problems defined on segmented interval boxes (i.e., $post(\mathcal{I}, a, f), \mathcal{I} \in B_I$). Additionally, we employ the parallelized computing by initial-set partition [13], a standard approach used in the reachability analysis of hybrid systems to obtain tighter bounds of reachable states. With the initial set partitioned into k subsets, the k sub-problems can be solved in parallel, which accelerates our approach with multiple cores.

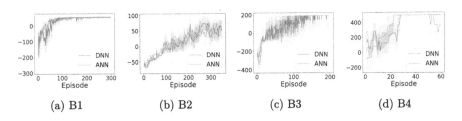

(a) B1   (b) B2   (c) B3   (d) B4

Fig. 7. Trend of cumulative rewards (y-axis) of the systems controlled by ANNs (orange) and DNNs (blue) trained by DDPG. (Color figure online)

Benchmarks. The benchmarks, as commonly adopted by most of the existing reachability analysis approaches such as Verisig 2.0 [32] and Polar [30], consist of seven reinforcement learning tasks with the dimensions ranging from 2 to 6. A reach-avoid property is defined for each task by specifying the goal region and unsafe region of the agent in the task. A trained DNN must guarantee that the reach-avoid property is satisfied when the agent is driven by the DNN.

For each task, we train four neural networks (two smaller networks chosen from [32] and two larger networks), thus 28 instances in total, with different activation functions and sizes of neurons. We also train the networks with different abstraction granularity levels to evaluate how abstraction granularity affects the efficiency. We use the well-known DDPG algorithm to train neural networks. Note that our approach makes no assumption on training algorithms and thus is applicable to other DRL algorithms. The detailed settings are provided in [56].

Experimental Setup. All experiments are conducted on a workstation equipped with a 32-core AMD Ryzen Threadripper CPU @ 3.6GHz and 256GB RAM, running Ubuntu 22.04.

6.2 Performance of Trained Neural Networks

We show that the extended abstract neural networks can be trained to achieve comparable performance against those conventional ones that have the same architectures and activation functions and are trained in the same approach. For each case, we train 5 times and record the cumulative reward during the training process with and without the abstraction layer. Figure 7 unfolds a

Table 1. Training time (s).

Task	ANN	DNN
B1	13.7	11.0
B2	7.4	6.6
B3	6.4	5.1
B4	5.8	3.2
B5	57.4	49.8
Tora	47.2	44.3
ACC	23.4	21.6

comparison of the trend of cumulative rewards during training between these two training approaches in B1-B4 (the other three are given in [56].) The solid lines and the shadows indicate the average reward and 95% confidence interval, respectively. The results show that an extended abstract neural network can make near-optimal decisions even under the constraint that it must yield the same action on each partitioned interval box. Importantly, the abstraction-based training incurs little and negligible time overhead only in several seconds, as shown in Table 1.

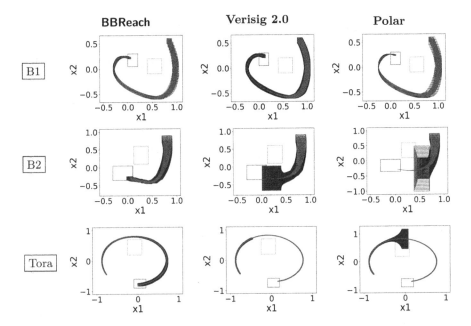

Fig. 8. Over-approximated reachable states (red box: over-approximated set; green lines: simulation trajectories; blue box: goal region; purple box: unsafe region). (Color figure online)

6.3   Tightness and Efficiency

We compare the tightness of the over-approximated reachable states by plotting the over-approximation sets computed by our approach and the state-of-the-art white-box tools including Polar [30] and Verisig 2.0 [32]. Because BBReach is designed for ANNs-controlled systems, while the white-box tools are for DNNs-controlled ones, the policy models for each task are different. To make the comparison as fair as possible, we use the same network architecture to train the ANN and DNN for the same task except that the ANN includes an additional abstraction layer. We also guarantee that all the trained systems can achieve the best cumulative reward for the same task. For instance, we initialize the neural networks with smaller weights as otherwise Verisig 2.0 would introduce larger over-approximation error (see Appendix B in [56]). In particular, we also simulated the trained systems and recorded trajectories as the baseline.

Figure 8 shows three representative cases. Verification succeeds if the system never enters the unsafe region (purple box) before reaching the goal region (blue box) which is also known as satisfying the reach-avoid property. All the four tools successfully verify the reach-avoid property in case B1, yet Verisig 2.0 is less tight than the other two. In case B2, BBReach outperforms the other two tools and succeeds in verifying the reach-avoid property. Both Verisig 2.0 and Polar terminate before reaching the goal region due to too large overestimation, and Polar outputs the over-approximation sets that intersect with the unsafe region.

Table 2. The verification results of reach-avoid properties and the time cost (s).

Task	Dim	Network	BBReach			Verisig 2.0					Polar			
			1C	20Cs	VR	1C	Impr.	20Cs	Impr.	VR	1C	Impr.	Impr.*	VR
B1	2	Tanh$_{2\times20}$	45.7	6.88	✓	45	0.98×	38	5.52×	✓	17	0.37×	2.47×	✓
		Tanh$_{3\times100}$	42.8	5.53	✓	413	9.65×	123	22.24×	✓	125	2.92×	22.60×	✓
		ReLU$_{2\times20}$	42.9	6.44	✓	—	—	—	—	✗^{c}	3	0.07×	0.47×	✓
		ReLU$_{3\times100}$	52.5	8.65	✓	—	—	—	—		—	—	—	✗^{b}
B2	2	Tanh$_{2\times20}$	10.0	1.19	✓	5.2	0.52×	4.1	3.45×	✗^{a}	5	0.50×	4.20×	✓
		Tanh$_{3\times100}$	10.8	1.36	✓	—	—	—	—	✗^{b}	—	—	—	✗^{b}
		ReLU$_{2\times20}$	8.6	1.30	✓	—	—	—	—	✗^{c}	3	0.35×	2.31×	✓
		ReLU$_{3\times100}$	12.4	1.42	✓	—	—	—	—		—	—	—	✗^{b}
B3	2	Tanh$_{2\times20}$	4.2	0.47	✓	36	8.57×	28	59.57×	✓	18	4.29×	39.29×	✓
		Tanh$_{3\times100}$	4.3	0.50	✓	357	83.02×	88	176.00×	✓	91	91.16×	182.00×	✓
		ReLU$_{2\times20}$	4.1	0.47	✓	—	—	—	—	✗^{c}	8	1.95×	17.02×	✓
		ReLU$_{3\times100}$	4.2	0.47	✓	—	—	—	—		14	3.33×	29.79×	✓
B4	3	Tanh$_{2\times20}$	1.3	0.32	✓	7	5.38×	5.1	15.94×	✓	5	3.85×	15.63×	✓
		Tanh$_{3\times100}$	1.0	0.24	✓	114	114.00×	31	129.17×	✓	27	27.00×	112.50×	✓
		ReLU$_{2\times20}$	1.9	0.48	✓	—	—	—	—	✗^{c}	2	1.05×	4.17×	✓
		ReLU$_{3\times100}$	1.8	0.43	✓	—	—	—	—		5	2.78×	11.63×	✓
B5	3	Tanh$_{3\times100}$	13.3	2.48	✓	157	11.80×	44	17.74×	✓	38	2.86×	15.32×	✓
		Tanh$_{4\times200}$	8.2	1.63	✓	1443	175.98×	191	117.18×	✓	157	19.15×	96.32×	✓
		ReLU$_{3\times100}$	5.8	1.08	✓	—	—	—	—	✗^{c}	7	1.21×	6.48×	✓
		ReLU$_{4\times200}$	13.5	2.50	✓	—	—	—	—		49	3.63×	19.60×	✓
Tora	4	Tanh$_{3\times20}$	133.2	8.61	✓	69	0.52×	46	5.34×	✓	45	0.34×	5.23×	✓
		Tanh$_{4\times100}$	112.3	9.78	✓	—	—	—	—	✗^{b}	✗^{b}	—	—	✗^{b}
		ReLU$_{3\times20}$	124.7	9.97	✓	—	—	—	—	✗^{c}	30	0.24×	3.01×	✓
		ReLU$_{4\times100}$	128.1	7.54	✓	—	—	—	—		53	0.41×	7.03×	✓
ACC	6	Tanh$_{3\times20}$	15.4	4.53	✓	113	7.34×	50	11.04×	✓	84	5.45×	18.54×	✓
		Tanh$_{4\times100}$	15.2	4.51	✓	2617	172.17×	375	83.15×	✓	677	44.54×	150.11×	✓
		ReLU$_{3\times20}$	15.2	4.45	✓	—	—	—	—	✗^{c}	26	1.71×	5.84×	✓
		ReLU$_{4\times100}$	18.4	5.49	✓	—	—	—	—		58	3.15×	10.56×	✓

Remarks. Improvement: time speedup of BBReach compared to Verisig or Polar (Verisig or Polar/BBReach). * denotes the comparison between BBReach with 20 cores (Cs) and Polar. Tanh/ReLU$_{n\times k}$: a neural network with the activation function Tanh/ReLU, n hidden layers, and k neurons per hidden layer. VR: verification result. ✓: the reach-avoid problem is successfully verified. ✗^{type}: the reach-avoid problem cannot be verified due to *type*: (a) large over-approximation error, (b) the calculation did not finish, (c) not applicable. —: no data available due to ✗^{b} or ✗^{c}

Nevertheless, the simulation results show the trained DNN-controlled system should satisfy the reach-avoid requirements. For Tora, BBReach significantly surpasses other tools. None of the two white-box tools finishes before reaching the goal region because of the huge over-approximation error. For instance, the resulting bound of action, upon Verisig 2.0's termination, reaches 10^7 which is too large to proceed, although the increase of reachable states by simulation is approximately in linear. The comparison results for B3, B4, B5, and ACC are similar as for B1. We refer to our technical report [56] for more detailed results.

Table 2 shows the verification results of all the 28 instances in column **VR**. BBReach successfully verifies all the instances, while Verisig 2.0 succeeds in 11

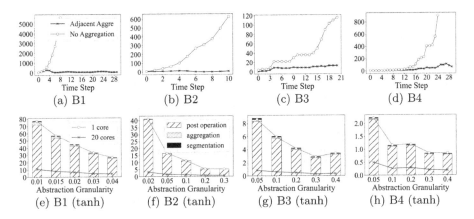

Fig. 9. Differential (Row 1) and decomposing (Row 2) analysis results. Y-axis in (a–d) indicates the number of interval boxes while in (e–h) the time overhead in seconds. Due to the space limitation, we use a scalar value g_1 to denote the n-dimensional abstraction granularity vector $\gamma = (g_1, ..., g_1)$.

instances and Polar in 24 instances. Verisig 2.0 reports 1 unknown case (marked by $\boldsymbol{X}^a$, indicating that over-approximated sets get outside of the goal region). Additionally, Verisig 2.0 and Polar report 1 case and 4 cases of terminating before reaching the goal region, respectively, denoted by $\boldsymbol{X}^b$. These also reflect that BBReach is tighter and introduces less overestimation than other tools.

Table 2 also shows the time cost. Note that Verisig 2.0 is not applicable to ReLU neural networks (marked by $\boldsymbol{X}^c$). BBReach costs much less time than Verisig 2.0 (up to 176× speedup) with parallelization enabled. Even with a single core, BBReach incurs less overhead than Verisig 2.0 in most cases. Compared to Polar, BBReach consumes more time in dealing with small-sized neural networks in B1, B2, and Tora because the finer-grained abstraction granularity is chosen in the three cases, which affects the performance (see Sect. 6.4). Nevertheless, BBReach consumes less time in all the remaining cases than Polar. In addition, BBReach outperforms Polar with up to 182× speedup (the latest release of Polar does not support parallelization), thanks to the parallel acceleration.

The efficiency advantage of BBReach becomes more notable with larger networks such as $\text{Tanh}_{4\times200}$, thanks to the black-box feature of our approach. BBReach consumes almost the same time even for larger neural networks as for small neural networks (e.g., $\text{Tanh}_{2\times20}$). In contrast, the time cost of both the white-box approach almost always increases significantly with larger neural networks. Moreover, Polar incurs more overhead to process the neural networks with the Tanh activation function compared to ReLU, while BBReach consumes similar times for both activation functions. Consequently, it is fair to conclude that BBReach is more efficient and scalable to large-sized neural networks with any activation functions. It is also evident that, via a decent design of neural networks, the reachability analysis for DNN-controlled systems is achievable while the planted decision-making neural networks are treated as black-box oracles, with significant rightness and efficiency outperformance over the white-box approaches.

6.4 Differential and Decomposing Analysis

Differential Analysis. To demonstrate the significance of the adjacent interval aggregation in Algorithm 2, we measure the growth rate of the number of interval boxes with adjacent aggregation, as well as with no aggregation. Figure 9(a-d) shows the comparison results on B1-B4 (the results for the other six benchmarks are similar and given in [56]. We observe that the number of interval boxes grows rapidly with no aggregation, which implies a dramatically increased verification overhead. With the adjacent interval aggregation, the number of interval boxes is extremely small and stable.

Decomposing Analysis. We evaluate how different abstraction granularity levels affect the performance of BBReach and its components. Abstraction granularity is a crucial hyper-parameter used in both training and calculation of over-approximation sets. To better understand the impact of abstraction granularity, for each benchmark, in addition to the default abstraction granularity levels (details can be found in [56]), we choose two finer and two coarser levels, respectively, to evaluate the verification efficiency on both Tanh and ReLU neural networks. We also measure the time consumed by each of the three steps, i.e., interval segmentation, post operation, and adjacent interval aggregation.

We present in Fig. 9(e-h) the results with the Tanh neural network in B1-B4 (the remaining results are similar and given in [56]). With a single core, as the abstraction granularity becomes coarse-grained, the verification time decreases; however, a fairly fine-grained abstraction granularity, e.g., (0.01, 0.01), could result in much higher verification overhead. We also observe that the post operation takes most of the verification time, while the overhead of the other two steps is negligible. Finally, as expected, the parallelization (with 20 cores) can significantly accelerate BBReach.

7 Related Work

Our work is a sequel of recently emerged approaches for the reachability analysis of DNN-controlled systems such as Verisig 2.0 [32], Polar [30], ReachNN* [31]. Besides these states of the art, NNV [58] introduces the star set analysis technique [57] to deal with the neural network and combines with the tool called CORA [5] for the reachability analysis of non-linear systems. JuliaReach [47] integrates the over-approximation of environment dynamics and DNNs together using Taylor models and zonotope. All these approaches treat DNNs as white boxes by over-approximating them with efficiently computable models such as Taylor models [13]. Due to the intrinsic complexity of DNNs, these white-box approaches are applicable to only a limited type of DNNs on small scales.

Our abstraction-based training method follows those machine learning methodologies which advocate a similar idea of pre-processing training data using either abstraction [2,34], fuzzing [12] or granulation [52] for various purposes of reducing the size of models, capturing uncertainties in input data and extracting abstract knowledge. Recent studies show that, rather than training on concrete datasets, training on symbolic datasets is helpful to build verification-friendly neural networks [20] and network-controlled systems [34]. The approach in [2] is focused on the training of finite-state systems, while the one in [34] needs to extend existing training methods to admit abstract states. Our design of abstract neural networks is more decent than the approach in [34] because we only need to insert abstraction layers into neural networks and do not impose any other changes to training algorithms.

There are several black-box but unsound verification approaches for DNN-controlled systems. For instance, Fan et al. proposed a hybrid approach of combining black-box simulation and white-box transition graph for a probabilistic verification result [22]. Xue et al. proposed a black-box model-checking approach for continuous-time dynamical systems based on Probably Approximate Correctness (PAC) learning [60]. Dong et al. built a discrete-time Markov chain from extracted trajectories of a DRL system and verified safety properties by probabilistic model checking [18]. However, these approaches are not sound and can only compute error probability and confidence with probably approximate correctness guarantees. The fundamental reason for the unsoundness is that only partial behaviors of systems can be modeled when conventional neural networks are treated as black-box oracles, i.e., fixing concrete system states and feeding them into the networks to determine the state transitions.

8   Conclusion and Future Work

We have presented an efficient and tight approach for the reachability analysis of DNN-controlled systems by bypassing the time-consuming and imprecise over-approximation of the DNNs in systems via abstraction-based training. Our method demonstrates the possibility of achieving sound but black-box reachability analysis through a decent abstraction-based training approach, breaking conventional intuitions that black-box methods only offer approximate correctness guarantees [22,60] and that over-approximating DNNs is inevitable for sound verification [30–32]. Compared to white-box approaches, our black-box approach offers several benefits, including significant efficiency improvements, improved tightness of computed overestimation sets, applicability and scalability to a wider range of extended abstract DNNs, regardless of their architectures, activation functions, and neuron size. Nevertheless, the reachability analysis part may suffer from state explosion in the worst case when the number of reachable states increases exponentially, as faced by all the related white-box approaches [30–32]. One possible solution is to coarsen the abstraction to reduce the size of abstract states, and learn an easy-to-verify linear policy for each coarsened abstract state. Such an approach has been successfully applied to reinforcement learning [4] and requires further investigation in the DNN-based setting.

Our work sheds light on a promising direction for studying efficient and sound formal verification approaches for DNN-controlled systems by treating

black-box-featured DNNs as black boxes. We believe that this first black-box
reachability analysis approach for DNN-controlled systems would stimulate more
future work, such as new abstraction methods, runtime verification and model-
checking of more complex safety and liveness properties.

Acknowledgment. The work has been supported by the National Key Project
(2020AAA0107800), NSFC Programs (62161146001, 62372176), Huawei Technologies
Co., Ltd., the Shanghai International Joint Lab (22510750100), the Shanghai Trusted
Industry Internet Software Collaborative Innovation Center, the Engineering and Physical Sciences Research Council (EP/T006579/1), the National Research Foundation
(NRF-RSS2022-009), Singapore, and the Shanghai Jiao Tong University Postdoc Schol-
arship.

References

1. Abadi, M., et al.: Tensorflow: a system for large-scale machine learning. In: OSDI,
   vol. 16, pp. 265–283. Savannah, GA, USA (2016)
2. Abel, D.: A theory of state abstraction for reinforcement learning. In: AAAI, vol.
   33, pp. 9876–9877 (2019)
3. Afzal, M., et al.: Veriabs: verification by abstraction and test generation. In: ASE,
   pp. 1138–1141. IEEE (2019)
4. Akrour, R., Veiga, F., Peters, J., Neumann, G.: Regularizing reinforcement learning
   with state abstraction. In: IROS, pp. 534–539. IEEE (2018)
5. Althoff, M.: An introduction to CORA 2015. In: Cyber-Physical Systems Virtual
   Organization (CPS-VO 2015), pp. 120–151 (2015)
6. Althoff, M., Magdici, S.: Set-based prediction of traffic participants on arbitrary
   road networks. IEEE Trans. Intell. Veh. **1**(2), 187–202 (2016)
7. Alur, R., et al.: The algorithmic analysis of hybrid systems. Theoret. Comput. Sci.
   138(1), 3–34 (1995)
8. Bacci, E., Parker, D.: Probabilistic guarantees for safe deep reinforcement learning.
   In: Bertrand, N., Jansen, N. (eds.) FORMATS 2020. LNCS, vol. 12288, pp. 231–
   248. Springer, Cham (2020). https://doi.org/10.1007/978-3-030-57628-8_14
9. Baier, C., Katoen, J.P.: Principles of Model Checking. MIT Press, Cambridge
   (2008)
10. Baluta, T., Chua, Z.L., Meel, K.S., Saxena, P.: Scalable quantitative verification
    for deep neural networks. In: ICSE, pp. 312–323. IEEE (2021)
11. Bertsekas, D.P., Rhodes, I.B.: On the minimax reachability of target sets and target
    tubes. Automatica **7**(2), 233–247 (1971)
12. Campos Souza, P.V.: Fuzzy neural networks and neuro-fuzzy networks: a review
    the main techniques and applications used in the literature. Appl. Soft Comput.
    92, 106275 (2020)
13. Chen, X., Ábrahám, E., Sankaranarayanan, S.: Taylor model flowpipe construction
    for non-linear hybrid systems. In: RTSS, pp. 183–192. IEEE (2012)
14. Chen, X., Ábrahám, E., Sankaranarayanan, S.: Flow*: an analyzer for non-linear
    hybrid systems. In: Sharygina, N., Veith, H. (eds.) CAV 2013. LNCS, vol. 8044, pp.
    258–263. Springer, Heidelberg (2013). https://doi.org/10.1007/978-3-642-39799-
    8_18
15. Christakis, M., et al.: Automated safety verification of programs invoking neural
    networks. In: Silva, A., Leino, K.R.M. (eds.) CAV 2021. LNCS, vol. 12759, pp.
    201–224. Springer, Cham (2021). https://doi.org/10.1007/978-3-030-81685-8_9

16. Collins, P., Bresolin, D., et al.: Computing the evolution of hybrid systems using rigorous function calculus. IFAC Proc. Vol. **45**(9), 284–290 (2012)
17. Cousot, P., Cousot, R.: Abstract interpretation: a unified lattice model for static analysis of programs by construction or approximation of fixpoints. In: POPL, pp. 238–252 (1977)
18. Dong, Y., Zhao, X., Huang, X.: Dependability analysis of deep reinforcement learning based robotics and autonomous systems through probabilistic model checking. In: IROS, pp. 5171–5178. IEEE (2022)
19. Dreossi, T., et al.: VerifAI: a toolkit for the formal design and analysis of artificial intelligence-based systems. In: Dillig, I., Tasiran, S. (eds.) CAV 2019. LNCS, vol. 11561, pp. 432–442. Springer, Cham (2019). https://doi.org/10.1007/978-3-030-25540-4_25
20. Drews, S., Albarghouthi, A., D'Antoni, L.: Proving data-poisoning robustness in decision trees. In: PLDI, pp. 1083–1097 (2020)
21. Dutta, S., Chen, X., Sankaranarayanan, S.: Reachability analysis for neural feedback systems using regressive polynomial rule inference. In: HSCC, pp. 157–168 (2019)
22. Fan, C., Qi, B., Mitra, S., Viswanathan, M.: DryVR: data-driven verification and compositional reasoning for automotive systems. In: Majumdar, R., Kunčak, V. (eds.) CAV 2017. LNCS, vol. 10426, pp. 441–461. Springer, Cham (2017). https://doi.org/10.1007/978-3-319-63387-9_22
23. Fan, J., Huang, C., Chen, X., Li, W., Zhu, Q.: ReachNN*: a tool for reachability analysis of neural-network controlled systems. In: Hung, D.V., Sokolsky, O. (eds.) ATVA 2020. LNCS, vol. 12302, pp. 537–542. Springer, Cham (2020). https://doi.org/10.1007/978-3-030-59152-6_30
24. Fang, X., Calinescu, R., Gerasimou, S., Alhwikem, F.: Fast parametric model checking through model fragmentation. In: ICSE, pp. 835–846. IEEE (2021)
25. Frehse, G.: SpaceEx: scalable verification of hybrid systems. In: Gopalakrishnan, G., Qadeer, S. (eds.) CAV 2011. LNCS, vol. 6806, pp. 379–395. Springer, Heidelberg (2011). https://doi.org/10.1007/978-3-642-22110-1_30
26. Gallestey, E., Hokayem, P.: Lecture notes in nonlinear systems and control (2019)
27. Gomes, L.: When will Google's self-driving car really be ready? It depends on where you live and what you mean by "ready" [news]. IEEE Spectr. **53**(5), 13–14 (2016)
28. Heo, K., Oh, H., Yang, H.: Resource-aware program analysis via online abstraction coarsening. In: ICSE, pp. 94–104. IEEE (2019)
29. Hildebrandt, C., Elbaum, S., Bezzo, N.: Blending kinematic and software models for tighter reachability analysis. In: ICSE(NIER), pp. 33–36 (2020)
30. Huang, C., Fan, J., Chen, X., Li, W., Zhu, Q.: POLAR: a polynomial arithmetic framework for verifying neural-network controlled systems. In: Bouajjani, A., Holík, L., Wu, Z. (eds.) Automated Technology for Verification and Analysis. ATVA 2022. LNCS, vol. 13505, pp. 414–430. Springer, Cham (2022). https://doi.org/10.1007/978-3-031-19992-9_27
31. Huang, C., Fan, J., Li, W., Chen, X., Zhu, Q.: ReachNN: reachability analysis of neural-network controlled systems. ACM Trans. Embed. Comput. Syst. **18**(5s), 1–22 (2019)
32. Ivanov, R., Carpenter, T., Weimer, J., Alur, R., Pappas, G., Lee, I.: Verisig 2.0: verification of neural network controllers using Taylor model preconditioning. In: Silva, A., Leino, K.R.M. (eds.) CAV 2021. LNCS, vol. 12759, pp. 249–262. Springer, Cham (2021). https://doi.org/10.1007/978-3-030-81685-8_11

33. Ivanov, R., Carpenter, T.J., Weimer, J., Alur, R., Pappas, G.J., Lee, I.: Verifying the safety of autonomous systems with neural network controllers. ACM Trans. Embed. Comput. Syst. **20**(1), 1–26 (2020)
34. Jin, P., Tian, J., Zhi, D., Wen, X., Zhang, M.: TRAINIFY: A CEGAR-driven training and verification framework for safe deep reinforcement learning. In: Shoham, S., Vizel, Y. (eds) Computer Aided Verification. CAV 2022. Lecture Notes in Computer Science, vol. 13371, pp. 193–218. Springer, Cham (2022). https://doi.org/10.1007/978-3-031-13185-1_10
35. LeCun, Y., Bengio, Y., Hinton, G.: Deep learning. Nature **521**(7553), 436–444 (2015)
36. Lillicrap, T.P., Hunt, J.J., Pritzel, A., et al.: Continuous control with deep reinforcement learning. In: ICLR, OpenReview.net (2016)
37. Limon, D., Bravo, J., Alamo, T., Camacho, E.: Robust MPC of constrained nonlinear systems based on interval arithmetic. IEE Proc. Control Theory App. **152**(3), 325–332 (2005)
38. Lygeros, J., Tomlin, C., Sastry, S.: Controllers for reachability specifications for hybrid systems. Automatica **35**(3), 349–370 (1999)
39. Makino, K., Berz, M.: Taylor models and other validated functional inclusion methods. Int. J. Pure Appl. Math. **6**, 239–316 (2003)
40. Meiss, J.D.: Differential dynamical systems, Mathematical modeling and computation, vol. 14. SIAM (2007)
41. Minsky, M.L.: Computation. Prentice-Hall Englewood Cliffs, Hoboken (1967)
42. Mnih, V., Kavukcuoglu, K., Silver, D., et al.: Human-level control through deep reinforcement learning. Nature **518**(7540), 529–533 (2015)
43. Moore, R.E., Kearfott, R.B., Cloud, M.J.: Introduction to interval analysis. SIAM (2009)
44. Park, S., Kim, J., Kim, G.: Time discretization-invariant safe action repetition for policy gradient methods. In: NeurIPS 2021, vol. 34, pp. 267–279 (2021)
45. Paszke, A., et al.: Pytorch: an imperative style, high-performance deep learning library. In: NeurIPS, vol. 32 (2019)
46. Pereira, A., Althoff, M.: Over approximative human arm occupancy prediction for collision avoidance. IEEE Trans. Autom. Sci. Eng. **15**(2), 818–831 (2017)
47. Schilling, C., Forets, M., Guadalupe, S.: Verification of neural-network control systems by integrating Taylor models and zonotopes. In: AAAI, vol. 36, pp. 8169–8177 (2022)
48. Schmidt, L., Kontes, G., Plinge, A., Mutschler, C.: Can you trust your autonomous car? Interpretable and verifiably safe reinforcement learning. In: IV, pp. 171–178. IEEE (2021)
49. Schürmann, B., Kochdumper, N., Althoff, M.: Reached model predictive control for disturbed nonlinear systems. In: CDC, pp. 3463–3470. IEEE (2018)
50. Scott, J., Raimondo, D., Marseglia, G., Braatz, R.: Constrained zonotopes: a new tool for set-based estimation and fault detection. Automatica **69**, 126–136 (2016)
51. Singh, S.P., Jaakkola, T., Jordan, M.I.: Reinforcement learning with soft state aggregation. NeurIPS **7**, 361–368 (1995)
52. Song, M., Jing, Y., Pedrycz, W.: Granular neural networks: a study of optimizing allocation of information granularity in input space. Appl. Soft Comput. **77**, 67–75 (2019)
53. Su, J., Chen, W.H.: Model-based fault diagnosis system verification using reachability analysis. IEEE Trans. Syst. Man Cybern. Syst. **49**(4), 742–751 (2017)
54. Sun, X., Khedr, H., Shoukry, Y.: Formal verification of neural network controlled autonomous systems. In: HSCC, pp. 147–156 (2019)

55. Szegedy, C., et al.: Intriguing properties of neural networks. In: ICLR (2014)
56. Tian, J., Zhi, D., Liu, S., Wang, P., Katz, G., Zhang, M.: Taming reachability analysis of DNN-controlled systems via abstraction-based training (2023)
57. Tran, H.-D., Bak, S., Xiang, W., Johnson, T.T.: Verification of deep convolutional neural networks using imagestars. In: Lahiri, S.K., Wang, C. (eds.) CAV 2020. LNCS, vol. 12224, pp. 18–42. Springer, Cham (2020). https://doi.org/10.1007/978-3-030-53288-8_2
58. Tran, H.-D., Yang, X., Manzanas Lopez, D., Musau, P., Nguyen, L.V., Xiang, W., Bak, S., Johnson, T.T.: NNV: the neural network verification tool for deep neural networks and learning-enabled cyber-physical systems. In: Lahiri, S.K., Wang, C. (eds.) CAV 2020. LNCS, vol. 12224, pp. 3–17. Springer, Cham (2020). https://doi.org/10.1007/978-3-030-53288-8_1
59. Wang, Z., Albarghouthi, A., Prakriya, G., Jha, S.: Interval universal approximation for neural networks. In: POPL, vol. 6, pp. 1–29. ACM (2022)
60. Xue, B., Zhang, M., Easwaran, A., Li, Q.: Pac model checking of black-box continuous-time dynamical systems. IEEE Trans. Comput. Aided Des. Integr. Circuits Syst. **39**(11), 3944–3955 (2020)
61. Zhang, Y., et al.: QVIP: an ILP-based formal verification approach for quantized neural networks. In: ASE, pp. 1–13. No. 80 (2022)

Verification of Neural Networks' Local Differential Classification Privacy

Roie Reshef[(⊠)], Anan Kabaha, Olga Seleznova, and Dana Drachsler-Cohen

Technion, Haifa, Israel
{sror,anan.kabaha}@campus.technion.ac.il, olga.s@technion.ac.il,
ddana@ee.technion.ac.il

Abstract. Neural networks are susceptible to privacy attacks. To date, no verifier can reason about the privacy of individuals participating in the training set. We propose a new privacy property, called *local differential classification privacy (LDCP)*, extending local robustness to a differential privacy setting suitable for black-box classifiers. Given a neighborhood of inputs, a classifier is LDCP if it classifies all inputs the same regardless of whether it is trained with the full dataset or whether any single entry is omitted. A naive algorithm is highly impractical because it involves training a very large number of networks and verifying local robustness of the given neighborhood separately for every network. We propose Sphynx, an algorithm that computes an abstraction of all networks, with a high probability, from a small set of networks, and verifies LDCP directly on the abstract network. The challenge is twofold: network parameters do not adhere to a known distribution probability, making it difficult to predict an abstraction, and predicting too large abstraction harms the verification. Our key idea is to transform the parameters into a distribution given by KDE, allowing to keep the over-approximation error small. To verify LDCP, we extend a MILP verifier to analyze an abstract network. Experimental results show that by training only 7% of the networks, Sphynx predicts an abstract network obtaining 93% verification accuracy and reducing the analysis time by $1.7 \cdot 10^4$x.

1 Introduction

Neural networks are successful in various tasks but are also vulnerable to attacks. One kind of attacks that is gaining a lot of attention is privacy attacks. Privacy attacks aim at revealing sensitive information about the network or its training set. For example, membership inference attacks recover entries in the training set [11,35,37,40,43,57,76], model inversion attacks reveal sensitive attributes of these entries [22,23], model extraction attacks recover the model's parameters [65], and property inference attacks infer global properties of the model [24]. Privacy attacks have been shown successful even against platforms providing a limited access to a model, including black-box access and a limited number of queries. Such restricted access is common for plat-

© The Author(s), under exclusive license to Springer Nature Switzerland AG 2024
R. Dimitrova et al. (Eds.): VMCAI 2024, LNCS 14500, pp. 98–123, 2024.
https://doi.org/10.1007/978-3-031-50521-8_5

forms providing machine-learning-as-a-service (e.g., Google's Vertex AI[1], Amazon's ML on AWS[2], and BigML[3]).

A common approach to mitigate privacy attacks is *differential privacy* (DP) [18]. DP has been adopted by numerous network training algorithms [1, 8,10,28,44]. Given a privacy level, a DP training algorithm generates the same network with a similar probability, regardless of whether a particular individual's entry is included in the dataset. However, DP is not an adequately suitable privacy criterion for black-box classifiers, for two reasons. First, it poses a too strong requirement: it requires that the training algorithm returns the same network (i.e., assign the same score to every class), whereas black-box classifiers are considered the same if they predict the same class (i.e., assign the *maximal* score to the same class). Second, DP can only be satisfied by randomized algorithms, adding noise to the computations. Consequently, the accuracy of the resulting network decreases. The amount of noise is often higher than necessary because the mathematical analysis of differentially private algorithms is highly challenging and thus practically not tight (e.g., it often relies on compositional theorems [18]). Thus, network designers often avoid adding noise to their networks. This raises the question: *What can a network designer provide as a privacy guarantee for individuals participating in the training set of a black-box classifier?*

We propose a new privacy property, called *local differential classification privacy (LDCP)*. Our property is designed for black-box classifiers, whose training algorithm is not necessarily DP. Conceptually, it extends the local robustness property, designed for adversarial example attacks [27,63], to a "deterministic differential privacy" setting. Local robustness requires that the network classifies all inputs in a given neighborhood the same. We extend this property by requiring that the network classifies all inputs in a given neighborhood the same *regardless of whether a particular individual's entry is included in the dataset*.

Proving that a network is LDCP is challenging, because it requires to check the local robustness of a very large number of networks: $|\mathcal{D}|+1$ networks, where $\mathcal{D}$ is the dataset, which is often large (e.g., $> 10k$ entries). To date, verification of local robustness [25,31,42,55,59,64,66,72], analyzing a single network, takes non-negligible time. A naive, accurate but highly unscalable algorithm checks LDCP by training every possible network (i.e., one for every possibility to omit a single entry from the dataset), checking that all inputs in the neighborhood are classified the same network-by-network, and verifying that all networks classify these inputs the same. However, this naive algorithm does not scale since training and analyzing thousands of networks is highly time-consuming.

We propose Sphynx (**S**afety **P**rivacy analyzer via **Hy**per-**N**etworks) for determining whether a network is LDCP at a given neighborhood (Fig. 1). Sphynx takes as inputs a network, its training set, its training algorithm, and a neighborhood. Instead of training $|\mathcal{D}|$ more networks, Sphynx computes a *hyper-network* abstracting these networks with a high probability (under several conditions). A

---

[1] https://cloud.google.com/vertex-ai.

[2] https://aws.amazon.com/machine-learning/.

[3] https://bigml.com/.

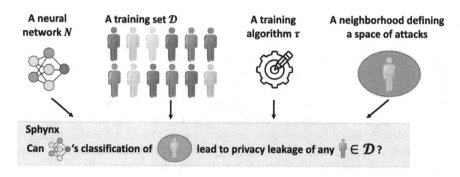

Fig. 1. Sphynx checks leakage of individuals' entries at a given neighborhood.

hyper-network abstracts a set of networks by associating to each network parameter an interval that contains the respective values in all networks. If Sphynx would train all networks, computing the hyper-network would be straightforward. However, Sphynx's goal is to reduce the high time overhead and thus it does not train all networks. Instead, it predicts a hyper-network from a small number of networks. Sphynx then checks LDCP at the given neighborhood directly on the hyper-network. These two ideas enable Sphynx to obtain a practical analysis time. The main challenges in predicting a hyper-network are: (1) network parameters do not adhere to a known probability distribution and (2) the inherent trade-off between a sound abstraction, where each parameter's interval covers all values of this parameter in every network (seen and unseen) and the ability to succeed in verifying LDCP given the hyper-network. Naturally, to obtain a sound abstraction, it is best to consider large intervals for each network parameter, e.g., by adding noise to the intervals (like in adaptive data analysis [5,15–17,21,29,53,69]). However, the larger the intervals the harder it is to verify LDCP, because the hyper-network abstracts many more (irrelevant) networks. To cope, Sphynx transforms the network parameters into a distribution given by kernel density estimation (KDE), allowing to predict the intervals without adding noise.

To predict a hyper-network, Sphynx executes an iterative algorithm. In every iteration, it samples a few entries from the dataset. For each entry, it trains a network given all the dataset except this entry. Then, given all trained networks, Sphynx predicts a hyper-network, i.e., it computes an interval for every network parameter. An interval is computed by transforming every parameter's observed values into a distribution given by KDE, using normalization and the Yeo-Johnson transformation [78]. Then, Sphynx estimates whether the hyper-network abstracts every network with a high probability, and if so, it terminates.

Given a hyper-network, Sphynx checks LDCP directly on the hyper-network. To this end, we extend a local robustness verifier [64], relying on mixed-integer linear programming (MILP), to analyze a hyper-network. Our extension replaces the equality constraints, capturing the network's affine computations, with inequality constraints, since our network parameters are associated with intervals and not real numbers. To mitigate an over-approximation error, we propose two approaches. The first approach relies on preprocessing to the network's inputs and the second one relies on having a lower bound on the inputs.

We evaluate Sphynx on data-sensitive datasets: Adult Census [13], Bank Marketing [41] and Default of Credit Card Clients [77]. We verify LDCP on three kinds of neighborhoods for checking safety to label-only membership attacks [11,35,40], adversarial example attacks in a DP setting (like [34,46]), and sensitive attributes (like [22,23]). We show that by training only 7% of the networks, Sphynx predicts a hyper-network abstracting an (unseen) network with a probability of at least 0.9. Our hyper-networks obtain 93% verification accuracy. Compared to the naive algorithm, Sphynx provides a significant speedup: it reduces the training time by 13.6x and the verification time by $1.7 \cdot 10^4$x.

2 Preliminaries

In this section, we provide the necessary background.

Neural Network Classifiers. We focus on binary classifiers, which are popular for data-sensitive tasks. As example, we describe the data-sensitive datasets used in our evaluation and their classifier's task (Sect. 6). Adult Census [13] consists of user records of the socioeconomic status of people in the US. The goal of the classifier is to predict whether a person's yearly income is higher or lower than 50K USD. Bank Marketing [41] consists of user records of direct marketing campaigns of a Portuguese banking institution. The goal of the classifier is to predict whether the client will subscribe to the product or not. Default of Credit Card Clients [77] consists of user records of demographic factors, credit data, history of payments, and bill statements of credit card clients in Taiwan. The goal of the classifier is to predict whether the default payment will be paid in the next month. We note that our definitions and algorithms easily extend to non-binary classifiers. A binary classifier N maps an input, a user record, $x \in \mathcal{X} \subseteq [0,1]^d$ to a real number $N(x) \in \mathbb{R}$. If $N(x) \geq 0$, we say the classification of x is 1 and write $\text{class}(N(x)) = 1$, otherwise, it is -1, i.e., $\text{class}(N(x)) = -1$. We focus on classifiers implemented by a fully-connected neural network. This network consists of an input layer followed by L layers. The input layer x_0 takes as input $x \in \mathcal{X}$ and passes it as is to the next layer (i.e., $x_{0,k} = x_k$). The next layers are functions, denoted $f_1, f_2, \ldots, f_L$, each taking as input the output of the preceding layer. The network's function is the composition of the layers: $N(x) = f_L(f_{L-1}(\cdots(f_1(x))))$. A layer m consists of neurons, denoted $x_{m,1}, \ldots, x_{m,k_m}$. Each neuron takes as input the outputs of all neurons in the preceding layer and outputs a real number. The output of layer m is the vector $(x_{m,1}, \ldots, x_{m,k_m})^T$ consisting of all its neurons' outputs. A neuron $x_{m,k}$ has a weight for each input $w_{m,k,k'} \in \mathbb{R}$ and a bias $b_{m,k} \in \mathbb{R}$. Its function is the composition of an affine computation, $\hat{x}_{m,k} = b_{m,k} + \sum_{k'=1}^{k_{m-1}} w_{m,k,k'} \cdot x_{m-1,k'}$, followed by an activation function computation, $x_{m,k} = \sigma(\hat{x}_{m,k})$. Activation functions are typically non-linear functions. In this work, we focus on the ReLU activation function, $\text{ReLU}(\hat{x}) = \max(0, \hat{x})$. We note that, while we focus on fully-connected networks, our approach can extend to other architectures, e.g., convolutional networks or residual networks. The weights and biases of a neural network are determined by a training process. A training algorithm $\mathcal{T}$ takes as inputs a network (typically, with random weights and biases) and a labelled training set $\mathcal{D} = \{(x_1, y_1), \ldots, (x_n, y_n)\} \subseteq \mathcal{X} \times \{-1, +1\}$. It returns a network with updated weights and biases. These parameters are computed with the goal of minimizing a given loss function, e.g., binary cross-entropy, capturing the level of inaccuracy of the network. The computation typically relies on iterative numerical optimization, e.g., stochastic gradient descent (SGD).

Differential Privacy (DP). DP focuses on algorithms defined over arrays (in our context, a dataset). At high-level, an algorithm is DP if for any two inputs differing in a single entry, it returns the same output with a similar probability. Formally, DP is a probabilistic privacy property requiring that the probability of returning different outputs is upper bounded by an expression defined by two parameters, denoted ϵ and δ [18]. Note that this requirement is too strong for classifiers providing only black-box access, which return only the input's classification. For such classifiers, it is sufficient to require that the classification is the same, and there is no need for the network's output (the score of every class) to be the same. To obtain the DP guarantee, DP algorithms add noise, drawn from some probability distribution (e.g., Laplace or Gaussian), to the input or their computations. That is, DP algorithms trade off their output's accuracy with privacy guarantees: the smaller the DP parameters (i.e., ϵ and δ) the more private the algorithm is, but its outputs are less accurate. The accuracy loss is especially severe in DP algorithms that involve loops in which every iteration adds noise. The loss is high because (1) many noise terms are added and (2) the mathematical analysis is not tight (it typically relies on compositional theorems [18]), leading to adding a higher amount of noise than necessary to meet the target privacy guarantee. Nevertheless, DP has been adopted by numerous network training algorithms [1,8,10,28,44]. For example, one algorithm adds noise to every gradient computed during training [1]. Consequently, the network's accuracy decreases significantly, discouraging network designers from employing DP. To cope, we propose a (non-probabilistic) privacy property that (1) only requires the network's classification to be the same and (2) can be checked even if the network has not been trained by a DP training algorithm.

Local Robustness. Local robustness has been introduced in response to adversarial example attacks [27,63]. In the context of network classifiers, an adversarial example attack is given an input and a space of perturbations and it returns a perturbed input that causes the network to misclassify. Ideally, to prove that a network is robust to adversarial attacks, one should prove that *for any valid input,* the network classifies the same under any valid perturbation. In practice, the safety property that has been widely-studied is *local robustness.* A network is locally robust at *a given input* if perturbing the input by a perturbation in a given space does not cause the network to change the classification. Formally, the space of allowed perturbations is captured by a *neighborhood* around the input.

Definition 1 (Local Robustness). Given a network N, an input $x \in \mathcal{X}$, a neighborhood $I(x) \subseteq \mathcal{X}$ containing x, and a label $y \in \{-1, +1\}$, the network N is *locally robust* at $I(x)$ *with respect to* y if $\forall x' \in I(x).\ \mathrm{class}(N(x')) = y$.

A well-studied definition of a neighborhood is the ϵ-*ball* with respect to the L_∞ norm [25,31,42,55,59,64,66,72]. Formally, given an input x and a bound on the perturbation amount $\epsilon \in \mathbb{R}^+$, the ϵ-ball is: $I_\epsilon^\infty(x) = \{x' \mid \|x' - x\|_\infty \leq \epsilon\}$. A different kind of a neighborhood captures *fairness* with respect to a given sensitive feature $i \in [d]$ (e.g., gender) [6,38,54,70]: $I_i^S(x) = \left\{ x' \mid \bigwedge_{j \in [d] \setminus \{i\}} x'_j = x_j \right\}$.

3   Local Differential Classification Privacy

In this section, we define the problem of verifying that a network is locally differentially classification private (LDCP) at a given neighborhood.

Local Differential Classification Privacy (LDCP). Our property is defined given a classifier N, trained on a dataset $\mathcal{D}$, and a neighborhood I, defining a space of attacks (perturbations). It considers N private with respect to an individual's entry (x', y') if N classifies all inputs in I the same, whether N has been trained with (x', y') or not. If there is discrepancy in the classification, the attacker may exploit it to infer information about x'. Namely, if the network designer cared about a single individual's entry (x', y'), our privacy property would be defined over two networks: the network trained with (x', y') and the network trained without (x', y'). Naturally, the network designer wishes to show that the classifier is private for every (x', y') participating in $\mathcal{D}$. Namely, our property is defined over $|\mathcal{D}| + 1$ networks. The main challenge in verifying this privacy property is that the training set size is typically very large (>10k entries). Formally, our property is an extension of local robustness to a differential privacy setting suitable for classifiers providing black-box access. It requires that the inputs in I are classified the same by *every* network trained on $\mathcal{D}$ or trained on $\mathcal{D}$ except for any single entry. Unlike DP, our definition is applicable to any network, even those trained without a probabilistic noise. We next formally define our property.

Definition 2 (Local Differential Classification Privacy). Given a network N trained on a dataset $\mathcal{D} \subseteq \mathcal{X} \times \{-1, +1\}$ using a training algorithm $\mathcal{T}$, a neighborhood $I(x) \subseteq \mathcal{X}$ and a label $y \in \{-1, +1\}$, the network N is *locally differentially classification private (LDCP)* if (1) N is locally robust at $I(x)$ with respect to y, and (2) for every $(x', y') \in \mathcal{D}$, the network of the same architecture as N trained on $\mathcal{D} \setminus \{(x', y')\}$ using $\mathcal{T}$ is locally robust at $I(x)$ with respect to y.

Our privacy property enables to prove safety against privacy attacks by defining a suitable set of neighborhoods. For example, several membership inference attacks [11,35,40] are label-only attacks. Namely, they assume the attacker has a set of inputs and they can query the network to obtain their classification. To prove safety against these attacks, given a set of inputs $X \subseteq \mathcal{X}$, one has to check LDCP for every neighborhood defined by $I_0^\infty(x)$ where $x \in X$. To prove safety against attacks aiming to reveal sensitive features (like [22,23]), given a set of inputs $X \subseteq \mathcal{X}$, one has to check that the network classifies the same regardless of the value of the sensitive feature. This can be checked by checking LDCP for every neighborhood defined by $I_i^S(x)$ where $x \in X$ and i is the sensitive feature.

Problem Definition. We address the problem of determining whether a network is locally differentially classification private (LDCP) at a given neighborhood, while minimizing the analysis time. A definite answer involves training and verifying $|\mathcal{D}| + 1$ networks (the network trained given all entries in the training set, and for each entry in the training set, the network trained given all entries except this entry). However, this results in a very long training time and verification time (even local robustness verification takes a non-negligible time). Instead, our problem is to provide an answer which is correct with a high probability. This problem is highly challenging. On the one hand, the fewer trained networks the lower the training and verification time. On the other hand, determining an answer, with a high probability, from a small number of networks is challenging, since network parameters do not adhere to a known probabilistic distribution. Thus, our problem is not naturally amenable to a probabilistic analysis.

Prior Work. Prior work has considered different aspects of our problem. As mentioned, several works adapt training algorithms to guarantee that the resulting network satisfies differential privacy [1,8,10,28,44]. However, these training algorithms tend to return networks of lower accuracy. A different line of research proposes algorithms for *machine unlearning* [9,26,36], in which an algorithm retrains a network "to forget" some entries of its training set. However, these approaches do not guarantee that the forgetting network is equivalent to the network obtained by training without these entries from the beginning. It is also more suitable for settings in which there are multiple entries to forget, unlike our differential privacy setting which focuses on omitting a single entry at a time. Several works propose verifiers to determine local robustness [19,30,31,45,48,56,74]. However, these analyze a single network and not a group of similar networks. A recent work proposes a proof transfer between similar networks [68] in order to reduce the verification time of similar networks. However, this work requires to explicitly have all networks, which is highly time consuming in our setting. Other works propose verifiers that check robustness to data poisoning or data bias [12,39,62]. These works consider an attacker that can manipulate or omit up to several entries from the dataset, similarly to LDCP allowing to omit up to a single entry. However, these verifiers either target patch attacks of image classifiers [62], which allows them to prove robustness without considering every possible network, or target decision trees [12,39], relying on predicates, which are unavailable for neural networks. Thus, neither is applicable to our setting.

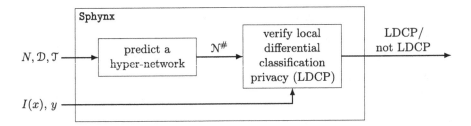

Fig. 2. Given a network classifier N, its training set $\mathcal{D}$ and its training algorithm $\mathcal{T}$, Sphynx predicts an abstract hyper-network $\mathcal{N}^{\#}$. It then checks whether $\mathcal{N}^{\#}$ is locally robust at $I(x)$ with respect to y to determine whether N is LDCP.

4   Our Approach

In this section, we present our approach for determining whether a network is locally differentially classification private (LDCP). Our key idea is to *predict an abstraction* of all concrete networks from a small number of networks. We call the abstraction a *hyper-network*. Thanks to this abstraction, Sphynx does not need to train all concrete networks and neither verify multiple concrete networks. Instead, Sphynx first predicts an abstract hyper-network, given a network N, its training set $\mathcal{D}$ and its training algorithm $\mathcal{T}$, and then verifies LDCP directly on the hyper-network. This is summarized in Fig. 2. We next define these terms.

Hyper-networks. A *hyper-network* abstracts a set of networks $\mathcal{N} = \{N_1, \ldots, N_K\}$ with the same architecture and in particular the same set of parameters, i.e., weights $\mathcal{W} = \{w_{1,1,1}, \ldots, w_{L,d_L,d_{L-1}}\}$ and biases $\mathcal{B} = \{b_{1,1}, \ldots, b_{L,d_L}\}$. The difference between the networks is the parameters' values. In our context, $\mathcal{N}$ consists of the network trained with the full dataset and the networks trained without any single entry: $\mathcal{N} = \{\mathcal{T}(N, \mathcal{D})\} \cup \{\mathcal{T}(N, \mathcal{D} \setminus \{(x', y')\}) \mid (x', y') \in \mathcal{D}\}$. A *hyper-network* is a network $\mathcal{N}^{\#}$ with the same architecture and set of parameters, but the domain of the parameters is not $\mathbb{R}$ but rather an abstract domain $\mathcal{A}$. As standard, we assume the abstract domain corresponds to a lattice and is equipped with a concretization function γ. We focus on a non-relational abstraction, where each parameter is abstracted independently. The motivation is twofold. First, non-relational domains are computationally lighter than relational domains. Second, the relation between the parameters is highly complex because it depends on a long series of optimization steps (e.g., SGD steps). Thus, while it is possible to bound these computations using simpler functions (e.g., polynomial functions), the over-approximation error would be too large to be useful in practice.

Formally, given a set of networks $\mathcal{N}$ with the same architecture and set of parameters $\mathcal{W} \cup \mathcal{B}$ and given an abstract domain $\mathcal{A}$ and a concretization function $\gamma : \mathcal{A} \to \mathcal{P}(\mathbb{R})$, a *hyper-network* is a network $\mathcal{N}^{\#}$ with the same architecture and set of parameters, where the parameters range over $\mathcal{A}$ and satisfy the following:

$$\forall N' \in \mathcal{N} : \left[\forall w_{m,k,k'} \in \mathcal{W} : w^{N'}_{m,k,k'} \in \gamma\left(w^{\#}_{m,k,k'}\right) \wedge \forall b_{m,k} \in \mathcal{B} : b^{N'}_{m,k} \in \gamma\left(b^{\#}_{m,k}\right) \right]$$

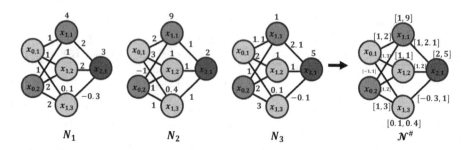

Fig. 3. Three networks, N_1, N_2, N_3, and their interval hyper-network $\mathcal{N}^\#$.

where $w_{m,k,k'}^{N'}$ and $b_{m,k}^{N'}$ are the values of the parameters in the network N' and $w_{m,k,k'}^\#$ and $b_{m,k}^\#$ are the values of the parameters in the hyper-network $\mathcal{N}^\#$.

Interval Abstraction. In this work, we focus on the interval domain. Namely, the abstract elements are intervals $[l, u]$, with the standard meaning: an interval $[l, u]$ abstracts all real numbers between l and u. We thus call our hyper-networks *interval hyper-networks*. Figure 3 shows an example of an interval hyper-network. An interval corresponding to a weight is shown next to the respective edge, and an interval corresponding to a bias is shown next to the respective neuron. For example, the neuron $x_{1,2}$ has two weights and a bias whose values are: $w_{1,2,1}^\# = [1,3]$, $w_{1,2,2}^\# = [1,2]$, and $b_{1,2}^\# = [1,1]$. Computing an interval hyper-network is straightforward if all concrete networks are known. However, computing all concrete networks defeats the purpose of having a hyper-network. Instead, Sphynx predicts an interval hyper-network with a high probability (Sect. 5.1).

Checking LDCP Given a Hyper-network. Given an interval hyper-network $\mathcal{N}^\#$ for a network N, a neighborhood $I(x)$ and a label y, Sphynx checks whether N is LDCP by checking whether $\mathcal{N}^\#$ is locally robust at $I(x)$ with respect to y. If $\mathcal{N}^\#$ is robust, then N is LDCP at $I(x)$, with a high probability. Otherwise, Sphynx determines that N is not LDCP at $I(x)$. Note that this is a conservative answer since N is either not LDCP or that the abstraction or the verification lose too much precision. Sphynx verifies local robustness of $\mathcal{N}^\#$ by extending a MILP verifier [64] checking local robustness of a neural network (Sect. 5.2).

5   Sphynx: Safety Privacy Analyzer via Hyper-networks

In this section, we present Sphynx, our system for verifying local differential classification privacy (LDCP). As described, it consists of two components, the first component predicts a hyper-network, while the second one verifies LDCP.

5.1   Prediction of an Interval Hyper-network

In this section, we introduce Sphynx's algorithm for predicting an interval hyper-network, called PredHyperNet. PredHyperNet takes as inputs a network N, its

---

Algorithm 1: PredHyperNet $(N, \mathcal{D}, \mathcal{T}, \alpha)$

Input: a network N, a training set $\mathcal{D}$, a training algorithm $\mathcal{T}$, an error bound α.
Output: an interval hyper-network.

1   nets $\leftarrow \{N\}$
2   entr $\leftarrow \emptyset$
3   $\mathcal{N}^{\#}_{prev} \leftarrow \bot$
4   $\mathcal{N}^{\#}_{curr} \leftarrow N^{\#}$
5   **while** $\Sigma_{w \in \mathcal{W} \cup \mathcal{B}} \mathbb{1} \left[1 - J([l^w_{curr}, u^w_{curr}], [l^w_{prev}, u^w_{prev}]) \leq R \right] < M \cdot |\mathcal{W} \cup \mathcal{B}|$ **do**
6      $\mathcal{N}^{\#}_{prev} \leftarrow \mathcal{N}^{\#}_{curr}$
7   **for** k *iterations* **do**
8         $(x, y) \leftarrow$ Random $(\mathcal{D} \setminus \text{entr})$
9   entr $\leftarrow$ entr $\cup \{(x, y)\}$
10   nets $\leftarrow$ nets $\cup \{\mathcal{T}(N, \mathcal{D} \setminus \{(x, y)\})\}$
11   **for** $w \in \mathcal{W} \cup \mathcal{B}$ **do**
12        $\mathcal{V}_w \leftarrow \{w^{N'} \mid N' \in \text{nets}\}$
13        $w^{\mathcal{N}^{\#}_{curr}} \leftarrow$ PredInt $\left(\mathcal{V}_w, \frac{\alpha}{|\mathcal{W} \cup \mathcal{B}|} \right)$

14 **return** $\mathcal{N}^{\#}_{curr}$

---

training set $\mathcal{D}$, its training algorithm $\mathcal{T}$, and a probability error bound α, where $\alpha \in (0, 1)$ is a small number. It returns an interval hyper-network $\mathcal{N}^{\#}$ which, with probability $1 - \alpha$ (under certain conditions), abstracts an (unseen) concrete network returned by $\mathcal{T}$ given N and $\mathcal{D} \setminus \{(x, y)\}$, where $(x, y) \in \mathcal{D}$. The main idea is to predict an abstraction for every network parameter from a small number of concrete networks. To minimize the number of networks, PredHyperNet executes iterations. An iteration trains k networks and predicts an interval hyper-network using all trained networks. If the intervals' distributions have not converged to the expected distributions, another iteration begins. We next provide details.

PredHyperNet's Algorithm. PredHyperNet (Algorithm 1) begins by initializing the set of trained networks nets to N and the set of entries entr, whose corresponding networks are in nets, to $\emptyset$. It initializes the previous hyper-network $\mathcal{N}^{\#}_{prev}$ to $\bot$ and the current hyper-network $\mathcal{N}^{\#}_{curr}$ to the interval abstraction of N, i.e., the interval of a network parameter $w \in \mathcal{W} \cup \mathcal{B}$ (a weight or a bias) is $[w^N, w^N]$. Then, while the stopping condition (Line 5), checking convergence using the Jaccard distance as described later, is false, it runs an iteration. An iteration trains k new networks (Line 7–Line 10). A training iteration samples an entry (x, y), adds it to entr, and runs the training algorithm $\mathcal{T}$ on (the architecture of) N and $\mathcal{D} \setminus \{(x, y)\}$. The trained network is added to nets. PredHyperNet then computes an interval hyper-network from nets (Line 11–Line 13). The computation is executed via PredInt (Line 13) independently on each network parameter w. PredInt's inputs are all observed values of w and a probability error bound α', which is α divided by the number of network parameters.

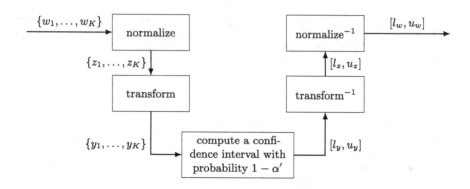

Fig. 4. The flow of `PredInt` for predicting an interval for a network parameter.

Interval Prediction: Overview `PredInt` predicts an interval for a parameter w from a (small) set of values $\mathcal{V}_w = \{w_1, \ldots, w_K\}$, obtained from the concrete networks, and given an error bound α'. If it used interval abstraction, w would be mapped to the interval defined by the minimum and maximum in $\mathcal{V}_w$. However, this approach cannot guarantee to abstract the unseen concrete networks, since we aim to rely on a number of concrete networks significantly smaller than $|\mathcal{D}|$. Instead, `PredInt` defines an interval by predicting the minimum and maximum of w, over all its values in the (seen and unseen) networks. There has been an extensive work on estimating statistics, with a high probability, of an unknown probability distribution (e.g., expectation [5,15–17,21,29,53,69]). However, `PredInt`'s goal is to estimate the *minimum* and *maximum* of unknown samples from an unknown probability distribution. The challenge is that, unlike other statistics, the minimum and maximum are highly sensitive to outliers. To cope, `PredInt` transforms the unknown probability distribution of w into a known one and then predicts the minimum and maximum. Before the transformation, `PredInt` normalizes the values in $\mathcal{V}_w$. Overall, `PredInt`'s operation is (Fig. 4): (1) it normalizes the values in $\mathcal{V}_w$, (2) it transforms the values into a known probability distribution, (3) it predicts the minimum and maximum by computing a confidence interval with a probability of $1 - \alpha'$, and (4) it inverses the transformation and normalization to fit the original scale. We note that executing normalization and transformation and then their inversion does not result in any information loss, because they are bijective functions. We next provide details.

Transformation and Normalization `PredInt` transforms the values in $\mathcal{V}_w$ to make them seem as if drawn from a known probability distribution. It employs the Yeo-Johnson transformation [78], transforming an unknown distribution into a Gaussian distribution. This transformation has the flexibility that the input random variables can have any real value. It has a parameter λ, whose value is determined using maximum likelihood estimation (MLE) (i.e., λ maximizes the likelihood that the transformed data is Gaussian). It is defined as follows:

$$T_\lambda(z) = \left\{ \begin{array}{ll} \frac{(1+z)^\lambda - 1}{\lambda}, & \lambda \neq 0, z \geq 0; \log(1+z), \quad \lambda = 0, z \geq 0 \\ -\frac{(1-z)^{2-\lambda} - 1}{2-\lambda}, & \lambda \neq 2, z < 0; -\log(1-z), \lambda = 2, z < 0 \end{array} \right\}$$

PredInt employs several adaptations to this transformation to fit our setting better. First, it requires $\lambda \in [0, 2]$. This is because if $\lambda < 0$, the transformed outputs have an upper bound $\frac{1}{-\lambda}$ and if $\lambda > 2$, they have a lower bound $-\frac{1}{\lambda-2}$. Since our goal is to predict the minimum and maximum values, we do not want the transformed outputs to artificially limit them. Second, the Yeo-Johnson transformation transforms into a Gaussian distribution. Instead, we transform to a distribution given by kernel density estimation (KDE) with a Laplace kernel, which is better suited for our setting (Sect. 6). We describe KDE in our extended version of the paper [51]. Third, the Yeo-Johnson transformation operates best when given values centered around zero. However, training algorithms produce parameters of different scales and centered around different values. Thus, before the transformation, PredInt normalizes the values in $\mathcal{V}_w$ to be centered around zero and have a similar scale as follows: $z_i \leftarrow \frac{w_i - \mu}{\Delta}, \forall i \in [K]$. Consequently, PredInt is invariant to translation and scaling and is thus more robust to the computations of the training algorithm $\mathcal{T}$. There are several possibilities to define the normalization's parameters, μ and Δ, each is based on a different norm, e.g., the L_1, L_2 or L_∞ norm. In our setting, a normalization based on the L_1 norm works best, since we use a Laplace kernel. Its center point is $\mu = \text{median}(w_1, \ldots, w_K)$ and its scale is the centered absolute first moment: $\Delta = \frac{1}{K} \sum_{i=1}^{K} |w_i - \mu|$.

Interval Prediction. After the transformation, PredInt has a cumulative distribution function (CDF) for the transformed values: $F_y(v) = \mathbb{P}\{y \leq v\}$. Given the CDF, we compute a confidence interval, defining the minimum and maximum. A confidence interval, parameterized by $\alpha' \in (0, 1)$, is an interval satisfying that the probability of an unknown sample being inside it is at least $1 - \alpha'$. It is defined by: $[l_y, u_y] = \left[F_y^{-1} \left(\frac{\alpha'}{2} \right), F_y^{-1} \left(1 - \frac{\alpha'}{2} \right) \right]$. Since we wish to compute an interval hyper-network $\mathcal{N}^\#$ abstracting an unseen network with probability $1 - \alpha$ and since there are $|\mathcal{W} \cup \mathcal{B}|$ parameters, we choose $\alpha' = \frac{\alpha}{|\mathcal{W} \cup \mathcal{B}|}$ (Line 13). By the union bound, we obtain confidence intervals guaranteeing that the probability that an unseen concrete network is covered by the hyper-network is at least $1 - \alpha$.

Stopping Condition. The goal of PredHyperNet's stopping condition is to identify when the distributions of the intervals computed for $\mathcal{N}_{curr}^\#$ have converged to their expected distributions. This is the case when the intervals have not changed significantly in $\mathcal{N}_{curr}^\#$ compared to $\mathcal{N}_{prev}^\#$. Formally, for each network parameter w, it compares the current interval to the previous interval by computing their Jaccard distance. Given the previous and current intervals of w, I_{prev} and I_{curr}, the Jaccard Index is: $J(I_{curr}, I_{prev}) = \frac{I_{curr} \sqcap I_{prev}}{I_{curr} \sqcup I_{prev}}$. For any two intervals, the Jaccard distance $1 - J(I_{curr}, I_{prev}) \in [0, 1]$, such that the smaller the distance, the more similar the intervals are. If the Jaccard distance is below a ratio R (a small number), we consider the interval of w as converged to the expected CDF. If at least $M \cdot 100\%$ of the hyper-network's intervals have converged, we consider that the hyper-network has converged, and thus `PredHyperNet` terminates.

5.2 Verification of a Hyper-network

In this section, we explain how `Sphynx` verifies that an interval hyper-network is locally robust, in order to show that the network is LDCP. Our verification extends a local robustness verifier [64], designed for checking local robustness of a (concrete) network, to check local robustness given an interval hyper-network. This verifier encodes the verification task as a mixed-integer linear program (MILP), which is submitted to a MILP solver, and provides a sound and complete analysis. We note that we experimented with extending incomplete analysis techniques to our setting (interval analysis and DeepPoly [59]) to obtain a faster analysis. However, the over-approximation error stemming both from these techniques and our hyper-network led to a low precision rate. The challenge with extending the verifier by Tjeng et al. [64] is that both the neuron values and the network parameters are variables, leading to quadratic constraints, which are computationally heavy for constraint solvers. Instead, `Sphynx` relaxes the quadratic constraints. To mitigate an over-approximation error, we propose two approaches. We begin with a short background and then describe our extension.

Background. The MILP encoding by Tjeng et al. [64] precisely captures the neurons' affine computation as linear constraints. The ReLU computations are more involved, because ReLU is non-linear. For each ReLU computation, a boolean variable is introduced along with four linear constraints. The neighborhood is expressed by constraining each input value by an interval, and the local robustness check is expressed by a linear constraint over the network's output neurons. This encoding has been shown to be sound and complete. To scale the analysis, every neuron is associated with a real-valued interval. This allows to identify ReLU neurons whose computation is linear, which is the case if the input's interval is either non-negative or non-positive. In this case, the boolean variable is not introduced for this ReLU computation and the encoding's complexity decreases.

Extension to Hyper-networks. To extend this verifier to analyze hyper-networks, we encode the affine computations differently because network parameters are associated with intervals and not real numbers. A naive extension replaces the parameter values in the linear constraints by variables, which are bounded by intervals. However, this leads to quadratic constraints and significantly increases the problem's complexity. To keep the constraints linear, one option is to introduce a fresh variable for every multiplication of a neuron's input and a network parameter. However, such variable would be bounded by the interval abstracting the multiplication of the two variables, which may lead to a very large over-approximation error. Instead, we rely on the following observation: if the input to every affine computation is non-negative, then the abstraction of the multiplication does not introduce an over-approximation error. This allows us to replace the constraint of each affine variable $\hat{x}$ (defined in Sect. 2), previously captured by an equality constraint, with two inequalities providing a lower and upper bound on $\hat{x}$. Formally, given lower and upper bounds of the weights and biases at the matrices l_W and u_W and the vectors l_b and u_b, the affine variable $\hat{x}$ is bounded by:

$$l_W \cdot x + l_b \leq \hat{x} \leq u_W \cdot x + u_b$$

To guarantee that the input to every affine computation x is non-negative our analysis requires (1) preprocessing of the network's inputs and (2) a non-negative lower bound to every activation function's output. The second requirement holds since the MILP encoding of ReLU explicitly requires a non-negative output. If preprocessing is inapplicable (i.e., $x \in X \not\subseteq [0,1]^d$) or the activation function may be negative (e.g., leaky ReLU), we propose another approach to mitigate the precision loss, given a lower bound $l_x \leq x$. Given lower and upper bounds for the weights and biases l_W, u_W, l_b, and u_b, we can bound the output by:

$$l_W \cdot x + l_b - (u_W - l_W) \cdot \max(0, -l_x) \leq \hat{x} \leq u_W \cdot x + u_b + (u_W - l_W) \cdot \max(0, -l_x)$$

We provide a proof in the extended version [51]. Note that each bound is penalized by $(u_W - l_W) \cdot \max(0, -l_x) \geq 0$, which is an over-approximation error term for the case where the lower bound l_x is negative.

5.3   Analysis of Sphynx

In this section, we discuss the correctness of Sphynx and its running time. Proofs are provided in the extended version [51].

Correctness. Our first lemma states the conditions under which PredInt computes an abstraction for the values of a single network parameter with a high probability.

Lemma 1. *Given a parameter w and an error bound α', if the observed values $w_1, \ldots, w_K$ are IID and suffice to predict the correct distribution, and if there exists $\lambda \in [0,2]$ such that the distribution of the transformed normalized values $y_1, \ldots, y_K$ is similar to a distribution given by KDE with a Laplace kernel, then PredInt computes a confidence interval containing an unseen value w_i, for $i \in \{K+1, \ldots, |\mathcal{D}|+1\}$, with a probability of $1 - \alpha'$.*

Note that our lemma does not make any assumption about the distribution of the observed values $w_1, \ldots, w_K$. Next, we state our theorem pertaining the correctness of PredHyperNet. The theorem states that when the stopping condition identifies correctly when the observed values have converged to the correct distribution, then the hyper-network abstracts an unseen network with the expected probability.

Theorem 1. *Given a network N, its training set $\mathcal{D}$, its training algorithm $\mathcal{T}$, and an error bound α, if R is close to 0 and M is close to 1, then PredHyperNet returns a hyper-network abstracting an unseen network with probability $1 - \alpha$.*

Our next theorem states that our verifier is sound and states when it is complete. Completeness means that if the hyper-network is locally robust at the given neighborhood, Sphynx is able to prove it.

Theorem 2. *Our extension to the MILP verifier provides a sound analysis. It is also complete if all inputs to the affine computations are non-negative.*

By Theorem 2 and Definition 2, if Sphynx determines that the hyper-network is locally robust, then the network is LDCP. Note that it may happen that the network is LDCP, but the hyper-network is not locally robust due to the abstraction's precision loss.

Running Time. The running time of Sphynx consists of the running time of PredHyperNet and the running time of the verifier. The running time of PredHyperNet consists of the networks' training time and the time to compute the hyper-networks (the running time of the stopping condition is negligible). The training time is the product of the number of networks and the execution time of the training algorithm $\mathcal{T}$. The time complexity of PredInt is $O(K^2)$, where $K = |\text{nets}|$, and thus the computation of a hyper-network is: $O\left(K^2 \cdot |\mathcal{W} \cup \mathcal{B}|\right)$. Since PredHyperNet runs $\frac{K}{k}$ iterations, overall, the running time is $O\left(|\mathcal{T}| \cdot K + \frac{K^3}{k} \cdot |\mathcal{W} \cup \mathcal{B}|\right)$. In practice, the second term is negligible compared to the first term ($|\mathcal{T}| \cdot K$). Thus, the fewer trained networks the faster PredHyperNet is. The running time of the verifier is similar to the running time of the MILP verifier [64], verifying local robustness of a single network, which is exponential in the number of ReLU neurons whose computation is non-linear (their input's interval contains negative and positive numbers). Namely, Sphynx reduces the verification time by a factor of $|\mathcal{D}| + 1$ compared to the naive algorithm that verifies robustness network-by-network.

6   Evaluation

We implemented Sphynx in Python[4]. Experiments ran on an Ubuntu 20.04 OS on a dual AMD EPYC 7713 server with 2TB RAM and 8 NVIDIA A100 GPUs. The hyper-parameters of PredHyperNet are: $\alpha = 0.1$, the number of trained networks in every iteration is $k = 400$, the thresholds of the stopping condition are $M = 0.9$ and $R = 0.1$. We evaluate Sphynx on the three data-sensitive datasets described in Sect. 2: Adult Census [13] (Adult), Bank Marketing [41] (Bank), and Default of Credit Card Clients [77] (Credit). We preprocessed the input values to range over $[0, 1]$ as follows. Continuous attributes were normalized to range over $[0, 1]$ and categorical attributes were transformed into two features ranging over $[0, 1]$: $\cos\left(\frac{\pi}{2} \cdot \frac{i}{m-1}\right)$ and $\sin\left(\frac{\pi}{2} \cdot \frac{i}{m-1}\right)$, where m is the number of categories and i is the category's index $i \in \{0, \ldots, m-1\}$. Binary attributes were transformed with a single feature: 0 for the first category and 1 for the second one. While one hot encoding is highly popular for categorical attributes, it has also

---

[4] Code is at: https://github.com/Robgy/Verification-of-Neural-Networks-Privacy.

been linked to reduced model accuracy when the number of categories is high [52, 75]. However, we note that the encoding is orthogonal to our algorithm. We consider three fully-connected networks: 2×50, 2×100, and 4×50, where the first number is the number of intermediate layers and the second one is the number of neurons in each intermediate layer. Our network sizes are comparable to or larger than the sizes of the networks analyzed by verifiers targeting these datasets [4, 38, 70]. All networks reached around 80% accuracy. The networks were trained over 10 epochs, using SGD with a batch size of 1024 and a learning rate of 0.1. We used L_1 regularization with a coefficient of 10^{-5}. All networks were trained with the same random seed, so Sphynx can identify the maximal privacy leakage (allowing different seeds may reduce the privacy leakage since it can be viewed as adding noise to the training process). We remind that Sphynx's challenge is intensified compared to local robustness verifiers: their goal is to prove robustness of a single network, whereas Sphynx's goal is to prove that privacy is preserved over a very large number of concrete networks: 32562 networks for Adult, 31649 networks for Bank and 21001 networks for Credit. Namely, the number of parameters that Sphynx reasons about is the number of parameters of all $|\mathcal{D}| + 1$ concrete networks. Every experiment is repeated 100 times, for every network, where each experiment randomly chooses dataset entries and trains their respective networks.

Performance of Sphynx. We begin by evaluating Sphynx's ability to verify LDCP. We consider three kinds of neighborhoods, each is defined given an input x, and the goal is to prove that all inputs in the neighborhood are classified as a label y:

1. *Membership*, $I(x) = \{x\}$: safety to label-only membership attacks [11, 35, 40].
2. *DP-Robustness*, $I_\epsilon^\infty(x) = \{x' \mid \|x' - x\|_\infty \le \epsilon\}$, where $\epsilon = 0.05$: safety to adversarial example attacks in a DP setting (similarly to [34, 46]).
3. *Sensitivity*, $I_i^S(x) = \left\{ x' \mid \bigwedge_{j \in [d] \setminus \{i\}} x'_j = x_j \right\}$, where the sensitive feature is $i = sex$ for Adult and Credit and $i = age$ for Bank: safety to attacks revealing sensitive attributes (like [22, 23]). We note that sensitivity is also known as *individual fairness* [30].

For each dataset, we pick 100 inputs for each of these neighborhoods. We compare Sphynx to the naive but most accurate algorithm that trains all concrete networks and verifies the neighborhoods' robustness network-by-network. Its verifier is the MILP verifier [64] on which Sphynx's verifier builds. We let both algorithms run on all networks and neighborhoods. Table 1 reports the confusion matrix of Sphynx compared to the ground truth (computed by the naive algorithm):

- True Positive (TP): the number of neighborhoods that are LDCP and that Sphynx returns they are LDCP.
- True Negative (TN): the number of neighborhoods that are not LDCP and Sphynx returns they are not LDCP.
- False Positive (FP): the number of neighborhoods that are not LDCP and Sphynx returns they are LDCP. A false positive may happen because of

Table 1. Sphynx's confusion matrix.

Dataset	Network	Membership				DP-Robustness				Sensitivity			
		TP	TN	FP	FN	TP	TN	FP	FN	TP	TN	FP	FN
Adult	2 × 50	93	0	0	7	75	21	0	4	85	10	0	5
	2 × 100	82	1	0	17	54	39	0	7	75	10	0	15
	4 × 50	93	0	0	7	11	86	3	0	1	97	1	1
Bank	2 × 50	100	0	0	0	100	0	0	0	100	0	0	0
	2 × 100	99	0	0	1	98	1	0	1	99	0	0	1
	4 × 50	81	0	0	19	22	62	9	7	2	71	8	19
Credit	2 × 50	93	0	0	7	91	0	0	9	92	0	0	8
	2 × 100	100	0	0	0	91	2	2	5	100	0	0	0
	4 × 50	91	0	0	9	2	95	3	0	0	96	4	0

the probabilistic abstraction which may miss concrete networks that are not locally robust at the given neighborhood.

- False Negative (FN): the number of neighborhoods that are LDCP but Sphynx returns they are not LDCP. A false negative may happen because the hypernetwork may abstract spurious networks that are not locally robust at the given neighborhood (an over-approximation error).

Results show that Sphynx's average accuracy is 93.3% (TP+TN). The FP rate is 1.1% and at most 9%. The FN rate (i.e., the over-approximation error) is 5.5%. Results further show how private the different networks are. All networks are very safe to label-only membership attacks. Although Sphynx has several false negative results, it still allows the user to infer that the networks are very safe to such attack. For DP-Robustness, results show that some networks are fairly robust (e.g., Bank 2 × 50 and 2 × 100), while others are not (e.g., Bank 4 × 50). For Sensitivity, results show that Sphynx enables the user to infer what networks are sensitive to the sex/age attribute (e.g., Credit 4 × 50) and what networks are not (e.g., Credit 2 × 100). An important characteristic of Sphynx's accuracy is that the false positive and false negative rates do not lead to inferring a wrong conclusion. For example, if the network is DP-robust, Sphynx proves LDCP (TP+FP) for significantly more DP-robustness neighborhoods than the number of DP-robustness neighborhoods for which it does not prove LDCP (TN+FN). Similarly, if the network is not DP-robust, Sphynx determines for significantly more DP-robustness neighborhoods that they are not LDCP (TN+FN) than the number of DP-robustness neighborhoods that it proves they are LDCP (TP+FP).

Table 2 compares the execution time of Sphynx to the naive algorithm. It shows the number of trained networks (which is the size of the dataset plus one for the naive algorithm, and $K = |nets|$ for Sphynx), the overall training time on a single GPU in hours and the verification time of a single neighborhood (in hours for the naive algorithm, and in *seconds* for Sphynx). Results show the two strengths of Sphynx: (1) it reduces the training time by 13.6x, because it

Table 2. Training and verification time of Sphynx and the naive algorithm.

Dataset	Network	Trained networks		GPU training time		Verification time	
		naive	Sphynx	naive	Sphynx	naive	Sphynx
		$\lvert \mathcal{D} \rvert + 1$	K	hours	hours	hours	seconds
Adult	2×50	32562	2436	44.64	3.33	2.73	0.35
	2×100	32562	2024	46.56	2.89	5.24	0.72
	4×50	32562	1464	52.37	2.35	0.33	0.87
Bank	2×50	31649	2364	41.04	3.06	2.73	0.35
	2×100	31649	2536	41.76	3.34	5.60	0.69
	4×50	31649	1996	49.41	3.11	1.3	1.19
Credit	2×50	21001	2724	18.72	2.42	2.1	0.35
	2×100	21001	2234	19.21	2.04	3.6	0.64
	4×50	21001	1816	22.67	1.96	0.08	0.75

Table 3. PredInt vs. the interval abstraction and several variants.

	Int. Abs	-Transform	-Normalize	-KDE	PredInt
Weight abstraction rate	19.60	55.25	22.89	48.10	70.61
Miscoverage	7.20	5.42	1.5×10^{-6}	2.13	2.49
Overcoverage	1	1.12	299539	1.62	2.80

requires to train only 7% of the networks and (2) it reduces the verification time by $1.7 \cdot 10^4$x. Namely, Sphynx reduces the execution time by four orders of magnitude compared to the naive algorithm. The cost is the minor decrease in Sphynx's accuracy ($<7\%$). That is, Sphynx trades off precision with scalability, like many local robustness verifiers do [2,7,25,42,49,56,58,59,71,72].

Ablation Study. We next study the importance of Sphynx's steps in predicting an interval hyper-network. Recall that PredInt predicts an interval for a given network parameter by running a normalization and the Yeo-Johnson transformation and transforming it into a distribution given by KDE. We compare PredInt to interval abstraction, mapping a set of values into the interval defined by their minimum and maximum, and to three variants of PredInt: (1) *-Transform*: does not use the normalization or the transformation and directly estimates the density with KDE, (2) *-Normalize*: does not use the normalization but uses the transformation, (3) *-KDE*: transforms into a Gaussian distribution (as common), and thus employs a normalization based on the L_2 norm. We run all approaches on the 2×100 network, trained for Adult. Table 3 reports the following:

– Weight abstraction rate: the average percentage of weights whose interval provides a (sound) abstraction. Note that this metric is not related to Lemma 1, guaranteeing that the value of a network parameter of a *single* network is inside its predicted interval with probability $1 - \alpha'$. This metric is more chal-

Table 4. `PredHyperNet` vs. the interval abstraction variant (using K networks).

Dataset	Network	Network abstraction rate		Overcoverage	
		Int. Abs	Sphynx	Int. Abs	Sphynx
Adult	2×50	77.04	94.55	1.00	4.52
	2×100	55.20	92.82	1.00	2.80
	4×50	58.24	92.22	1.00	2.83
Bank	2×50	80.06	93.89	1.00	2.54
	2×100	73.21	90.13	1.00	3.21
	4×50	74.80	90.47	1.00	1.93
Credit	2×50	84.07	95.98	1.00	15.07
	2×100	67.16	90.59	1.00	3.68
	4×50	73.14	93.19	1.00	3.53

lenging: it measures how many values of a given network parameter, over $|\mathcal{D}| + 1$ networks, are inside the corresponding predicted interval.

- Miscoverage: measures how much the predicted intervals need to expand to be an abstraction. It is the average over all intervals' miscoverage. The interval miscoverage is the ratio of the size of the difference between the optimal interval and the predicted interval and the size of the predicted interval.
- Overcoverage: measures how much wider than necessary the predicted intervals are. It is the geometric average over all intervals' overcoverage. The interval overcoverage is the ratio of the size of the join of the predicted interval and the optimal interval and the size of the optimal interval.

Results show that the weight abstraction rate of the interval abstraction is very low and `PredInt` has a 3.6x higher rate. Results further show that `PredInt` obtains a very low miscoverage, by less than 2.9x compared to the interval abstraction. As expected, these results come with a cost: a higher overcoverage. An exception is the interval abstraction which, by definition, does not have an overcoverage. Results further show that the combination of normalization, transformation, and KDE improve the weight abstraction rate of `PredInt`. Next, we study how well `PredHyperNet`'s hyper-networks abstract an (unseen) concrete network with a probability ≥ 0.9. We compare to a variant that replaces `PredInt` by the interval abstraction, computed using the K concrete networks reported in Table 2. Table 4 reports the network abstraction rate (the average percentage of concrete networks abstracted by the hyper-network) and overcoverage. Results show that `PredHyperNet` obtains a very high network abstraction rate and always above $1 - \alpha = 0.9$. In contrast, the variant obtains lower network abstraction rate with a very large variance. As before, the cost is the overapproximation error.

7   Related Work

Network Abstraction. Our key idea is to abstract a set of concrete networks (seen and unseen) into an interval hyper-network. Several works rely on network abstraction to expedite verification of a single network. The goal of the abstraction is to generate a smaller network, which can be analyzed faster, and that preserves soundness with respect to the original network. One work proposes an abstraction-refinement approach that abstracts a network by splitting neurons into four types and then merging neurons of the same type [20]. Another work merges neurons similarly but chooses the neurons to merge by clustering [3]. Other works abstract a neural network by abstracting its network parameters with intervals [47] or other abstract domains [61]. In contrast, Sphynx abstracts a large set of networks, seen and unseen, and proves robustness for all of them.

Robustness Verification. Sphynx extends a MILP verifier [64] to verify local robustness given a hyper-network. There are many local robustness verifiers. Existing verifiers leverage various techniques, e.g., over-approximation [2,49,71], linear relaxation [7,25,42,56,58,59,72], simplex [20,31,32], mixed-integer linear programming [33,60,64], and duality [14,50]. A different line of works verifies robustness to small perturbations to the network parameters [67,73]. These works assume the parameters' perturbations are confined in a small L_∞ ϵ-ball and compute a lower and upper bounds on the network parameters by linearly bounding their computations. In contrast, our network parameters are not confined in an ϵ-ball, and our analysis is complete if the inputs are (or processed to be) non-negative.

Adaptive Data Analysis. PredHyperNet relies on an iterative algorithm to predict the minimum and maximum of every network parameter. Adaptive data analysis deals with estimating statistics based on data that is obtained iteratively. Existing works focus on statistical queries computing the expectation of functions [15–17,21,29,53,69] or low-sensitivity queries [5].

8   Conclusion

We propose a privacy property for neural networks, called local differential classification privacy (LDCP), extending local robustness to the setting of differential privacy for black-box classifiers. We then present Sphynx, a verifier for determining whether a network is LDCP at a given neighborhood. Instead of training all networks and verifying local robustness network-by-network, Sphynx predicts an interval hyper-network, providing an abstraction with a high probability, from a small number of networks. To predict the intervals, Sphynx transforms the observed parameter values into a distribution given by KDE, using the Yeo-Johnson transformation. Sphynx then verifies LDCP at a neighborhood directly on the hyper-network, by extending a local robustness MILP verifier. To mitigate an over-approximation error, we rely on preprocessing to the network's inputs or on a lower bound for them. We evaluate Sphynx on data-sensitive datasets and show that by training only 7% of the networks, Sphynx predicts a hyper-network abstracting any concrete network with a probability of at least 0.9, obtaining 93% verification accuracy and reducing the verification time by $1.7 \cdot 10^4$x.

Acknowledgements. We thank the anonymous reviewers for their feedback. This research was supported by the Israel Science Foundation (grant No. 2605/20).

References

1. Abadi, M., et al.: Deep learning with differential privacy. In: Weippl, E.R., Katzenbeisser, S., Kruegel, C., Myers, A.C., Halevi, S. (eds.) Proceedings of the ACM SIGSAC Conference on Computer and Communications Security. ACM (2016)
2. Anderson, G., Pailoor, S., Dillig, I., Chaudhuri, S.: Optimization and abstraction: a synergistic approach for analyzing neural network robustness. In: McKinley, K.S., Fisher, K. (eds.) Proceedings of the 40th ACM SIGPLAN Conference on Programming Language Design and Implementation, PLDI (2019)
3. Ashok, P., Hashemi, V., Křetínský, J., Mohr, S.: DeepAbstract: neural network abstraction for accelerating verification. In: Hung, D.V., Sokolsky, O. (eds.) ATVA 2020. LNCS, vol. 12302, pp. 92–107. Springer, Cham (2020). https://doi.org/10. 1007/978-3-030-59152-6_5
4. Baharlouei, S., Nouiehed, M., Beirami, A., Razaviyayn, M.: Rényi fair inference. In: 8th International Conference on Learning Representations, ICLR. OpenReview.net (2020)
5. Bassily, R., Nissim, K., Smith, A.D., Steinke, T., Stemmer, U., Ullman, J.R.: Algorithmic stability for adaptive data analysis. In: Wichs, D., Mansour, Y. (eds.) Proceedings of the 48th Annual ACM SIGACT Symposium on Theory of Computing, STOC. ACM (2016)
6. Bastani, O., Zhang, X., Solar-Lezama, A.: Probabilistic verification of fairness properties via concentration. Proc. ACM Program. Lang. 3(OOPSLA) (2019)
7. Boopathy, A., Weng, T., Chen, P., Liu, S., Daniel, L.: CNN-CERT: an efficient framework for certifying robustness of convolutional neural networks. In: The Thirty-Third AAAI Conference on Artificial Intelligence, AAAI, The Thirty-First Innovative Applications of Artificial Intelligence Conference, IAAI, The Ninth AAAI Symposium on Educational Advances in Artificial Intelligence, EAAI. AAAI Press (2019)
8. Bu, Z., Dong, J., Long, Q., Su, W.J.: Deep learning with gaussian differential privacy. CoRR abs/1911.11607 (2019)
9. Cao, Y., Yang, J.: Towards making systems forget with machine unlearning. In: IEEE Symposium on Security and Privacy, SP. IEEE Computer Society (2015)
10. Chamikara, M.A.P., Bertók, P., Khalil, I., Liu, D., Camtepe, S., Atiquzzaman, M.: Local differential privacy for deep learning. IEEE Internet Things J. (2020)
11. Choquette-Choo, C.A., Tramèr, F., Carlini, N., Papernot, N.: Label-only membership inference attacks. In: Meila, M., Zhang, T. (eds.) Proceedings of the 38th International Conference on Machine Learning, ICML. Proceedings of Machine Learning Research, PMLR (2021)
12. Drews, S., Albarghouthi, A., D'Antoni, L.: Proving data-poisoning robustness in decision trees. In: Donaldson, A.F., Torlak, E. (eds.) Proceedings of the 41st ACM SIGPLAN International Conference on Programming Language Design and Implementation, PLDI. ACM (2020)

13. Dua, D., Graff, C.: UCI machine learning repository (2017). http://archive.ics.uci.edu/ml
14. Dvijotham, K., Stanforth, R., Gowal, S., Mann, T.A., Kohli, P.: A dual approach to scalable verification of deep networks. In: Globerson, A., Silva, R. (eds.) Proceedings of the Thirty-Fourth Conference on Uncertainty in Artificial Intelligence, UAI. AUAI Press (2018)
15. Dwork, C., Feldman, V., Hardt, M., Pitassi, T., Reingold, O., Roth, A.: Generalization in adaptive data analysis and holdout reuse. In: Cortes, C., Lawrence, N.D., Lee, D.D., Sugiyama, M., Garnett, R. (eds.) Advances in Neural Information Processing Systems 28: Annual Conference on Neural Information Processing Systems (2015)
16. Dwork, C., Feldman, V., Hardt, M., Pitassi, T., Reingold, O., Roth, A.: The reusable holdout: preserving validity in adaptive data analysis. Science (2015)
17. Dwork, C., Feldman, V., Hardt, M., Pitassi, T., Reingold, O., Roth, A.L.: Preserving statistical validity in adaptive data analysis. In: Servedio, R.A., Rubinfeld, R. (eds.) Proceedings of the Forty-Seventh Annual ACM on Symposium on Theory of Computing, STOC. ACM (2015)
18. Dwork, C., Roth, A.: The algorithmic foundations of differential privacy. Found. Trends Theor. Comput. Sci. 9(3–4) (2014)
19. Ehlers, R.: Formal verification of piece-wise linear feed-forward neural networks. In: D'Souza, D., Narayan Kumar, K. (eds.) ATVA 2017. LNCS, vol. 10482, pp. 269–286. Springer, Cham (2017). https://doi.org/10.1007/978-3-319-68167-2_19
20. Elboher, Y.Y., Gottschlich, J., Katz, G.: An abstraction-based framework for neural network verification. In: Lahiri, S.K., Wang, C. (eds.) CAV 2020. LNCS, vol. 12224, pp. 43–65. Springer, Cham (2020). https://doi.org/10.1007/978-3-030-53288-8_3
21. Feldman, V., Steinke, T.: Generalization for adaptively-chosen estimators via stable median. In: Kale, S., Shamir, O. (eds.) Proceedings of the 30th Conference on Learning Theory, COLT. Proceedings of Machine Learning Research, PMLR (2017)
22. Fredrikson, M., Jha, S., Ristenpart, T.: Model inversion attacks that exploit confidence information and basic countermeasures. In: Ray, I., Li, N., Kruegel, C. (eds.) Proceedings of the 22nd ACM SIGSAC Conference on Computer and Communications Security. ACM (2015)
23. Fredrikson, M., Lantz, E., Jha, S., Lin, S.M., Page, D., Ristenpart, T.: Privacy in pharmacogenetics: An end-to-end case study of personalized warfarin dosing. In: Fu, K., Jung, J. (eds.) Proceedings of the 23rd USENIX Security Symposium. USENIX Association (2014)
24. Ganju, K., Wang, Q., Yang, W., Gunter, C.A., Borisov, N.: Property inference attacks on fully connected neural networks using permutation invariant representations. In: Lie, D., Mannan, M., Backes, M., Wang, X. (eds.) Proceedings of the ACM SIGSAC Conference on Computer and Communications Security, CCS. ACM (2018)
25. Gehr, T., Mirman, M., Drachsler-Cohen, D., Tsankov, P., Chaudhuri, S., Vechev, M.T.: AI2: safety and robustness certification of neural networks with abstract interpretation. In: IEEE Symposium on Security and Privacy, SP. IEEE Computer Society (2018)
26. Goel, S., Prabhu, A., Kumaraguru, P.: Evaluating inexact unlearning requires revisiting forgetting. CoRR abs/2201.06640 (2022)
27. Goodfellow, I.J., Shlens, J., Szegedy, C.: Explaining and harnessing adversarial examples. In: Bengio, Y., LeCun, Y. (eds.) 3rd International Conference on Learning Representations, ICLR, Conference Track Proceedings (2015)

28. Ha, T., Dang, T.K., Dang, T.T., Truong, T.A., Nguyen, M.T.: Differential privacy in deep learning: an overview. In: Lê, L., Dang, T.K., Minh, Q.T., Toulouse, M., Draheim, D., Küng, J. (eds.) International Conference on Advanced Computing and Applications, ACOMP. IEEE Computer Society (2019)

29. Hardt, M., Ullman, J.R.: Preventing false discovery in interactive data analysis is hard. In: 55th IEEE Annual Symposium on Foundations of Computer Science, FOCS. IEEE Computer Society (2014)

30. John, P.G., Vijaykeerthy, D., Saha, D.: Verifying individual fairness in machine learning models. In: Adams, R.P., Gogate, V. (eds.) Proceedings of the Thirty-Sixth Conference on Uncertainty in Artificial Intelligence, UAI. Proceedings of Machine Learning Research, AUAI Press (2020)

31. Katz, G., Barrett, C., Dill, D.L., Julian, K., Kochenderfer, M.J.: Reluplex: an efficient SMT solver for verifying deep neural networks. In: Majumdar, R., Kunčak, V. (eds.) CAV 2017. LNCS, vol. 10426, pp. 97–117. Springer, Cham (2017). https://doi.org/10.1007/978-3-319-63387-9_5

32. Katz, G., et al.: The marabou framework for verification and analysis of deep neural networks. In: Dillig, I., Tasiran, S. (eds.) CAV 2019. LNCS, vol. 11561, pp. 443–452. Springer, Cham (2019). https://doi.org/10.1007/978-3-030-25540-4_26

33. Lazarus, C., Kochenderfer, M.J.: A mixed integer programming approach for verifying properties of binarized neural networks. In: Espinoza, H., et al. (eds.) Proceedings of the Workshop on Artificial Intelligence Safety co-located with the Thirtieth International Joint Conference on Artificial Intelligence (IJCAI). CEUR Workshop Proceedings, CEUR-WS.org (2021)

34. Lécuyer, M., Atlidakis, V., Geambasu, R., Hsu, D., Jana, S.: Certified robustness to adversarial examples with differential privacy. In: 2019 IEEE Symposium on Security and Privacy, SP. IEEE (2019)

35. Li, Z., Zhang, Y.: Membership leakage in label-only exposures. In: Kim, Y., Kim, J., Vigna, G., Shi, E. (eds.) CCS: ACM SIGSAC Conference on Computer and Communications Security. ACM (2021)

36. Liu, Y., Ma, Z., Liu, X., Ma, J.: Learn to forget: Machine unlearning via neuron masking. IEEE Trans. Dependable and Secure Comput. (2022)

37. Long, Y., et al.: Understanding membership inferences on well-generalized learning models. CoRR abs/1802.04889 (2018)

38. Mazzucato, D., Urban, C.: Reduced products of abstract domains for fairness certification of neural networks. In: Drăgoi, C., Mukherjee, S., Namjoshi, K. (eds.) SAS 2021. LNCS, vol. 12913, pp. 308–322. Springer, Cham (2021). https://doi.org/10.1007/978-3-030-88806-0_15

39. Meyer, A.P., Albarghouthi, A., D'Antoni, L.: Certifying robustness to programmable data bias in decision trees. In: Ranzato, M., Beygelzimer, A., Dauphin, Y.N., Liang, P., Vaughan, J.W. (eds.) Advances in Neural Information Processing Systems 34: Annual Conference on Neural Information Processing Systems NeurIPS (2021)

40. Monreale, A., Naretto, F., Rizzo, S.: Agnostic label-only membership inference attack. In: Li, S., Manulis, M., Miyaji, A. (eds.) Network and System Security. Lecture Notes in Computer Science, vol. 13983, pp. 249–264. Springer, Cham (2023). https://doi.org/10.1007/978-3-031-39828-5_14

41. Moro, S., Cortez, P., Rita, P.: A data-driven approach to predict the success of bank telemarketing. Decis. Support Syst. **62**, 22–31 (2014)

42. Müller, C., Serre, F., Singh, G., Püschel, M., Vechev, M.T.: Scaling polyhedral neural network verification on GPUs. In: Smola, A., Dimakis, A., Stoica, I. (eds.) Proceedings of Machine Learning and Systems, MLSys. mlsys.org (2021)

43. Nasr, M., Shokri, R., Houmansadr, A.: Comprehensive privacy analysis of deep learning: passive and active white-box inference attacks against centralized and federated learning. In: IEEE Symposium on Security and Privacy, SP. IEEE (2019)
44. Nasr, M., Shokri, R., Houmansadr, A.: Improving deep learning with differential privacy using gradient encoding and denoising. CoRR abs/2007.11524 (2020)
45. Pham, L.H., Sun, J.: Verifying neural networks against backdoor attacks. In: Shoham, S., Vizel, Y. (eds.) Computer Aided Verification - 34th International Conference, CAV, Proceedings. Springer, Part I. Lecture Notes in Computer Science (2022). https://doi.org/10.1007/978-3-031-13185-1_9
46. Phan, N., Thai, M.T., Hu, H., Jin, R., Sun, T., Dou, D.: Scalable differential privacy with certified robustness in adversarial learning. In: Proceedings of the 37th International Conference on Machine Learning, ICML. Proceedings of Machine Learning Research, vol. 119. PMLR (2020)
47. Prabhakar, P., Afzal, Z.R.: Abstraction based output range analysis for neural networks. In: Wallach, H.M., Larochelle, H., Beygelzimer, A., d'Alché-Buc, F., Fox, E.B., Garnett, R. (eds.) Advances in Neural Information Processing Systems 32: Annual Conference on Neural Information Processing Systems NeurIPS (2019)
48. Pulina, L., Tacchella, A.: Challenging SMT solvers to verify neural networks. AI Commun. $25(2)$, 117–135 (2012)
49. Qin, C., et al.: Verification of non-linear specifications for neural networks. In: 7th International Conference on Learning Representations, ICLR. OpenReview.net (2019)
50. Raghunathan, A., Steinhardt, J., Liang, P.: Certified defenses against adversarial examples. In: 6th International Conference on Learning Representations, ICLR, Conference Track Proceedings. OpenReview.net (2018)
51. Reshef, R., Kabaha, A., Seleznova, O., Drachsler-Cohen, D.: Verification of neural networks local differential classification privacy. CoRR abs/2310.20299 (2023)
52. Rodríguez, P., Bautista, M.Á., Gonzàlez, J., Escalera, S.: Beyond one-hot encoding: lower dimensional target embedding. Image Vis. Comput. 75, 21–31 (2018)
53. Rogers, R., Roth, A., Smith, A.D., Srebro, N., Thakkar, O., Woodworth, B.E.: Guaranteed validity for empirical approaches to adaptive data analysis. In: Chiappa, S., Calandra, R. (eds.) The 23rd International Conference on Artificial Intelligence and Statistics, AISTATS. Proceedings of Machine Learning Research, PMLR (2020)
54. Ruoss, A., Balunovic, M., Fischer, M., Vechev, M.T.: Learning certified individually fair representations. In: Larochelle, H., Ranzato, M., Hadsell, R., Balcan, M., Lin, H. (eds.) Advances in Neural Information Processing Systems 33: Annual Conference on Neural Information Processing Systems, NeurIPS (2020)
55. Ryou, W., Chen, J., Balunovic, M., Singh, G., Dan, A., Vechev, M.: Scalable polyhedral verification of recurrent neural networks. In: Silva, A., Leino, K.R.M. (eds.) CAV 2021. LNCS, vol. 12759, pp. 225–248. Springer, Cham (2021). https://doi.org/10.1007/978-3-030-81685-8_10
56. Salman, H., Yang, G., Zhang, H., Hsieh, C., Zhang, P.: A convex relaxation barrier to tight robustness verification of neural networks. In: Wallach, H.M., Larochelle, H., Beygelzimer, A., d'Alché-Buc, F., Fox, E.B., Garnett, R. (eds.) Advances in Neural Information Processing Systems 32: Annual Conference on Neural Information Processing Systems, NeurIPS (2019)
57. Shokri, R., Stronati, M., Song, C., Shmatikov, V.: Membership inference attacks against machine learning models. In: IEEE Symposium on Security and Privacy, SP. IEEE Computer Society (2017)

58. Singh, G., Ganvir, R., Püschel, M., Vechev, M.T.: Beyond the single neuron convex barrier for neural network certification. In: Wallach, H.M., Larochelle, H., Beygelzimer, A., d'Alché-Buc, F., Fox, E.B., Garnett, R. (eds.) Advances in Neural Information Processing Systems 32: Annual Conference on Neural Information Processing Systems, NeurIPS (2019)

59. Singh, G., Gehr, T., Püschel, M., Vechev, M.T.: An abstract domain for certifying neural networks. Proc. ACM Program. Lang. (POPL) (2019)

60. Singh, G., Gehr, T., Püschel, M., Vechev, M.T.: Boosting robustness certification of neural networks. In: 7th International Conference on Learning Representations, ICLR. OpenReview.net (2019)

61. Sotoudeh, M., Thakur, A.V.: Abstract neural networks. In: Pichardie, D., Sighireanu, M. (eds.) SAS 2020. LNCS, vol. 12389, pp. 65–88. Springer, Cham (2020). https://doi.org/10.1007/978-3-030-65474-0_4

62. Sun, Y., Usman, M., Gopinath, D., Pasareanu, C.S.: VPN: verification of poisoning in neural networks. In: Isac, O., Ivanov, R., Katz, G., Narodytska, N., Nenzi, L. (eds.) Software Verification and Formal Methods for ML-Enabled Autonomous Systems. Lecture Notes in Computer Science, vol. 13466, pp. 3–14. Springer, Cham (2022). https://doi.org/10.1007/978-3-031-21222-2_1

63. Szegedy, C., et al.: Intriguing properties of neural networks. In: Bengio, Y., LeCun, Y. (eds.) 2nd International Conference on Learning Representations, ICLR, Conference Track Proceedings (2014)

64. Tjeng, V., Xiao, K.Y., Tedrake, R.: Evaluating robustness of neural networks with mixed integer programming. In: 7th International Conference on Learning Representations, ICLR. OpenReview.net (2019)

65. Tramèr, F., Zhang, F., Juels, A., Reiter, M.K., Ristenpart, T.: Stealing machine learning models via prediction APIs. In: Holz, T., Savage, S. (eds.) 25th USENIX Security Symposium, USENIX Security 16. USENIX Association (2016)

66. Tran, H.-D., Bak, S., Xiang, W., Johnson, T.T.: Verification of deep convolutional neural networks using ImageStars. In: Lahiri, S.K., Wang, C. (eds.) CAV 2020. LNCS, vol. 12224, pp. 18–42. Springer, Cham (2020). https://doi.org/10.1007/978-3-030-53288-8_2

67. Tsai, Y., Hsu, C., Yu, C., Chen, P.: Formalizing generalization and adversarial robustness of neural networks to weight perturbations. In: Ranzato, M., Beygelzimer, A., Dauphin, Y.N., Liang, P., Vaughan, J.W. (eds.) Advances in Neural Information Processing Systems 34: Annual Conference on Neural Information Processing Systems, NeurIPS (2021)
68. Ugare, S., Singh, G., Misailovic, S.: Proof transfer for fast certification of multiple approximate neural networks. Proc. ACM Program. Lang. **6**(OOPSLA1) (2022)
69. Ullman, J.R., Smith, A.D., Nissim, K., Stemmer, U., Steinke, T.: The limits of post-selection generalization. In: Bengio, S., Wallach, H.M., Larochelle, H., Grauman, K., Cesa-Bianchi, N., Garnett, R. (eds.) Advances in Neural Information Processing Systems 31: Annual Conference on Neural Information Processing Systems, NeurIPS (2018)
70. Urban, C., Christakis, M., Wüstholz, V., Zhang, F.: Perfectly parallel fairness certification of neural networks. Proc. ACM Program. Lang. **4**(OOPSLA) (2019)
71. Wang, S., Pei, K., Whitehouse, J., Yang, J., Jana, S.: Efficient formal safety analysis of neural networks. In: Bengio, S., Wallach, H.M., Larochelle, H., Grauman, K., Cesa-Bianchi, N., Garnett, R. (eds.) Advances in Neural Information Processing Systems 31: Annual Conference on Neural Information Processing Systems, NeurIPS (2018)
72. Wang, S., et al.: Beta-crown: efficient bound propagation with per-neuron split constraints for neural network robustness verification. In: Ranzato, M., Beygelzimer, A., Dauphin, Y.N., Liang, P., Vaughan, J.W. (eds.) Advances in Neural Information Processing Systems 34: Annual Conference on Neural Information Processing Systems, NeurIPS (2021)
73. Weng, T., Zhao, P., Liu, S., Chen, P., Lin, X., Daniel, L.: Towards certificated model robustness against weight perturbations. In: The Thirty-Fourth AAAI Conference on Artificial Intelligence, AAAI. AAAI Press (2020)
74. Wu, H., et al.: Parallelization techniques for verifying neural networks. In: Formal Methods in Computer Aided Design, FMCAD. IEEE (2020)
75. Ye, A.: Stop one-hot encoding your categorical variables (2020). https://medium.com/analytics-vidhya/stop-one-hot-encoding-your-categorical-variables-bbb0fba89809
76. Ye, J., Maddi, A., Murakonda, S.K., Bindschaedler, V., Shokri, R.: Enhanced membership inference attacks against machine learning models. In: Yin, H., Stavrou, A., Cremers, C., Shi, E. (eds.) Proceedings of the ACM SIGSAC Conference on Computer and Communications Security, CCS. ACM (2022)
77. Yeh, I.C., Lien, C.H.: The comparisons of data mining techniques for the predictive accuracy of probability of default of credit card clients. Expert Syst. Appl. **36**, 2473–2480 (2009)
78. Yeo, I.K., Johnson, R.A.: A new family of power transformations to improve normality or symmetry. Biometrika **87**, 954–959 (2000)

AGNES: Abstraction-Guided Framework for Deep Neural Networks Security

Akshay Dhonthi[1,2]($\boxtimes$), Marcello Eiermann[3], Ernst Moritz Hahn[2], and Vahid Hashemi[1]

[1] AUDI AG, Auto-Union-Straße 1, 85057 Ingolstadt, Germany
akshay.dhonthirameshbabu@audi.de
[2] Formal Methods and Tools, University of Twente, Enschede, Netherlands
[3] Information and Computing Sciences, Utrecht University, Utrecht, Netherlands

Abstract. Deep Neural Networks (DNNs) are becoming widespread, particularly in safety-critical areas. One prominent application is image recognition in autonomous driving, where the correct classification of objects, such as traffic signs, is essential for safe driving. Unfortunately, DNNs are prone to *backdoors*, meaning that they concentrate on attributes of the image that should be irrelevant for their correct classification. Backdoors are integrated into a DNN during training, either with malicious intent (such as a manipulated training process, because of which a yellow sticker always leads to a traffic sign being recognised as a stop sign) or unintentional (such as a rural background leading to any traffic sign being recognised as "animal crossing", because of biased training data).

In this paper, we introduce AGNES, a tool to detect backdoors in DNNs for image recognition. We discuss the principle approach on which AGNES is based. Afterwards, we show that our tool performs better than many state-of-the-art methods for multiple relevant case studies.

Keywords: Neural network analysis · Security testing · Backdoor detection

1 Introduction

Deep Neural Networks (DNNs) are widely used nowadays in safety-critical application such as automotive [4,8], avionics [7] and medical [5] industries. In particular, in the automative domain, the safety-critical applications range from automated driver assistance systems (ADAS) such as Automated Emergency Braking Systems (AEBS) up to fully Automated Driving (AD) vehicles. In such applications, machines take over more and more critical decisions. This leads to two types of risks: *safety* and *security* [24]. The former describes functional safety, i.e., protecting the environment and other road users from the machines. In contrast, the latter describes the protection of machines from the environment

This research was funded in part by the EU under project 864075 CAESAR, the project Audi Verifiable AI, and the BMWi funded KARLI project (grant 19A21031C).

and other external interventions. In this work, we describe the tool AGNES, which focuses on security risks in DNNs. We specifically focus on classification functions where, given an image, the DNN predicts to what class the image belongs. This is an important task particularly in safety-critical domains such as the traffic sign detection function in an autonomous system.

A security risk occurs when a DNN does not predict the true class of an image, but instead focuses on regions in the image which are not intended to influence the classification. Such behaviour results if the dataset used to train a DNN is intentionally manipulated by adding triggers to a small part of the dataset or when the DNN learns from regions in the training images that are irrelevant for decision-making. We call such behaviours *backdoors* of the network [10,17]. Further, with *trigger*, we denote the part of the image which activates the backdoor. A trigger may be as simple as a small patch in a specific location of the image, or it can also be a nearly invisible watermark spread over the whole image [10]. It could also be very complex: with inappropriate training, a DNN might focus on the background of an image to guess the traffic sign to be detected rather than on the traffic sign itself [23,28]. Additionally, proving the presence of a backdoor, describing its triggers, and removing backdoors are highly complex tasks. This is particularly true if one is unaware of the exact type of possible triggers.

Identification of backdoors has been an interest for a while now, and several techniques have been developed in recent years [12,22]. Backdoor detection can be performed by synthesising the trigger and then using it to analyse the performance of DNNs. Apart from techniques that synthesise triggers, other defence methods directly defend from backdoors either during runtime [19] or post-hoc analysis [11,14]. However, in this work, we focus on techniques based on trigger-synthesis because we can visualise the backdoors. Mainly, the techniques that are worth mentioning are Neural cleanse (NC) [27], Artificial Brain Stimulation (ABS) [16], ABS+Exray [18], and K-Arm [25].

NC is the first proposed trigger-synthesis based defence method. Here, potential trigger patterns are obtained for each class via trigger optimisation [15]. The ABS technique is based on stimulating/changing every neuron output in the hidden layers and propagating this change up to the output layer to see if it impacts the network output. When the neurons are stimulated, the neurons that change the network's output are considered *compromised neuron candidates* (CNCs) and are later used to generate triggers. The ABS+Exray method uses a trigger inversion technique to generate triggers and then analyses if the triggers contain features that are not considered natural distinctive features between the victim (the original class) and the target class. The K-Arm optimization method, which is built on top of ABS, propose a Reinforcement Learning based method to detect backdoors. We integrate ABS in our tool because it does not require pre-knowledge of the trigger type. This helps to synthesise triggers by utilising only benign images, making it more interesting to automotive applications where the trigger is unknown.

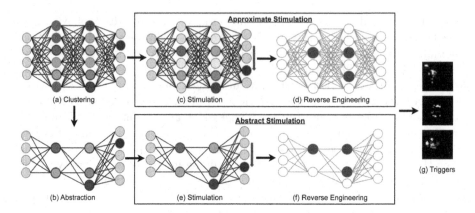

Fig. 1. AGNES Framework. The colours on neurons represent various clusters. The neurons with the dark blue outline are cluster representatives of each cluster. The neurons with red outlines undergo stimulation while the rest are skipped. The red neurons in the reverse engineering step are the compromised neurons. (Color figure online)

However, as mentioned in [6], the ABS technique is slow because the stimulation of neurons is time-consuming. ABS analyses all the features in the hidden layers sequentially to identify backdoors. Doing so is inefficient due to the large number of features in the hidden layers. Also, ABS generates an intermediate list of neurons called compromised neuron candidates, which contain many false positives. This means that most of these neurons may not lead to identifying triggers [25]. To address these issues, we propose a way to abstract the inner parameters of a network such that the identification of compromised neurons in the network is fast and robust. We apply our method over several model architectures, trigger types, and trigger parameters (such as trigger size and transparency) to evaluate its performance.

We depict the principle workflow of our tool AGNES in Fig. 1. It can find the most critical neurons quickly and outperforms state-of-the-art (SOTA) techniques in identification of some types of triggers. Our tool works with the DNN as a black-box in so far as that it only needs the trained network and a small benign dataset, but not the training data or further information about the nature of potential triggers. AGNES abstracts the neural network by clustering the neurons in each layer and abstracts each cluster to its *cluster representatives* (CR) (neuron that is closest to the cluster centre) based on DeepAbstract [2] technique. We propose two methods: The first one is *Abstract Stimulation Method* (AbsSM), where we stimulate the abstract network. The second one is the *Approximate Stimulation* (AproxSM), where we cluster each hidden layer based on their activation values and then stimulate only the cluster representatives for analysis. We evaluate both techniques on several model architectures, trigger types, and trigger parameters such as trigger size and transparency. In many cases (cf. Sect. 5), AGNES performs better in terms of time complexity and accuracy than other techniques.

Overall, our contributions in this paper are as follows:

- We introduce the tool AGNES, which improves SOTA backdoor identification techniques in terms of performance and runtime.
- We propose two methods, namely AbsSM and AproxSM for stimulation analysis. The former is a precise method which abstracts the network for analysis while the latter is an approximate method in which instead of abstracting the network, a smaller set of neurons for analysis is extracted. The proper method selection for analysis is done automatically by AGNES.
- We perform an extensive evaluation of our tool on SOTA model architectures for traffic sign classification problem while adhering to the latest machine learning frameworks such as TensorFlow 2.0 [1] and PyTorch [21].

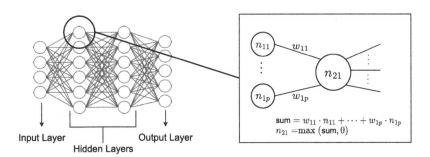

Fig. 2. The left image depicts the DNN structure, and the right image depicts the computation of a single neuron output. We show, as an example, computation for neuron n_{21}.

2   Preliminaries

In this section, we introduce the key components of AGNES.

2.1   Deep Neural Networks

In Fig. 2, we provide an example of a simple DNN. The left part provides a high-level overview of the network, depicting the connections between the neurons. The input layer takes the shape of an image, where *shape* represents the total number of elements and dimensions. The image can have the shape $[l, w, c]$, where l and w are the length and width of the image, and c represent the channels typically 3 (R, G, B). The output layer is the shape of the number of classes to which an image can be classified. The hidden layers can be but not limited to convolutional, activation, pooling, or fully connected layers. The right part details how a single neuron works in a fully connected layer. The neuron detailed receives input values from its predecessor neurons, weighted with the strength of their connections. These values are summed up, resulting in a real value. On this sum, we apply the *activation function* of this neuron, which is usually the same for all neurons of the layer. In this case, we use *ReLU* (maximum of 0 and the weighted input sum), which is the most commonly used activation function nowadays.

DNNs can approximate arbitrary functions [9] and perform a wide range of tasks. Their advantage is that they can be trained on a large set of examples to provide the correct output on inputs sufficiently similar to the ones they have been trained with. Their main disadvantage is that they do not have any easily understandable structure, consisting mostly of a large number of real numbers representing the weights connecting the individual neurons. This is also why it is easy for them to have hidden backdoors that are hard to find in a DNN [10].

2.2 Abstraction

For this work, we use DeepAbstract [2] as the abstraction technique, but any abstraction method would be suitable. DeepAbstract clusters the neurons in a layer via the k-means function. For each cluster, we chose a representative: the neuron closest to the centroids of the respective cluster. This abstraction technique only works for networks with Fully Connected (FC) layers. Therefore, we propose an approximate technique, AproxSM, in Sect. 3 to handle non-FC networks such as convolutional networks. Later after identifying clusters, all the neurons in a cluster are dissolved into the respective CR by redirecting all the weights into the CR. Since the activation values of all the neurons in a cluster are similar, the loss of information flow resulting from summing all the connections in the cluster to the CR will be low.

We obtain the optimal number of clusters by making sure the loss in accuracy of the DNN is minimal after clustering. The method is illustrated in Fig. 1(b), where the original network is abstracted using only CR. Neurons belonging to the same class mostly behave alike. Therefore, our intuition is that stimulating the CR neurons can alone identify the neurons responsible for triggering backdoors.

2.3 Stimulation

We utilise the stimulation technique from ABS, where each neuron is systematically stimulated to check whether the output of the network changes. The stimulation is done as follows: We increase or decrease the neuron output under analysis in a hidden layer by adding *stimulation value*. Positive/negative stimulation values will increase/decrease the neuron output, respectively. While doing so, we keep the rest of the neuron outputs in that layer as they are. A neuron is then marked as a *compromised neuron candidate* (CNC) if, under some stimulation value, the output class of the network will shift when stimulated neruon value is forward propagated to the output layer.

We compute stimulation outputs for images from different classes, and analyse whether the output shift is consistent. This means that we identify the neurons such that for a specific stimulation value, the network output remains the same regardless of the class of the image. Such a neuron is deemed compromised as it completely influences the network's output. Our goal is to identify such neurons with high precision. ABS performs this task systematically for all the neurons in the hidden layers, which can be slow for large models. Instead, in AGNES, we restrict the stimulation analysis only to the CR. Doing so can be efficient with respect to time. This process is further explained in Sect. 3.

3   Methodology

We propose two methods: stimulation over the abstract network (AbsSM) and over the original network (AproxSM). The difference between both techniques concerns the neurons and the network used for stimulation, as depicted in Fig. 1. The former technique stimulates all the neurons but works with the abstract network. However, the latter technique stimulates the CR on the original network and ignores the rest of the neurons for stimulation analysis. This section will first describe the clustering and abstraction procedure and, later, the two stimulation techniques.

3.1   Identifying Cluster Representatives

Algorithm 1 depicts our approach for identifying CR. The inputs to the algorithm are the trained DNN and the dataset. In line 1, we choose whether to use abstraction. This selection is done automatically in AGNES which we discuss in Sect. 4. For stimulation on the abstract network, we utilise DeepAbstract. DeepAbstract, as explained in Sect. 2.2, clusters neurons in each hidden layer via the K-means approach and then systematically removes neurons by abstracting them into the CR. One limitation of DeepAbstract is that it only supports FC layers while image classification models are built based on *convolutional neural network* CNN. Therefore, we provide a way to work with convolutional neural networks by converting all CNN layers to FC layers, as depicted in line 2. This

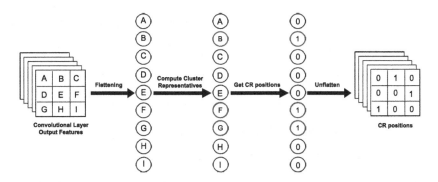

Fig. 3. Obtaining CR positions for one of the features in a convolutional layer. The red mark represent the CR. (Color figure online)

is possible and indeed a straightforward transformation because a convolutional layer is just a sparsely connected FC layer [20]. Nevertheless, our tool is agnostic to the abstraction technique and therefore, DeepAbstract can easily be replaced with a different method that supports other complex layers. The next step, as in lines 3–6, is to run the abstraction function from DeepAbstract. We then return the abstract network and end the algorithm.

However, abstracting large models with CNN layers would be difficult due to their multidimensional structure. FC layers are one-dimensional and can directly be clustered, whereas the outputs from CNN layers are two-dimensional features. For very large networks, converting FC to CNN layers would drastically increase the number of connections, making it difficult to subsume them into CR during abstraction. Therefore, we propose AproxSM in which, instead of abstracting the network, we first obtain the positions of CR, which is just the information of the location of the CR neuron. And then, we stimulate all the CR neurons using the position of CR. All the other neurons are not stimulated. To obtain CR in line 11, we flatten the multidimensional layer output to one-dimensional. Figure 3 depicts the procedure to get the CR positions. We compute CR via k-means and then store their positions. We then unflatten the CR position array to get the shape of the original CNN layer and then use it for stimulation.

3.2 Stimulation Techniques

This section explains how to identify CNC via stimulation. We follow a similar approach as in ABS [16]. The difference is that ABS stimulates all the neurons in the network sequentially, whereas we only stimulate the CR. Doing so is straightforward for stimulation over the abstract network because it contains only CR; therefore, we directly run the stimulation function on neurons in the abstract network. Our intuition is that since our clustering is based on the activation values obtained from the benign dataset, the neurons activated in the presence of a trigger called *trojan neurons* may have lower values on benign images. However, these trojan neurons would have similar activation values, and they would belong to one cluster. We analysed a trojan model by comparing activation values of benign and trojan images and found out that our intuition is, in fact, true. Therefore, amplifying the value of that trojan neuron would emulate the trigger behaviour. Note that there will also be benign CR, but these neurons will be filtered out during stimulation as they will not affect the network output.

ABS for CNN layers stimulates each output feature one after the other by setting all the neurons in that feature to the stimulation value (as seen from the source code). This change can be too strong because all the neurons in one feature can, in reality, never be that high and, therefore, may have a high impact on the network output. To address this issue, instead of stimulating all the neurons in a feature, we utilise the positions of the CR and stimulate only those neurons. Figure 4 depicts our idea for stimulation where we apply stimulation values to the CR using the CR position matrix. This replaces the neuron outputs for only the CR while keeping the neuron values of the rest as it is. Intuitively, this kind of stimulation would focus only on critical regions in the intermediate feature and,

---

Algorithm 1. Computing Cluster Representatives

Input: $\mathcal{N}$: Trained DNN,
$X_{test} = \{x_1, \cdots, x_T\}$: benign test data,
abstract: if *TRUE* would abstract the network, otherwise would reshape the layer outputs,
Output: $\hat{\mathcal{N}}$: Abstract Network, and CR.
1: **if** *abstract* = *TRUE* **then**
2:    Convert all convolutional layers to FC layers
3:    **for** layer in hidden layers **do**
4:       Obtain clusters and CR.
5:       Abstract the layer based on DeepAbstract and identify CR.
6:    **end for**
7:    **return** DNN $\hat{\mathcal{N}}$.
8: **else**
9:    **for** layer in hidden layers **do**
10:       Flatten the outputs of all neurons in the layer
11:       Perform K-means clustering on the layer and extract CR
12:    **end for**
13:    **return** CR
14: **end if**

---

therefore, can produce better results than ABS. Another advantage is that the stimulation runtime would also decrease because we can skip the features that do not have any CR. We can call this an approximate method because we stimulate the same neurons as in AbsSM. But unlike AproxSM, the forward propagation of the stimulation value is on all the neurons including non-CR neurons which may slightly affect the output.

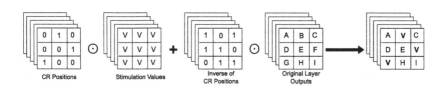

Fig. 4. AproxSM method. Here, $\odot$ is the Hadamard product operation. v is the stimulation value.

4   Tool Architecture

This section gives an overview of the tool, its functionalities, advantages and limitations. Figure 5 depicts the tool architecture. AGNES runs on Python and supports state-of-the-art machine learning frameworks such as TensorFlow and PyTorch. We utilise the *ONNX* library [3] to convert Pytorch and TensorFlow models to an *onnx file*, which is a generic model that can be loaded into either framework. Therefore, our tool is compatible with most of the SOTA model architectures. Additionally, ONNX helps us integrate other abstraction and stimulation techniques with different library and framework requirements.

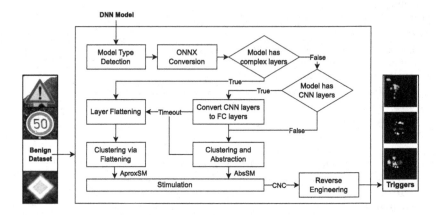

Fig. 5. Tool Framework.

The input to our tool is a trained model and a benign test dataset. The models can be of either TensorFlow or PyTorch frameworks; we automatically convert them to the required framework based on the function to be run. We have tested the tool's capabilities on several classification model architectures such as *LeNet, AlexNet, VGG, and ResNet*. More information on these architectures can be found in Sect. 5. We first identify the suitable stimulation method based on the type of model. Abstract models can be stimulated if the model has only fully connected (FC) and convolutional layers (Conv). This limitation can, however, be lifted by using a different abstraction technique that supports other layer types. We switch to the AproxSM method if the model contains any other complex layers, i.e., layers apart from FC and Conv. We also set a timeout for stimulation over the abstract network so that we switch to the AproxSM method if it takes too long to run abstraction.

The output from AGNES are CNCs and the range of stimulation values for compromised neurons at which these neurons trigger the trojan behaviour. Next, we use this to run a reverse engineering function that generates masked images that closely represent the trigger. We utilise the reverse engineering function from ABS. However, our tool is agnostic of the type of reverse engineering. Note that we focus on the traffic sign classification problem in this paper, but the tool is not limited to this application. The method works on any dataset and applications where a DNN is built. Finally, we compute the percentage of wrong predictions, i.e., the *Attack Success Rate* (ASR) on the images that contain the reverse-engineered masks.

5 Experiments

In this section, we show experimental results to validate our method. Our evaluation is based on the following model architectures:

- *FC1:* Flatten, FC(864, 400, 160), FC(43),
- *FC2:* Flatten, FC(864, 864, 512, 256), FC(43),
- *FC3:* Flatten, FC(864, 864, 864, 512, 512), FC(43),
- *NN1:* Conv(8, 16, 32, 16, 8), Flatten, FC(43),
- *LeNet:* Conv(6), MaxPool, Conv(16), MaxPool, Flatten, FC(400, 120), Dropout(0.5), FC(80), Dropout(0.5), FC(43),
- *AlexNet:* Conv(9), MaxPool, Conv(32), MaxPool, Conv(48, 64, 96), MaxPool, Flatten, FC(864, 400), Dropout(0.5), FC(160), Dropout(0.5), FC(43),
- *VGG:* Conv(16), MaxPool, Conv(32), MaxPool, Conv(64), MaxPool, Conv (64), MaxPool, Conv(128), MaxPool, Conv(64), MaxPool, FC(1024, 1024), FC(43)

Here, we represent the type of hidden layer and the respective number of neurons or kernels in the brackets. For example, FC(864, 400) means that the network has two fully connected layers, with the number of neurons in the hidden layers being 864 and 400, respectively. The Flatten layer converts a multidimensional layer to one dimension. Conv denotes convolutional layers, and the number in the bracket represents the number of features/kernels trained in that layer. MaxPool layers pool neurons by keeping the max value out of the neurons in the pooling window. A Dropout(0.5) layer randomly drops out fifty per cent of the neurons. Notice that all these models end with FC(43) because the models are built for the dataset with 43 classes.

We evaluate AbsSM with *FC1, FC2, FC3* models and evaluate AproxSM with *NN1, LeNet, AlexNet, VGG* models. The reason for this setup is that *FC1, FC2, FC3* are compatible with DeepAbstract, which we use in the AbsSM. We train all the models on the GTSRB dataset [26] for traffic sign classification. The GTSRB dataset has over 50, 000 images from German traffic signs that belong to 43 different classes. We train all these models on various trigger types as shown in Fig. 6 where RP, BP, and LRP stand for Red Pixel, Blue Pixel, and Long Red Pixel triggers, respectively. The numbers 0.2, 0.4, and 0.6 denote the percentage of trigger transparency; a lower number means high transparency. Note that the trigger is enlarged for visualisation; however, we set the trigger size to be 2.5% of the original image size in the experiments. We replace 20% of training images with the trigger and change their labels to the target label to prepare the trojan dataset. Our target class is set to 'stop sign', meaning the output from the trojan model in the presence of a trigger will always output 'stop sign'. Table 1 depicts the performance and attack success rates of all these model architectures.

We showcase via experimental evaluations that our methods, AbsSM and AproxSM, perform better in identifying the number of backdoors than other methods. We evaluate these backdoors by reverse engineering them and computing their ASR. Both our methods reduce runtime, especially for large models. We also show that our methods perform better in identifying backdoors even when the trigger is transparent compared to state-of-the-art methods.

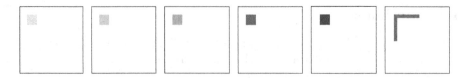

Fig. 6. Trojan Triggers. The first four images are Red Pixel (RP) Triggers at transparency 0.2, 0.4, 0.6, 1.0 respectively. Followed by a Blue Pixel (BP) and a Long Red Pixel (LRP) triggers. (Color figure online)

5.1 Backdoor Identification and Runtime Reduction

We evaluate our model's performance using three main metrics: number of identified backdoors, ASR, and runtime. We count the number of identified backdoors by reverse engineering all the CNCs and checking whether the generated trigger can misclassify at least 80% of the classes. We measure the attack success rate based on the percentage of images with triggers misclassified by the model. Finally, we measure the runtime, which consists of the time to build the abstraction and identify backdoors.

Table 1. Accuracy and ASR of the trained trojan models.

	FC1	FC2	FC3	NN1	LeNet	AlexNet	VGG
Validation Accuracy	0.8055	0.8693	0.9371	0,9064	0,9198	0,9506	0.9453
Attack Success Rate	0.9464	0.9997	0.9998	0.9840	0.9792	0.9864	0.9989
Trainable Parameters	3,072K	3,987K	4,877K	80K	123K	1,264K	1,893K

Table 2 depicts the model's performance based on ASR and the number of backdoors. All the results in the table are averaged over at least 10 runs for each model architecture, and the trigger used is RP with no transparency. We separately show the performance of our two methods and compare them to other methods, such as ABS and NC. For AbsSM, we get the abstraction rates 35%, 45%, 50% for models *FC1, FC2, FC3* respectively by setting the drop in accuracy on test dataset for the abstract model to 5%. On the other hand, the number of CR for the models used in AproxSM denoted as *Clustering Rate* is set to 10%, which means that 90% of the neurons are not simulated. We see some improvements in detecting the number of backdoors with respect to other methods, but overall, the performance is consistent with the SOTA. With this setup, although ASR is 100% with our tool and ABS, we see improvements in the number of identified backdoors.

Table 2 also depicts the difference in runtime (in sec) with respect to other methods. We show abstraction and stimulation time separately for our tool. We see a great reduction in the stimulation runtime for all AbsSM models. This is because the abstract model we use for stimulation is smaller than the original model. For AproxSM, the runtime reduction depends on whether cluster representatives are spread out throughout the layer or concentrated at a few features. If all the CR are concentrated at a few features, then after unflattening the neurons as shown in Fig. 3, the rest of the features without CR need not be stimulated. We do see such behaviour in some models. This behaviour is because several features that do not propagate large values and do not influence the network output will not become CR. On the other hand, due to the K-means clustering function being slow, the abstraction runtime is high for models such as *NN1, Lenet, VGG*. The two factors that increase the K-means clustering time are the number of images (5000) in the dataset and the number of neurons in the model. We can reduce the clustering runtime by reducing the dataset's size or replacing the clustering function.

Table 2. Method Performance.

	Model	Attack Success Rate			# of Backdoors			Runtime			
		Ours	ABS	NC	Ours	ABS	NC	Ours		ABS	NC
								Abst	Stim	Stim	Stim
AbsSM	*FC1*	1.0	1.0	1.0	**10**	7	1	27	**120**	301	600
	FC2	1.0	1.0	0.94	**8**	6	1	42	**131**	175	781
	FC3	1.0	1.0	1.0	**5**	4	1	63	**121**	211	906
AproxSM	*NN1*	1.0	1.0	0.31	2	2	0	1684	**156**	174	2,186
	LeNet	1.0	1.0	0.98	12	12	2	99	133	134	632
	AlexNet	1.0	1.0	0.07	**5**	2	0	446	234	236	1,284
	VGG	1.0	0.84	0.84	**2**	1	1	681	**330**	342	1,412

Overall, we see improvements in average performance in identifying backdoors while reducing runtime. This justifies our initial claim that abstracting the network would focus on the important neurons, and stimulating them would lead to identifying more backdoors compared to SOTA. We have some limitations with runtime due to the slow clustering method. However, the function-agnostic architecture of our tool simplifies the limitation as we can use other clustering or abstraction methods.

5.2   Performance on Various Triggers

Table 3 depicts the method performance concerning the number of backdoors identified when different kinds of triggers are applied. The numbers in the brackets represent the trigger transparency as explained before in Fig. 6. As we can see, our technique outperforms backdoor identification when the trigger transparency is high. Since our technique clusters all the neurons with similar values, the neurons representing these transparent triggers could be grouped. Stimulating this cluster would mean we are enhancing the influence of these neurons while other neuron values remain the same. Due to this reason, our technique performs better for transparent triggers compared to other methods.

Table 3. Number of identified backdoors comparing with ABS.

	Model	RP(0.2)		RP(0.4)		RP(0.6)		BP(0.2)		LRP(0.2)	
		Ours	ABS	Ours	ABS	Ours	ABS	Ours	ABS	Ours	ABS
AbsSM	FC1	**14**	6	12	12	**10**	8	7	7	3	4
	FC2	**4**	3	2	4	13	14	2	2	9	8
	FC3	**5**	2	**11**	4	3	3	1	0	6	5
AproxSM	NN1	**2**	1	1	1	**2**	0	0	0	2	0
	LeNet	**7**	5	**9**	8	**8**	6	7	7	6	6
	AlexNet	**2**	0	1	1	2	3	**6**	2	5	3
	VGG	0	1	1	1	**2**	1	**1**	0	**1**	0

Another factor we consider is the influence of the clustering rate on the network's performance. We do this experiment for only AproxSM because, for AbsSM, we fix the abstraction percentage by setting the drop in accuracy to 5%. Note that the clustering rate would not affect the model performance but instead, the number of CRs obtained. Table 4 depicts the change in runtime and backdoor identification at different clustering rates. The runtime is linearly dependent on the clustering rate. The reason is that a lower clustering rate implies a lower number of clusters to be made and in turn, less time required for the computation of the k-means. On the other hand, the number of backdoors detected varies for each model, and it is important to set the correct clustering rate. We could improve the way we set the clustering rate by considering different clustering methods such as DBSCAN [13], which does not require the clustering rate to be defined manually but instead optimally finds the best clustering parameter. We leave the exploration of other clustering techniques as future work.

All these experiments justify our approach. Moreover, developing this as a tool capable of evaluating large SOTA model architectures built on different frameworks makes analysis much easier and systematic. We want to add more abstraction and stimulation techniques to our tool in the future to improve its performance.

Table 4. Performance at different clustering rates.

	Model	Clustering Runtime				Number of Backdoors			
		10%	20%	30%	40%	10%	20%	30%	40%
AproxSM	NN1	3506	4273	6367	8634	1	1	4	0
	LeNet	106	194	290	389	7	8	5	9
	AlexNet	487	1165	1747	2344	2	2	1	1
	VGG	715	1444	2208	3003	0	1	1	0

6   Conclusion

In this paper, we have introduced the tool AGNES for finding backdoors in DNNs. We have provided experimental evidence that the two analysis methods implemented are superior in terms of performance to state-of-the-art methods. The two methods are complementary in that the best choice of the method depends on the model under consideration.

In the future, we will improve our tool in several dimensions further. As one of the next steps, we will consider different abstraction methods, which will speed up the computation of abstract DNNs to be used for analysis. We will also plan to further reduce the runtime by using parallel computation approaches.

References

1. Abadi, M.: Tensorflow: learning functions at scale. In: Proceedings of the 21st ACM SIGPLAN International Conference on Functional Programming, p. 1 (2016)
2. Ashok, P., Hashemi, V., Křetínský, J., Mohr, S.: DeepAbstract: neural network abstraction for accelerating verification. In: Hung, D.V., Sokolsky, O. (eds.) ATVA 2020. LNCS, vol. 12302, pp. 92–107. Springer, Cham (2020). https://doi.org/10.1007/978-3-030-59152-6_5
3. Bai, J., Lu, F., Zhang, K.: ONNX: open neural network exchange, github (online) (2023). https://github.com/onnx/onnx
4. Chen, C., Seff, A., Kornhauser, A., Xiao, J.: Deepdriving: learning affordance for direct perception in autonomous driving. In: Proceedings of the IEEE International Conference on Computer Vision, pp. 2722–2730 (2015)
5. Chen, X., et al.: Recent advances and clinical applications of deep learning in medical image analysis. Med. Image Anal. **79**, 102444 (2022)
6. Dhonthi, A., Hahn, E.M., Hashemi, V.: Backdoor mitigation in deep neural networks via strategic retraining. In: Chechik, M., Katoen, J.P., Leucker, M. (eds.) FM 2023. LNCS, vol. 14000, pp. 635–647. Springer, Cham (2023). https://doi.org/10.1007/978-3-031-27481-7_37
7. Dmitriev, K., Schumann, J., Holzapfel, F.: Toward certification of machine-learning systems for low criticality airborne applications. In: 2021 IEEE/AIAA 40th Digital Avionics Systems Conference (DASC), pp. 1–7. IEEE (2021)
8. Faster, R.: Towards real-time object detection with region proposal networks. In: Advances in Neural Information Processing Systems, vol. 9199, no. 10.5555, pp. 2969239–2969250 (2015)
9. Goodfellow, I., Bengio, Y., Courville, A.: Deep Learning. MIT Press, Cambridge (2016)
10. Gu, T., Dolan-Gavitt, B., Garg, S.: Badnets: identifying vulnerabilities in the machine learning model supply chain. In: Proceedings of Machine Learning and Computer Security Workshop (2017)
11. Huang, S., Peng, W., Jia, Z., Tu, Z.: One-pixel signature: characterizing CNN models for backdoor detection. In: Vedaldi, A., Bischof, H., Brox, T., Frahm, J.-M. (eds.) ECCV 2020. LNCS, vol. 12372, pp. 326–341. Springer, Cham (2020). https://doi.org/10.1007/978-3-030-58583-9_20
12. Huang, X., et al.: A survey of safety and trustworthiness of deep neural networks: verification, testing, adversarial attack and defence, and interpretability. Comput. Sci. Rev. **37**, 100270 (2020)

13. Khan, K., Rehman, S.U., Aziz, K., Fong, S., Sarasvady, S.: DBSCAN: past, present and future. In: The Fifth International Conference on the Applications of Digital Information and Web Technologies (ICADIWT 2014), pp. 232–238. IEEE (2014)

14. Kolouri, S., Saha, A., Pirsiavash, H., Hoffmann, H.: Universal litmus patterns: revealing backdoor attacks in CNNs. In: Proceedings of the IEEE/CVF Conference on Computer Vision and Pattern Recognition, pp. 301–310 (2020)

15. Li, Y., Jiang, Y., Li, Z., Xia, S.T.: Backdoor learning: a survey. IEEE Trans. Neural Netw. Learn. Syst. 1–18 (2022). https://doi.org/10.1109/TNNLS.2022.3182979. https://ieeexplore.ieee.org/document/9802938/

16. Liu, Y., Lee, W.C., Tao, G., Ma, S., Aafer, Y., Zhang, X.: ABS: scanning neural networks for back-doors by artificial brain stimulation. In: Proceedings of the 2019 ACM SIGSAC Conference on Computer and Communications Security, pp. 1265–1282 (2019)

17. Liu, Y., et al.: Trojaning attack on neural networks. In: 25th Annual Network and Distributed System Security Symposium (NDSS 2018). Internet Soc (2018)

18. Liu, Y., Shen, G., Tao, G., Wang, Z., Ma, S., Zhang, X.: Complex backdoor detection by symmetric feature differencing. In: Proceedings of the IEEE/CVF Conference on Computer Vision and Pattern Recognition, pp. 15003–15013 (2022)

19. Ma, S., Liu, Y., Tao, G., Lee, W.C., Zhang, X.: NIC: detecting adversarial samples with neural network invariant checking. In: 26th Annual Network and Distributed System Security Symposium (NDSS 2019). Internet Soc (2019)

20. Ma, W., Lu, J.: An equivalence of fully connected layer and convolutional layer. arXiv preprint arXiv:1712.01252 (2017)

21. Paszke, A., et al.: Pytorch: an imperative style, high-performance deep learning library. In: Advances in Neural Information Processing Systems, vol. 32 (2019)

22. Räuker, T., Ho, A., Casper, S., Hadfield-Menell, D.: Toward transparent AI: a survey on interpreting the inner structures of deep neural networks. In: 2023 IEEE Conference on Secure and Trustworthy Machine Learning (SaTML), pp. 464–483. IEEE (2023)

23. Saha, A., Subramanya, A., Pirsiavash, H.: Hidden trigger backdoor attacks. In: Proceedings of the AAAI Conference on Artificial Intelligence, vol. 34, pp. 11957–11965 (2020)

24. Salay, R., Queiroz, R., Czarnecki, K.: An analysis of ISO 26262: using machine learning safely in automotive software. arXiv preprint arXiv:1709.02435 (2017)

25. Shen, G., et al.: Backdoor scanning for deep neural networks through k-arm optimization. In: International Conference on Machine Learning, pp. 9525–9536. PMLR (2021)

26. Stallkamp, J., Schlipsing, M., Salmen, J., Igel, C.: Man vs. computer: benchmarking machine learning algorithms for traffic sign recognition. Neural Netw. **32**, 323–332 (2012)

27. Wang, B., et al.: Neural cleanse: identifying and mitigating backdoor attacks in neural networks. In: 2019 IEEE Symposium on Security and Privacy (SP), pp. 707–723. IEEE (2019)

28. Zhao, S., Ma, X., Zheng, X., Bailey, J., Chen, J., Jiang, Y.G.: Clean-label backdoor attacks on video recognition models. In: Proceedings of the IEEE/CVF Conference on Computer Vision and Pattern Recognition, pp. 14443–14452 (2020)

Probabilistic and Quantum Programs

Guaranteed Inference for Probabilistic Programs: A Parallelisable, Small-Step Operational Approach

Michele Boreale$^{(\boxtimes)}$ and Luisa Collodi

Dipartimento di Statistica, Informatica, Applicazioni "G. Parenti", Università degli Studi di Firenze, Florence, Italy
{michele.boreale,luisa.collodi}@unifi.it

Abstract. We put forward an approach to the semantics of probabilistic programs centered on an action-based language equipped with a small-step operational semantics. This approach provides benefits in terms of both clarity and effective implementation. Discrete and continuous distributions can be freely mixed, unbounded loops are allowed. In measure-theoretic terms, a product of Markov kernels is used to formalize the small-step operational semantics. This approach directly leads to an exact sampling algorithm that can be efficiently SIMD-parallelized. An observational semantics is also introduced based on a probability space of infinite sequences, along with a finite approximation theorem. Preliminary experiments with a proof-of-concept implementation based on TensorFlow show that our approach compares favourably to state-of-the-art tools for probabilistic programming and inference.

Keywords: Probabilistic programming · Operational semantics · Measure theory · Monte Carlo simulation · SIMD parallelism

1 Introduction

In the field of Probabilistic Programming (PP), the denotational approach dating back to Kozen's work is well established, see [4,12,16] and references therein. Equivalent operational approaches based on sequential composition and/or 'big step' semantics also have been considered, see e.g. [22]. Here we depart from this tradition and consider an action based probabilistic programming language with a 'small-step' operational semantics: we want to argue that this brings benefits both in terms of clarity of presentation and in pragmatical terms of effective implementation. The language we consider (Sect. 3) is a calculus à la Milner's CCS [18], where an action can be either a random sampling or an assignment. Continuous and discrete distributions can be freely mixed, conditionals and unbounded loops are allowed. In order to account for conditioning/observation, we also consider a fail statement indicating that the current computation must be rejected.

Work partially supported by the project SERICS (PE00000014) under the NRRP MUR program funded by the EU - NextGenerationEU.

R. Dimitrova et al. (Eds.): VMCAI 2024, LNCS 14500, pp. 141–162, 2024.
https://doi.org/10.1007/978-3-031-50521-8_7

The small-step operational semantics is formalized in terms of a (self) *product* of a Markov kernel (Sects. 2, 3), and directly leads to a sampling method, as we discuss below. On top of the operational semantics, an observational semantics (Sect. 4) is introduced, based on a probability space of infinite sequences of states. To this purpose, a standard cylindrical sigma-field construction [2, Ch.2] is leveraged. In fact, we are more flexible than this, and define the observational semantics in terms of expectation of measurable functions over the considered probability space, conditioned on non-failure. The main result here is an approximation theorem, which provides lower and upper bounds of the exact semantics (expectation), based on the semantics of the program truncated at a chosen finite execution length t: the finite semantics is effectively computable, or at least relatively easy to estimate with guarantees.

Indeed, one of the benefits and motivations of our approach is that the small-step semantics directly translates into an *exact* sampling algorithm, that can be efficiently parallelized (Sect. 5). The returned independent samples can be used for *Monte Carlo (MC) inference*, with formal statistical guarantees, expressed in terms of confidence intervals computed via exponential tail inequalities, such as Hoeffding's [14]. Specifically, we exploit a form of *Single Instruction Multiple Data (SIMD)* parallelism that arises naturally from the definition of our Markov kernel: the basic idea is that, on each transition, the kernel is applied to a whole vector of independent store-program pairs at once. Here SIMD should be contrasted with *MIMD* (M = Multiple), a form of thread-level parallelism that is a commonplace in MC methods[1]. Being able to exploit SIMD parallelism is practically relevant: indeed, modern CPUs and programming languages offer extensive support for massive SIMD parallelism in terms of *vectorization*, that is instructions operating on whole arrays at once, that can result in huge gains in performance. If desired, the proposed sampling scheme can be effectively combined with Importance Sampling (IS) to alleviate the problem of rejections, arising in the presence of conditioning/fail actions.

Preliminary experiments conducted with a proof-of-concept implementation based on TensorFlow [1], and comparison with some state-of-the-art tools, show promising results (Sect. 6). Some concluding remarks are drawn in the final section (Sect. 7). Due to lack of space, proofs and additional technical material will be made available in an online full version of the paper. All in all, we can summarize our two main contributions as follows: (1) A simple yet rigorous, measure-theoretic small-step operational semantics, that directly translates into an effective, SIMD-parallellisable inference algorithm. (2) A clean observational semantics and a finite approximation theorem, which provides a formal basis for MC estimation with guarantees.

Related Work. An introduction to and survey of recent literature on PP is the book [4]; see also the review article [12]. With few notable exceptions that we will review below, most work on the semantics of probabilistic programs still follows

---

[1] E.g. when multiple threads execute independent chains in a Markov Chain Monte Carlo method.

the denotational approach initiated by Kozen [16]. This include, among others, the work of Borgström, Gordon et al. see e.g. [5], and the work of Staton see e.g. [24,25], who consider a measure-theoretic and/or categorical point of view. In this line of work emphasis is, for instance, on providing conditions under which a *density* for a program-induced random variable exists. We do not consider such aspects in our framework, as a (cumulative) *distribution*, which always exists, is all that is needed.

As mentioned earlier, a few works depart from the traditional denotational setting, and are closer in spirit to ours. The work of Aditya et al. on Markov Chain Monte Carlo (MCMC) for R2 [22] considers a big-step sampling semantics. It is unclear how a big-step semantics would translate into a SIMD-parallel algorithm. Also, no approximation results in terms of finite execution paths is provided. In [8] the authors apply program analysis to IS. Here we too consider IS, but with a complementary concern: integrating IS into a clean measure-theoretic semantics, rather than devising syntax-driven transformations, as they do.

The work of Jasen et al. on model checking PP [15] also follows an operational approach. The main difference from us is that [15] only considers discrete distributions, for which techniques of discrete-space Markov chains [3] apply; they also consider nondeterminism, which is not a concern in our work. Similar remarks apply to Gretz et al. [13], who adopt a nondeterministic Dijkstra guarded command language extended with a discrete probabilistic choice (pGCL), basically corresponding to sampling from Bernoulli distributions. Again, the distributions definable by the programs are discrete, and discrete-space Markov Decision Processes (MDPs) are sufficient for operational modeling of the programs. They propose reasoning techniques based on probabilistic versions of weakest preconditions and invariants: these techniques allow in some cases to compute expected values of programs exactly, but they are not as generally applicable as the guaranteed MC estimation, based on finite approximation, we consider here.

Several works describe tools and algorithms for inference in PP, based on MCMC or variations thereof, like R2 [22] or WebPPL [11], among the many. In Sect. 6 we compare our sampling algorithm with these tools mainly in terms of efficiency. Not being MCMC an exact sampling technique (as the drawn samples are not truly independent), these algorithms cannot offer formal guarantees of accuracy. On the side of formal guarantees, symbolic tools like Hakaru [21] and Psi-solver [9] are based on formal manipulations of integrals, that can return explicit, exact answers in a limited set of cases. More similar in spirit to ours is the work of Sankaranarayanan et al. [23], who provide formal guarantees by only analyzing a subset of adequately chosen execution paths of a program. The analysed programs are loop-free queries. We leave an experimental comparison with [23] for future work.

It is common for programming languages, including the above mentioned webPPL and R2, to provide programmers with vectorization primitives. This language-level vectorization should not be confused with leveraging vectorization at the level of the *inference algorithm*, which is our concern here. Indeed, on nontrivial probabilistic programs and inference tasks, the difference in terms of execution time and accuracy between our approach and others shows up clearly, sometimes dramatically; see the experiments in Sect. 6.

2   Preliminaries on Measure Theory

We review a few basic concepts of measure theory following closely the presentation in the first two chapters of [2], which is a reference for whatever is not explicitly described below. Given a nonempty set Ω, a *sigma-field* $\mathcal{F}$ on Ω is a collection of subsets of Ω that contains Ω, and is closed under complement and under countable disjoint union. The pair $(\Omega, \mathcal{F})$ is called a *measurable space*. A (total) function $f : \Omega_1 \to \Omega_2$ is *measurable* w.r.t. the sigma-fields $(\Omega_1, \mathcal{F}_1)$ and $(\Omega_2, \mathcal{F}_2)$ if whenever $A \in \mathcal{F}_2$ then $f^{-1}(A) \in \mathcal{F}_1$. We let $\overline{\mathbb{R}} = \mathbb{R} \cup \{-\infty, +\infty\}$ be the set of extended reals, assuming the standard arithmetic for $\pm\infty$ (cf. [2, Sect. 1.5.2]), and $\overline{\mathbb{R}}^+$ the set of nonnegative reals including $+\infty$. The *Borel sigma-field* $\mathcal{F}$ on $\Omega = \overline{\mathbb{R}}^m$ is the minimal sigma-field that contains all rectangles of the form $[a_1, b_1] \times \cdots \times [a_n, b_n]$, with $a_i, b_i \in \overline{\mathbb{R}}$. With the above notation, f is said to be *Borel measurable* if $\mathcal{F}_2$ is a Borel sigma-field. Throughout the paper, **"measurable" means "Borel measurable"**. We shall mostly work with measurable spaces $(\Omega, \mathcal{F})$ where $\Omega = \overline{\mathbb{R}}^m$ for some $m \geq 1$ and $\mathcal{F}$ is the Borel sigma-field over Ω. On functions, Borel measurability is preserved by composition and other elementary operations on functions; continuous real functions are Borel measurable.

A *measure* over a measurable space $(\Omega, \mathcal{F})$ is a function $\mu : \mathcal{F} \to \overline{\mathbb{R}}^+$ that is countably additive, that is $\mu(\cup_{j \geq 1} A_j) = \sum_{j \geq 1} \mu(A_j)$ whenever A_j's are pairwise disjoint sets in $\mathcal{F}$. The *Lebesgue integral* of a Borel measurable function f w.r.t. a measure μ [2, Ch.1.5], both defined over a measurable space $(\Omega, \mathcal{F})$, is denoted by $\int_\Omega \mu(d\omega) f(\omega)$, with the subscript Ω omitted when clear from the context. When μ is the standard Lebesgue measure, we may omit μ and write the integral as $\int_\Omega d\omega f(\omega)$. For $A \in \mathcal{F}$, $\int_A \mu(d\omega) f(\omega)$ denotes $\int_\Omega \mu(d\omega) f(\omega) 1_A(\omega)$, where $1_A(\cdot)$ is the indicator function of the set A. We let δ_v denote Dirac's measure concentrated on v: for each set A in an appropriate sigma-field, $\delta_v(A) = 1$ if $v \in A$, $\delta_v(A) = 0$ otherwise. Otherwise said, $\delta_v(A) = 1_A(v)$. Another measure that arises (in connections with discrete distributions) is the counting measure, $\mu_C(A) := |A|$. In particular, for a nonnegative f, we have the equality $\int_A \mu_C(d\omega) f(\omega) = \sum_{\omega \in A} f(\omega)$.

Let h be a nonnegative measurable function and defined on $(\Omega, \mathcal{F})$, and μ be a measure on $\mathcal{F}$. The function $\nu(A) := \int_A \mu(d\omega) h(\omega)$ $(A \in \mathcal{F})$ defines a new measure on $\mathcal{F}$. The function h is called a *density* of ν w.r.t. μ. A basic fact about densities is that, for each measurable f, $\int \nu(d\omega) f(\omega) = \int \mu(d\omega) h(\omega) \cdot f(\omega)$, in the sense that if one of the integrals exists so does the other, and the two are equal: this is called *chain rule* for densities; see [2, Ch.2.2,Pr.4].

A *probability measure* is a measure μ defined on Ω such that $\int \mu(du) = 1$. For a given nonnegative measurable function f defined over Ω, its *expectation* w.r.t. a probability measure ν is just its integral: $\mathrm{E}_\nu[f] = \int \nu(d\omega) f(\omega)$. The following definition is central.

Definition 1 (Markov kernel). *Let* $(\Omega_1, \mathcal{F}_1)$ *and* $(\Omega_2, \mathcal{F}_2)$ *be measurable spaces. A function* $K : \Omega_1 \times \mathcal{F}_2 \longrightarrow \overline{\mathbb{R}}^+$ *is a* Markov kernel *from* Ω_1 *to* Ω_2 *if it satisfies the following properties:*

1. *for each* $\omega \in \Omega_1$*, the function* $K(\omega, \cdot) : \mathcal{F}_2 \to \overline{\mathbb{R}}^+$ *is a probability measure on* $(\Omega_2, \mathcal{F}_2)$*;*
2. *for each* $A \in \mathcal{F}_2$*, the function* $K(\cdot, A) : \Omega_1 \to \overline{\mathbb{R}}^+$ *is measurable.*

We will mostly be concerned with the case $\Omega_1 = \Omega_2$, $\mathcal{F}_1 = \mathcal{F}_2$. The following is a standard result about the construction of finite product of measures over a space $\Omega^t = \Omega \times \cdots \times \Omega$ (t times) for $t \geq 1$ an integer. The formulation below is a specialization of [2, Th.2.6.7] to Markov kernels and nonnegative functions. In particular, part (a) gives a way to construct a measure on the product space Ω^t, starting from an initial measure μ^1 and $t-1$ Markov kernels. The product space is, intuitively, the sample space of the paths of length t of a Markov chain. In particular, a path of length $t = 1$ consists of just an initial state — no transition has been fired. Part (b) is a generalization of Fubini theorem, which allows one to express an integral over the product space w.r.t. the measure of part (a) in terms of iterated integrals over the component spaces. Below, we will let ω^t range over Ω^t.

Theorem 1 (product of measures). *Let* $t \geq 1$ *be an integer. Let* μ^1 *be a probability measure on* Ω *and* $K_2, ..., K_t$ *be* $t-1$ *(not necessarily distinct) Markov kernels from* Ω *to* Ω*.*

(a) There is a unique probability measure μ^t *defined on* $(\Omega^t, \mathcal{F}^t)$ *such that for every* $A_1 \times \cdots \times A_t \in \mathcal{F}^t$ *we have:*

$$\mu^t(A_1 \times \cdots \times A_t) = \int_{A_1} \mu^1(d\omega_1) \int_{A_2} K_2(\omega_1)(d\omega_2) \cdots \int_{A_t} K_t(\omega_{t-1})(d\omega_t) \,. \tag{1}$$

(b) (Fubini) Let f *be a nonnegative measurable function defined on* Ω^t*. Then, letting* $\omega^t = (\omega_1, ..., \omega_t)$*, we have*

$$\int \mu^t(\omega^t) f(\omega^t) = \int \mu^1(d\omega_1) \int K_2(\omega_1)(d\omega_2) \cdots \int K_t(\omega_{t-1})(d\omega_t) f(\omega^t) \,. \tag{2}$$

In   particular,   on the right-hand side, for each $j = 1, ..., t - 1$ *and* $(\omega_1, ..., \omega_{j-1})$*, the function* $\omega_j \mapsto \int K_{j+1}(\omega_j)(d\omega_{j+1}) \cdots \int K_t(\omega_{t-1})(d\omega_t) f(\omega^t)$ *is measurable over* Ω*.*

It is customary to denote the measure μ^t defined by part (a) of the theorem also as $\mu^1 \otimes K_2 \otimes \cdots \otimes K_t$.

3 Probabilistic Programs

When writing programs, one will have to rely on a repertoire of basic distributions and functions. We introduce the necessary terminology below. Then we introduce the syntax and operational semantics of the language, given in terms of a Markov kernel.

Basic Elements. Both continuous and discrete basic distributions will be considered. Each basic distribution is assumed to admit a density w.r.t. a suitable measure. This enables efficient sampling, but does not imply that measures corresponding to whole programs (cf. Sect. 4) have a density. A crucial point for the expressiveness of the language is that a basic density may depend on *parameters*, whose value at runtime is determined by the state of the program. To ensure that the resulting programs define measurable functions, it is important that the dependence between the parameters and the density be in turn of measurable type. We formalize this with the concept of parametric density, introduced below. In the definition, $v \in \overline{\mathbb{R}}^m$ represent the parameters, and $G(v, \cdot)$ a density over the reals, determined by the parameters v.

Definition 2 (parametric density). *Let $G : \overline{\mathbb{R}}^m \times \overline{\mathbb{R}} \to \overline{\mathbb{R}}^+$ be a function. Assume G is measurable over $\overline{\mathbb{R}}^{m+1}$. We say G is a parametric density w.r.t. the measure μ_G over $\overline{\mathbb{R}}$ if the function $(v, A) \mapsto \int_A \mu_G(dr) G(v, r)$ is a Markov kernel from $\overline{\mathbb{R}}^m$ to $\overline{\mathbb{R}}$.*

Example 1. *Consider $\rho_U(v, r) = 1_{(0,1)}(r)$ (the argument v is ignored): this is the density of the uniform distribution on $(0, 1)$ w.r.t. the Lebesgue measure. For $h = (h_1, h_2)$ measurable functions, with h_2 positive, consider $\rho_{G,h}(v, r) := \frac{1}{h_2(v)\sqrt{2\pi}} \exp\left(-\frac{1}{2}\left(\frac{r - h_1(v)}{h_2(v)}\right)^2\right)$: this is a Normal (Gaussian) distribution of mean $h_1(v)$ and standard deviation $h_2(v)$ w.r.t. the Lebesgue measure. For an example of discrete distribution, consider a measurable function h which takes values in $[0, 1]$; then $\rho_{B,h}(v, r) = (1 - h(v)) \cdot 1_{\{0\}}(r) + h(v) \cdot 1_{\{1\}}(r)$ is the density of a Bernoulli distribution with success probability $h(v)$ w.r.t. the counting measure.*

Let $m \geq 1$ be fixed throughout the following definitions: m will represent the number of variables in the program. For the actual syntax of our language, we fix three sets of functions.

- *Update functions*: a countable set of measurable functions $g : \overline{\mathbb{R}}^m \to \overline{\mathbb{R}}$.
- *Parametric densities*: a countable set of parametric densities $G : \overline{\mathbb{R}}^m \times \overline{\mathbb{R}} \to \overline{\mathbb{R}}^+$. For simplicity, we will stick to the case where μ_G is either the standard Lebesgue measure (for continuous distributions) or the counting measure (for discrete ones).
- *Predicates*: a countable set of measurable functions $\phi : \overline{\mathbb{R}}^m \to \{0, 1\}$. By $\overline{\phi}$ we denote the predicate $1 - \phi$. An Iverson bracket style notation will be often employed, e.g.: $[x_1 \geq 1]$ is the predicate that on input v yields 1 if $v_1 \geq 1$, 0 otherwise.

Syntax. Fix a tuple of m distinct variables (symbols), $x = (x_1, ..., x_m)$. An *action* α is either an assignment or a random extraction[2]

$$\alpha ::= x_i = g \mid x_i \sim G \tag{3}$$

---

[2] Variables x_i are only introduced as syntactic sugar: they will not get instantiated, and might be dispensed with using a terser notation, like @$i = g$ and @$i \sim G$.

for any $1 \leq i \leq m$. Moreover, we fix a countable set of statement variables ranged over by $Y, Y', \ldots$. The syntax of *program statements* $S, S', \ldots$ is given by the following grammar:

$$S ::= Y \mid \text{nil} \mid \text{fail} \mid \alpha.S \mid \text{if } \phi \text{ then } S_1 \text{ else } S_2 \mid \text{rec } Y.S \qquad (4)$$

with the convention that rec Y binds Y in the last clause above, and that different occurrences of rec bind different variables. We shall only consider *closed* program statements with *guarded* recursion, a standard restriction in process calculi [18]: in each recursive subterm rec $Y.S$, all occurrences of Y in S are within the scope of some action $\alpha.(\cdot)$ operator. We let $\mathcal{P}$ denote the set of closed and guarded program statements. Note that $\mathcal{P}$ is, by construction, a countable set. In this syntax, nil stands for (correct) termination and fail for failure. The other clauses are self-explanatory. We shall use the following abbreviation

$$\text{obs}(\phi).S := \text{if } \phi \text{ then } S \text{ else fail} .$$

Example 2 (random walk). *The following program S models a random walk via a an unbounded loop. The initial value of variables, (i, r, y) in this example, is always assumed to be zero. First, r is drawn at random[3] in $(0, 1)$. Then, at each iteration y is updated according to a Normal distribution ρ_G of mean y and s.d. $2r$, until $|y| \geq 1$.*

$$S = r \sim \rho_U. \text{rec} Y. (\text{if} |y| < 1 \text{ then } (y \sim \rho_G(y, 2 \cdot r).i = i + 1.Y) \text{ else nil}) . \qquad (5)$$

As a variation of the above, one might be interested in observing only those computations that terminate in at least 3 iterations:

$$S' = r \sim \rho_U. \text{rec} Y. (\text{if} |y| < 1 \text{ then } (y \sim \rho_G(y, 2 \cdot r).i = i + 1.Y) \text{ else obs}(i \geq 3)) . \qquad (6)$$

Operational Semantics. The operational semantics of $\mathcal{P}$ will be given by a Markov kernel $\kappa(\cdot, \cdot)$. Some additional notation is in order. First, we assume a bijection between $\mathcal{P}$ and $\mathbb{N}$, so that without loss of generality we can regard $\mathcal{P}$ as a subset of $\mathbb{R}$. In particular, we will consider a set of *state-program pairs* $\overline{\mathbb{R}}^m \times \mathcal{P} \subseteq \overline{\mathbb{R}}^{m+1}$. Henceforth, we fix our state space and sigma-field as follows:

$$\Omega := \overline{\mathbb{R}}^{m+1} \qquad \qquad \mathcal{F} := \text{Borel sigma-field over } \overline{\mathbb{R}}^{m+1}$$

while reserving the symbol $\mathcal{F}_k$ for the Borel sigma-field over $\overline{\mathbb{R}}^k$, for any $k \geq 1$. For any $S \in \mathcal{P}$ and $A \in \mathcal{F}$, we let $A_S := \{v \in \overline{\mathbb{R}}^m : (v, S) \in A\}$ be the *section* of A at S. Note that $A_S \in \mathcal{F}_m$, as sections of measurable sets are measurable, see [2, Th.2.6.2,proof(1)].

For $v = (v_1, \ldots, v_m) \in \overline{\mathbb{R}}^m$, $r \in \overline{\mathbb{R}}$ and $1 \leq i \leq m$, we let $v[r @ i] := (v_1, \ldots, r, \ldots, v_m)$ denote the tuple where v_i has been replaced by r. Moreover, we let $v_{-i} \in \overline{\mathbb{R}}^{m-1}$ be the vector obtained from v by removing its i-th

---

[3] Technically, Definition 2 requires that we specify a density $\rho_G((c, 2r), \cdot)$ also when $r \leq 0$: this can be fixed arbitrarily for our purposes.

component, and let $A_{(v_{-i}, S)} = \{r \in \overline{\mathbb{R}} : (v[r @ i], S) \in A\} \subseteq \overline{\mathbb{R}}$ be the section of A at (v_{-i}, S), which is a measurable set in $\mathcal{F}_1$.

Now we define a function $\kappa : \Omega \times \mathcal{F} \to \mathbb{R}^+$. Due to the presence of rec, a plain inductive definition on the structure of S would not go through. Nevertheless, it is still possible to give a syntax-driven definition, as follows. The syntactic *depth* of S, written $\mathrm{dp}(S)$, is inductively defined by plain induction on S, as expected: $\mathrm{dp}(Y) = \mathrm{dp}(\mathsf{nil}) = \mathrm{dp}(\mathsf{fail}) = 0$, $\mathrm{dp}(\alpha.S) = 1 + \mathrm{dp}(S)$, $\mathrm{dp}(\mathsf{if}\ \phi\ \mathsf{then}\ S_1\ \mathsf{else}\ S_2) = 1 + \max\{\mathrm{dp}(S_1), \mathrm{dp}(S_2)\}$, and $\mathrm{dp}(\mathsf{rec}\,Y.S) = 1 + \mathrm{dp}(S)$. Let the *unfolding* of S, written $\mathrm{u}(S)$, be the term obtained from S by replacing each subterm $\mathsf{rec}\,Y.S'$ that is *not* in the scope of a rec $(\cdot)$ or $\alpha.(\cdot)$, by $S'[\mathsf{rec}\,Y.S'/Y]$. As an example, if $S = \mathsf{rec}\,Y.\mathsf{rec}\,Y'.\mathsf{if}\ \phi\ \mathsf{then}\ \alpha.Y\ \mathsf{else}\ \alpha'.Y'$ then

$$\mathrm{u}\,(S) = \mathsf{rec}\,Y'.\mathsf{if}\ \phi\ \mathsf{then}\ \alpha.S\ \mathsf{else}\ \alpha'.Y'\,.$$

We call S *stable* if $S = \mathrm{u}(S)$. Define the k-th unfolding of S, $\mathrm{u}^{(k)}(S)$, for $k \geq 0$, by induction on k as expected. We let the *stability depth* be $\mathrm{sdp}(S) := \min\{k \geq 0 : \mathrm{u}^{(k)}(S)$ is stable $\}$. This parameter is well defined and finite for programs $S \in \mathcal{P}$. In particular, if Y occurs in S, then $\mathrm{sdp}(S[\mathsf{rec}\,Y.S/Y]) < \mathrm{sdp}(\mathsf{rec}\,Y.S)$ for each term $\mathsf{rec}\,Y.S \in \mathcal{P}$. In what follows, we shall refer to the induction on the pairs $\mathrm{p}(S) := (\mathrm{sdp}(S), \mathrm{dp}(S))$ ordered lexicographically as to *structural induction on S*. Note that in particular $\mathrm{p}(S[\mathsf{rec}\,Y.S/Y]) < \mathrm{p}(\mathsf{rec}\,Y.S)$. Below, we shall write $\kappa(\omega, A)$ as $\kappa(\omega)(A)$, for any $A \in \mathcal{F}$.

Definition 3 (Markov kernel over Ω). *We let $\kappa : \Omega \times \mathcal{F} \to \overline{\mathbb{R}}^+$ be the function defined by structural induction on S as follows. Below, $v \in \overline{\mathbb{R}}^m$, $r \in \overline{\mathbb{R}}$ and $A \in \mathcal{F}$.*

$$\kappa(\omega)(A) = \delta_\omega(A)\ (\omega \notin \overline{\mathbb{R}}^m \times \mathcal{P})$$
$$\kappa(v, \mathsf{nil})(A) = \delta_{(v, \mathsf{nil})}(A)$$
$$\kappa(v, \mathsf{fail})(A) = \delta_{(v, \mathsf{fail})}(A)$$
$$\kappa(v, x_i = g.S)(A) = \delta_{(v[g(v) @ i], S)}(A)$$
$$\kappa(v, x_i \sim G.S)(A) = \int_{A_{(v_{-i}, S)}} \mu_G(dr)G(v, r)$$
$$\kappa(v, \mathsf{if}\ \phi\ \mathsf{then}\ S_1\ \mathsf{else}\ S_2)(A) = \phi(v)\kappa(v, S_1)(A) + \overline{\phi}(v)\kappa(v, S_2)(A)$$
$$\kappa(v, \mathsf{rec}\,Y.S)(A) = \kappa(v, S[\mathsf{rec}\,Y.S/Y])(A)\,.$$

Note that the first clause ensures that $\kappa(\cdot)(A)$ is defined on all elements of Ω, even those that do not represent valid state-program pairs.

Lemma 1. *The function κ is a Markov kernel from Ω to Ω.*

The proof that κ is a Markov kernel is a bit laborious but not difficult. We show separately that for each fixed $\omega \in \Omega$, the function $A \mapsto \kappa(\omega)(A)$ is a probability measure on $\mathcal{F}$, and that for each fixed $A \in \mathcal{F}$, the function $\omega \mapsto \kappa(\omega)(A)$ is measurable on Ω. This allow us to conclude, relying on some basic facts from measure theory (e.g. continuity implies measurability).

Example 3. *Consider the program S (random walk) in (5). Here we have $m = 3$ variables, (i, r, y). Assume for simplicity that $A = R \times \mathcal{P}$, for some rectangle $R = [a_1, b_1] \times [a_2, b_2] \times [a_3, b_3] \subseteq [0,1]^3$. Also call S_1, S_2, S_3 the continuations of S after the first, second and third action, respectively. Let $\rho_G((c, s), r)$ be the Normal parametric density of mean c and standard deviation s, and assume $s \in (0,1)$ below. Then one can check:*

$$\kappa(((0,0,0), S)(A) = \delta_1([a_1, b_1])$$
$$\kappa(((1,0,0), S_1)(A) = b_2 - a_2$$
$$\kappa(((1, s, 0), S_2)(A) = \int_{a_3}^{b_3} \rho_G((0, 2\, s), r) dr .$$

4   Observable Semantics

For any $t \geq 1$, we call Ω^t the set of *paths of length t*. Consider now the set of paths of infinite length, Ω^∞, that is the set of infinite sequences $\tilde{\omega} = (\omega_1, \omega_2, ...)$ with $\omega_i \in \overline{\mathbb{R}}^{m+1}$. For any $\omega^t \in \Omega^t$ and $\tilde{\omega} \in \Omega^\infty$, we identify the pair $(\omega^t, \tilde{\omega})$ with the element of Ω^∞ in which the prefix ω^t is followed by $\tilde{\omega}$. For $t \geq 1$ and a measurable $B^t \subseteq \Omega^t$, we let $B_t := B^t \times \Omega^\infty \subseteq \Omega^\infty$ be the *measurable cylinder* generated by B^t. We let $\mathcal{C}$ be the minimal sigma-field over Ω^∞ generated by all measurable cylinders. Under the same assumptions of Theorem 1 on the measure μ^1 and on the kernels $K_2, K_3, ...$ there exists a unique measure μ^∞ on $\mathcal{C}$ such that for each $t \geq 1$ and each measurable cylinder B_t, it holds that $\mu^\infty(B_t) = \mu^t(B^t)$: see [2, Th.2.7.2], also known as the *Ionescu-Tulcea theorem*. In the definition below, we let $0 = (0, ..., 0)$ (m times) and consider $\delta_{(0,S)}$, the Dirac's measure on Ω that concentrates all the probability mass in $(0, S)$.

Definition 4 (probability measure induced by S). *Let $S \in \mathcal{P}$. For each integer $t \geq 1$, we let μ_S^t be the probability measure over Ω^t uniquely defined by Theorem 1(a) by letting $\mu^1 = \delta_{(0,S)}$ and $K_2 = \cdots = K_t = \kappa$. We let μ_S^∞ be the unique probability measure on $\mathcal{C}$ induced by μ_1 and $K_2 = \cdots = K_t = \cdots = \kappa$, as determined by the Tulcea-Ionescu theorem.*

In other words, $\mu_S^t = \delta_{(0,S)} \otimes \kappa \otimes \cdots \otimes \kappa$ ($t - 1$ times κ). By convention, if $t = 1$, $\mu_S^t = \delta_{(0,S)}$. The measure μ_S^∞ can be informally interpreted as the limit of the measures μ_S^t and represents the semantics of S. We will be interested in the following events of Ω^t and Ω^∞, for $t \geq 1$. We start from the following three sets that form a partition of Ω: $\mathsf{L} := \overline{\mathbb{R}}^m \times (\overline{\mathbb{R}} \setminus \{\text{nil}, \text{fail}\})$, $\mathsf{T} := \overline{\mathbb{R}}^m \times \{\text{nil}\}$, $\mathsf{F} := \overline{\mathbb{R}}^m \times \{\text{fail}\}$.

- $\mathsf{F}_t := \mathsf{L}^{t-1} \times \mathsf{F} \times \Omega^\infty$, the paths that *fail* at time t;
- $\mathsf{T}_t := \mathsf{L}^{t-1} \times \mathsf{T} \times \Omega^\infty$, the paths that *terminate* at time t;
- $\mathsf{F}_\infty := \bigcup_{t \geq 1} \mathsf{F}_t$, the paths that *eventually fail*;
- $\mathsf{T}_\infty := \bigcup_{t \geq 1} \mathsf{T}_t$, the paths that *eventually terminate*.

We are interested in evaluating a program *only on non failed paths*: that is, we will work by conditioning on the event $(\mathsf{F}_\infty)^c$. Recall that the *support* of an (extended) real valued function f is the set $\text{supp}(f) := \{z : f(z) \neq 0\}$. In what follows, *we shall concentrate on nonnegative measurable functions f to avoid unnecessary complications with the existence of integrals*. The general case can be dealt with by the usual trick of decomposing f as $f = f^+ - f^-$, where for each z one defines $f^+(z):= \max(0, f(z))$ and $f^-(z):= -\min(0, f(z))$, and then dealing separately with f^+ and f^-.

Definition 5 (observable semantics). *Let f be a nonnegative measurable function defined on Ω^∞. We let the* unnormalized *semantics of S and f be $[S]f := \mathbb{E}_{\mu_S^\infty}[f](= \int \mu_S^\infty(d\tilde{\omega})f(\tilde{\omega}))$. Let the support of f be contained in $(F_\infty)^c$, then we let*

$$[\![S]\!]f := \frac{[S]f}{[S]1_{(F_\infty)^c}} \tag{7}$$

provided the denominator is not zero; otherwise $[\![S]\!]f$ is undefined.

We are mainly interested in $[\![S]\!]f$ in cases where, informally speaking, the value of f does not depend on what happens after termination, that is it is determined by the first terminated state: we call these functions *termination based*, and define them formally below. We also define *lifting*, a natural way of defining termination based functions from functions on Ω that only look at the value of variables in (correctly) terminated states.

Definition 6 (termination based f, lifting). *Let $f : \Omega^\infty \to \overline{\mathbb{R}}^+$ be a measurable function. We say f is* termination based *if $\mathrm{supp}(f) \subseteq T_\infty$ and for each $t \geq 1$ and $\omega^t \in L^{t-1} \times T$, we have that f is constant on $\{\omega^t\} \times \Omega^\infty$. For any such f and $t \geq 1$, we let $f_t : \Omega^t \to \overline{\mathbb{R}}^+$ be defined by $f_t(\omega^t):=f(\omega^t, *^\infty)$, for an arbitrarily fixed $* \in L$.*

Given a nonnegative measurable g on Ω with $\mathrm{supp}(g) \subseteq \mathbb{R}^m \times \{nil\}$, we define the lifting of g *to Ω^∞ as the function f defined as follows: for $\tilde{\omega} = (\omega_1, \omega_2, ...)$, $f(\tilde{\omega}):=g(\omega_t)$ if $\tilde{\omega} \in T_t$ for some $t \geq 1$; otherwise, $f(\tilde{\omega}):=0$. The lifting of g to Ω^t is f_t, where f is the lifting of g to Ω^∞.*

Most functions of interest can be defined via lifting.

Example 4. *$f = 1_{T_\infty}$, the indicator function of the termination event, is termination based, and is just the lifting of the predicate $g(v, S) = [S = nil]$ on Ω. Another example are lifting of functions g that just return the final values of the program's variables, or functions thereof, in terminated states. Outside the class of lifted functions, a function that, for instance, counts the number of times a given event $E \subseteq \Omega$ is observed until termination, is termination based.*

Informally speaking, the functions f_t approximate f over paths of length t. It is not difficult to check that, for any t, f_t is measurable over Ω^t, starting from the obvious case where f is the indicator function. The next result shows how to approximate $[\![S]\!]f$ with quantities defined *only* in terms of f_t and μ_S^t, which is the basis for the MC approximation algorithm in the next section. Formally, for $t \geq 1$ and a measurable function $h : \Omega^t \to \overline{\mathbb{R}}^+$, we let

$$[S]^t h := \mathbb{E}_{\mu_S^t}[h] \left(= \int \mu_S^t(d\omega^t)h(\omega^t)\right).$$

We also use the abbreviation $\mathsf{T}^{\leq t} := \cup_{j=1}^{t} \mathsf{L}^{j-1} \times \mathsf{T} \times \Omega^{t-j} \subseteq \Omega^t$. The intuitive content of the result is as follows. At a given time t, the probability mass over paths of length t is divided among terminated ones ($\mathsf{T}^{\leq t}$), live ones (L^t) and failed ones $((\mathsf{T}^{\leq t} \cup \mathsf{L}^t)^c)$. As t grows, the live probability mass can be distributed over terminated and failed paths. Then considering the numerator in (7), at any given time t we obtain an upper bound by moving all the live probability mass to terminated paths, while taking into account a bound M on f; and a lower bound by zeroing the live probability mass. We can argue similarly for the denominator.

Theorem 2 (finite approximation). *Assume that, for some* $t \geq 1$, $\mu_S^t(\mathsf{T}^{\leq t}) > 0$. *Then for any termination based function* f, *we have that* $[\![S]\!]f$ *is well defined. Moreover, given a uniform upper bound* $f_t \leq M$ ($M \in \overline{\mathbb{R}}^+$), *for each* t *large enough:*

$$\frac{[S]^t f_t}{[S]^t 1_{\mathsf{T}^{\leq t}} + [S]^t 1_{\mathsf{L}^t}} \leq [\![S]\!]f \leq \frac{[S]^t f_t + M \cdot [S]^t 1_{\mathsf{L}^t}}{[S]^t 1_{\mathsf{T}^{\leq t}}}. \tag{8}$$

In particular, if for some t *we have* $[S]^t 1_{\mathsf{L}^t} = 0$, *then*

$$[\![S]\!]f = \frac{[S]^t f_t}{[S]^t 1_{\mathsf{T}^{\leq t}}}. \tag{9}$$

When f is an indicator function, $f = 1_A$, we can of course take $M = 1$ in the theorem above.

Example 5. *Let us consider again the program* S *on the variables* (i, r, y) *in* (5). *Let* f *be the lifting of the indicator function* g *defined on* Ω *that characterizes the final states of the program in which* $i \geq 3$, *formally:* $g((a, b, c), z) = 1$ *if* $a \geq 3$ *and* $z = nil$, *and* 0 *elsewhere. So* $[\![S]\!]f$ *is the probability that the execution of* S *terminates with at least 3 iterations. Fix* $t = 120$. *Drawing a large number of independent samples[4] from the program, we can estimate the expected values in* (8) *as arithmetic means, and can check that in this case the bounds yield:*

$$0.45 \leq [\![S]\!]f \leq 0.56.$$

Also, $[S]^t 1_{\mathsf{T}^{\leq t}} \geq 0.93$, *that is the probability of terminating within time* $t = 120$ *is at least 0.93. Further details on the actual computation of these figures are postponed until Sect. 6.*

Remark 1. Concerning the probability of termination, that is the $\mu_S^\infty(\mathsf{T}_\infty) = [\![S]\!]f$ with $f = 1_{\mathsf{T}_\infty}$, we note only the *lower* bound in (8) can be useful. Indeed, in this case $f_t = 1_{\mathsf{T}^{\leq t}}$, so that the upper bound's numerator in (8) is always at least as large as the denominator.

---

[4] For this example, we have used 10^5 samples. In fact, the stated bounds hold with very high probability: this will be made rigorous in the next section in terms of confidence intervals.

Remark 2 (from expectations to distributions). While our semantics is given in terms of expectations, it is fairly general. In particular, for any nonnegative measurable h defined on Ω^∞, all the relevant information about h is conveyed by its *cumulative distribution function,* conditioned on non failure, defined for any $r \in \overline{\mathbb{R}}$ as

$$F(r) := \mu_S^\infty \left(h^{-1}([-\infty, r]) \cap (\mathsf{F}_\infty)^c \right) / \mu_S^\infty \left((\mathsf{F}_\infty)^c \right).$$

Now $F(r)$ can be recovered by letting $f = 1_{h \leq r} \cdot 1_{(\mathsf{F}_\infty)^c}$ in Definition 5, that is $F(r) = [\![S]\!](1_{h \leq r} \cdot 1_{(\mathsf{F}_\infty)^c})$. To see this, note that this f is just the indicator function of $h^{-1}([-\infty, r]) \cap (\mathsf{F}_\infty)^c$. We also note that this f is *not* termination based, so Theorem 2 as is does not apply to it. We think that at least for finite (rec-free) S, an extension of the theorem to such f is easy, but leave the details for future work.

5   Inference via Vectorized Monte Carlo Sampling

Monte Carlo Estimation. It is convenient to extend our notation for the measure μ_S^t to the case of an arbitrary initial vector $v \in \overline{\mathbb{R}}^m$, by letting $\mu_{S,v}^t := \delta_{(v,S)} \otimes \kappa \otimes \cdots \otimes \kappa$ ($t-1$ times κ), and $[\![S]\!]_v^t f := \int \mu_{S,v}^t(\omega^t) f(\omega^t)$.

Proposition 7. *Let $f : \Omega^t \to \overline{\mathbb{R}}^+$ be a nonnegative measurable function and $t \geq 1$. Then, for $\omega_1 := (v, S) \in \Omega$*

$$[\![S]\!]_v^t f = \int \kappa(\omega_1)(d\omega_2) \cdots \int \kappa(\omega_{t-1})(d\omega_t) f(\omega_1, \omega_2, ..., \omega_t). \tag{10}$$

Moreover, the function $v \mapsto [\![S]\!]_v^t f$ is measurable.

By virtue of Proposition 7, any expectation (integral) involved in (7) or in (8) can be expressed as $t-1$ iterated expectations (integrals). We shall sometimes abbreviate this as follows, letting $\omega_1 = (0, S)$:

$$[\![S]\!]^t f = \mathrm{E}_{\mu_S^t}[f] = \mathrm{E}_{\omega_2 \sim \kappa(\omega_1), \omega_3 \sim \kappa(\omega_2), ..., \omega_t \sim \kappa(\omega_{t-1})}[f(\omega_1, ..., \omega_t)]. \tag{11}$$

Equation (11) is the basis for implementing a Monte Carlo sampling inference algorithm. Indeed, the Law of Large Numbers implies that expectation can be approximated by arithmetic mean (see further below for a rigorous statement):

$$[\![S]\!]^t f \approx \frac{1}{N} \sum_{i=1}^N f(Z_i) \tag{12}$$

where $Z_1, Z_2, ..., Z_N$ are N i.i.d. random variables, with Z_i distributed according to μ_S^t. Explicitly, for each i, as prescribed by (11):

$$Z_i = (\omega_{i,1}, ...\omega_{i,t}), \text{ where: } \omega_{i,1} = (0, S), \omega_{i,2} \sim \kappa(\omega_{i,1}), ..., \omega_{i,t} \sim \kappa(\omega_{i,t-1}). \tag{13}$$

If f is of bounded variation, that is $f(\omega^t) \leq M < +\infty$ for each $\omega^t \in \Omega^t$, the quality of the approximation (12) can be controlled via the Hoeffding inequality [14], which implies that, for each $\epsilon > 0$

$$\Pr\left(\left|[S]^t f - \tfrac{1}{N} \sum_{i=1}^{N} f(Z_i)\right| > \epsilon\right) \leq \delta := 2e^{-2\frac{N\epsilon^2}{M^2}}. \tag{14}$$

From confidence intervals of width ϵ for expectations $[S]^t f$, it is easy to build confidence intervals for the ratios representing the lower and upper bounds in (8) via simple algebraic manipulations. The resulting bounds tend to be quite loose in some situations, e.g. when f returns small values. To obtain less naive bounds, consider the subset of the N samples that are in $\mathsf{T}^{\leq t} \cup \mathsf{L}^t$, say $\tilde{Z}_1, ..., \tilde{Z}_{N_a}$ for some $0 \leq N_a \leq N$; we call these the *accepted* samples. Out of these, there will be $0 \leq N_t \leq N_a$ samples in $\mathsf{T}^{\leq t}$ (terminated), and $N_a - N_t$ in L^t (live). Assuming $N_a > 0$, for $\epsilon > 0$, a tighter confidence interval is

$$\frac{\sum_{i=1}^{N_a} f_t(\tilde{Z}_i)}{N_a} - \epsilon \leq [S]f \leq \frac{\sum_{i=1}^{N_a} f_t(\tilde{Z}_i)}{N_t} + \epsilon + M \cdot \left(\frac{N_a}{\max\{0, N_t - \epsilon N_a\}} - 1\right) \tag{15}$$

where, by applying Hoeffding's inequality, one can prove that the both the above bounds hold with confidence at least $1 - (2\delta_a + \delta_t)$, with $\delta_a = e^{-2N_a \epsilon^2 / M^2}$ and $\delta_t = e^{-2N_a \epsilon^2}$. The convention $r/0 = +\infty$ for $r \geq 0$ applies above. When it is known that $\mu_S^t(\mathsf{L}^t) = 0$, the term $M \cdot (\cdots)$ can be omitted, and we can set $N_t = N_a$. Let us also mention that there is life beyond Hoeffding: in particular, empirical versions of Bernstein inequality [17,20] are convenient when the samples exhibit a low empirical variance.

A Vectorized Algorithm. While the above is standard, we argue that, thanks to the uniform formulation of the Markov kernel $\kappa(\cdot)$, (12) can be efficiently implemented in vectorized form. This is outlined in Fig. 1, where we use the following notation and conventions. We use a pair of arrays (V, P) to store N (independent) instances of pairs (program store, control pointer). In particular, $V \in \overline{\mathbb{R}}^{N \times m}$ and $P \in \mathcal{P}^{N \times 1}$ store, respectively: N program's stores $v \in \overline{\mathbb{R}}^m$; and N control pointers, each represented as a statement $S \in \mathcal{P}$. The basic operations we rely upon are the following.

1. *Boolean masking*: for any two arrays $W \in \overline{\mathbb{R}}^{K \times m}$ and $\beta \in \overline{\mathbb{R}}^{K \times 1}$, we let $W[\beta]$ denote the sub-array of W formed by all rows of W at indices i s.t. $\beta(i) \neq 0$.
2. *Vectorization*: for any function $h : \overline{\mathbb{R}}^m \to \overline{\mathbb{R}}$, possibly a predicate, and array $W \in \overline{\mathbb{R}}^{K \times m}$, we let $h(W)$ be the $K \times m$ array obtained by applying h elementwise on W, as expected.

Vectorization also applies to random sampling. Denote by $D_G(v)$ the distribution on $\overline{\mathbb{R}}$ induced by the parametric density $G(v, r)$, taking parameters $v \in \overline{\mathbb{R}}^m$: in $R \sim D_G(W)$, R is a $K \times 1$ array obtained by K independent extractions, drawn according to the distributions $D_G(w)$ for w in W. Modern CPUs offer extensive support for performing boolean masked assignments such as $W[\beta] \leftarrow W'[\beta]$

as *single instructions* (to some extent, depending on the size of W): this effectively leverages SIMD parallelism, eliminating levels of looping and iterations, and potentially leading to significant runtime speed-up. Vectorization is also supported by many high-level programming languages[5]. Note that, in the **for** loop of algorithm $\tilde{\kappa}$ at line 3, the number of iterations is bounded by $|set(P)|$, which is of the order of the syntactic size of S and independent of the number of samples N to be taken. An optimization of this scheme is possible by considering a suitable restricted language of program statements; this will be explored in Sect. 6.

$\tilde{\kappa}(V, P)$
1: $V', P' \leftarrow$ empty arrays of the same shape as V, P
2: $B \leftarrow \{ (S, 1_{\{S\}}(P)) : S \in set(P) \}$
3: **for** $(S, \beta) \in B$ **do**
4:     **switch** S **do**
5:         **case** nil or fail
6:             $V'[\beta] \leftarrow V[\beta], P'[\beta] \leftarrow S$
7:         **case** $x_i = g.S'$
8:             $V'[\beta, i] \leftarrow g(V[\beta]), P'[\beta] \leftarrow S'$
9:         **case** $x_i \sim G.S'$
10:            $V'[\beta, i] \leftarrow R, P'[\beta] \leftarrow S'$, where $R \sim D_G(V[\beta])$
11:        **case** if ϕ then S_1 else S_2
12:            $\beta_1 = \phi(V), \beta_2 = 1 - \beta_1$
13:            $V'[\beta \cdot \beta_j], P'[\beta \cdot \beta_j] \leftarrow \tilde{\kappa}(V[\beta \cdot \beta_j], S_j), j = 1, 2$
14:        **case** rec $Y.S'$
15:            $V'[\beta], P'[\beta] \leftarrow \tilde{\kappa}(V[\beta], S'[S/Y])$
16: **end for**
17: **return** (V', P')

$\mathcal{A}(S, h, t, N)$
1: $V \leftarrow N \times m$ array, filled with 0
2: $P \leftarrow N \times 1$ array, filled with S
3: **do** $t - 1$ **times**
4:     $V, P \leftarrow \tilde{\kappa}(V, P)$
5: **end do**
6: **return** $\frac{1}{N} \sum h(V, P)$

Fig. 1. Vectorized Monte-Carlo algorithm for estimating $[S]^t f$, with $f : \Omega^t \to \overline{\mathbb{R}}$. **Left:** Algorithm $\tilde{\kappa}$: given $(V, P) = ((v_j)_{j=1}^N, (S_j)_{j=1}^N)$, the returned (V', P') stores N independent drawings from $\kappa(v_1, S_1), ..., \kappa(v_N, S_N)$. **Right:** Algorithm $\mathcal{A}$, Monte Carlo estimation (12) of $[S]^t f$. Here we assume f is the lifting to Ω^t of a function h defined on Ω. **Notation.** $set(P) \subseteq \mathcal{P}$ is the set of elements that appear in P; $V[\beta, i]$ is the i-th column of $V[\beta]$; $\sum h(V, P) := \sum_{j=1}^N h(v_j, S_j)$, where v_j, S_j denote the j-th row of V, P, respectively. Broadcast of individual values to whole arrays is automatically performed as needed: $W \leftarrow w$ means filling the array W with a specific value w, and so on. $\beta \cdot \beta'$ denotes the element-wise product of β and β'.

Importance Sampling. Consider the program $S = x_1 \sim \rho_U(0,1).x_2 \sim \rho_G(x_1, 1).\text{obs}(|x_2 - x_1| > 3).\text{nil}$. The obs statement demands that x_2 fall more than three standard deviations away from x_1. Suppose we draw N samples from this program, relying in the algorithm $\mathcal{A}$ in the previous section. Recalling that $\text{obs}(\phi).\text{nil} = \text{if } \phi \text{ then nil else fail}$, a large fraction of the resulting computations— on average about 99.7%—will end into fail, that is will be *rejected*. If the purpose of this procedure is obtaining samples of the terminated states of the program, this is huge waste of computational resources. According to the principles of *importance sampling* (IS), it is better if in the second sampling step in S one draws directly from the distribution $\rho_G((x_1, 1), \cdot)$ *restricted* to the set $A_{x_1} =$

---

[5] For instance, in Python via the Numpy or TensorFlow packages.

$(-\infty, x_1 - 3) \cup (x_1 + 3, +\infty)$: call $\tilde{\rho}_G((x_1, 1), \cdot)$ this new distribution. This change of sampling distribution must be compensated by weighing each sample with an *importance weight* $W(x_1) = \int_{A_{x_1}} \mu(dr) \rho_G((x_1, 1), r)$. Formally, we apply Proposition 7 with $t = 3$, take into account the chain rule for densities and, assuming f is the lifting of some g, compute as follows (in the second step, we take into account that $g(r_1, r_2, \mathsf{fail}) = 0$; in the third step, the definition of the restricted density $\tilde{\rho}_G(r_1, 1; r_2)$):

$$
\begin{aligned}
[S]^t f &= \int \mu(dr_1) \rho_U(0, 1; r_1) \cdot \int \mu(dr_2) \rho_G((r_1, 1), r_2) \cdot [|r_2 - r_1| > 3] \cdot g(r_1, r_2, \mathsf{nil}) + \\
&\quad \int \mu(dr_1) \rho_U(0, 1; r_1) \cdot \int \mu(dr_2) \rho_G((r_1, 1), r_2) \cdot [|r_2 - r_1| \leq 3] \cdot g(r_1, r_2, \mathsf{fail}) \\
&= \int \mu(dr_1) \rho_U(0, 1; r_1) \cdot \int \mu(dr_2) \left(\frac{\rho_G(r_1, 1; r_2) \cdot [|r_2 - r_1| > 3]}{W(r_1)} \right) W(r_1) g(r_1, r_2, \mathsf{nil}) \\
&= \int \mu(dr_1) \rho_U(0, 1; r_1) \cdot \int \mu(dr_2) \tilde{\rho}_G((r_1, 1), r_2) W(r_1) g(r_1, r_2, \mathsf{nil}) \\
&= [\tilde{S}]^t \tilde{f}
\end{aligned}
$$
$$(16)$$

where $\tilde{S} = x_1 \sim \rho_U(0, 1). x_2 \sim \tilde{\rho}_G(x_1, 1). \mathsf{nil}$ and $\tilde{f}$ is the lifting of $W \cdot g$. Now, provided we know how to draw efficiently samples from $\tilde{\rho}_G$ and how to compute $W(x_1)$ (which is in fact easy), evaluating $[\tilde{S}]^t \tilde{f}$ by applying $\mathcal{A}$ to $\tilde{S}$ and $\tilde{f}$ will be much more efficient than applying $\mathcal{A}$ directly to S and f. In this example, when applying $\mathcal{A}$ to $\tilde{S}$ we will have no rejections. Moreover, from the point of view of the application of Hoeffding's bound (14), we can replace M with the much smaller $M \cdot \sup_{x_1} W(x_1)$, obtaining to a significative sharpening of the bound. In fact, in this case $W(x_1)$ is a constant, $W(x_1) \approx 0.003$. The above example can be generalized to a systematic procedure, more details will be made available in an online version of the paper. Let us remark here that applying this procedure can be costly, as, in the general case, the number of terms that may appear in an expansion like (16) can be exponential in the size of S. In our experiments, IS turned out to be less convenient than a pure vectorized Monte Carlo estimation.

6   Implementation and Evaluation

We will specialize the sampling algorithm $\mathcal{A}$ of Sect. 5 to a sublanguage of $\mathcal{P}$, and present an implementation based on TensorFlow (TF) [1], an open source library enabling efficient and SIMD-parallel manipulation of tensors (multidimensional arrays). We will then evaluate this implementation on a few probabilistic programs taken from the literature. For loop-free examples, we will also compare our results with those obtained by applying two state-of-the-art tools for probabilistic programming.

6.1   TSI: A TensorFlow Based Implementation

We focus on a sublanguage $\mathcal{P}_0 \subseteq \mathcal{P}$, and present a TF-based implementation of a specialized version of algorithm $\mathcal{A}$. The syntax of $\mathcal{P}_0$ is obtained by imposing that every rec subterm has the following form:

$$\mathsf{rec}\, Y.(\mathsf{if}\, \phi\, \mathsf{then}\, S_1\, \mathsf{else}\, S_2)$$

such that S_1 does not contain rec nor nil, $S_1 \neq$ fail and S_2 does not contain Y. So nested rec's are not allowed, and exiting a loop is possible only either via a fail statement inside S_1 or by making ϕ false, in the last case proceeding then to S_2. Accordingly, in the rest of the section we will also use the following abbreviation, for any rec-free statement $S_1 \neq$ fail where every nil is guarded by an action prefix, and S_2 does not contain Y:

$$\text{while } \phi S_1, S_2 := \text{rec } Y.(\text{if } \phi \text{ then } S_1[Y/\text{nil}] \text{ else } S_2)) \tag{17}$$

The sublanguage $\mathcal{P}_0$ has been tailored so as to be mapped easily into TF, as explained below; yet it is still expressive enough to allow for the description of nearly all the examples in the literature we have examined. Our translation leverages the following linguistic features of TF.

1. Vectorized **assignment** xi=g(x), where xi,x can be either scalar or tensor variables, or tuples thereof. Functions and predicates on scalars are automatically extended to tensors as expected. Likewise, vectorized **sampling** from a distribution with parameters x will be written G(x).sample(): when the parameter x is a tensor, this will yield a tensor of independent samples, one for each component of x.
2. Vectorized **if-then-else** expression tf.where(cond,a,b). Here cond is a tensor of booleans that acts as a mask: the result of tf.where is a tensor of the same shape as a, where the value of each component is the corresponding element from tensor a (if the corresponding element from cond is True) or from tensor b (otherwise). These operations are executed independently on the elements of the involved tensors cond,a,b, in a vectorized fashion.
3. Vectorized **while-loop** expression tf.while_loop(fcond,fbody,vars). Here both fcond and fbody are functions that can be called with the vars tensor variables as arguments: fcond returns a boolean scalar, and fbody returns a tuple of tensors that can be assigned to variables vars. The semantics dictates that the assignment vars=fbody(vars) be repeated as long as fcond(vars) yields True; upon termination, the current value of vars is returned. Therefore, although the termination condition fcond(vars) is global, different instances of fbody are independently executed on the elements of tensors vars, in a vectorized fashion.

An outline of our translation of $\mathcal{P}_0$ into TF is presented in Fig. 2. A statement $S \in \mathcal{P}_0$ is translated into a TF-Python function f_S, introduced by the declaration $\mathbb{T}(S)$. When invoked, f_S will return N independent samples, each drawn according to (13). In more detail, the command x,mask,t=f_S(x,mask,t) executes N independent instances of the initial statement S, applying successive transitions, until each instance has reached at least a prescribed time (=path length) of t_0. The tuple of tensors x = x1,...,xm, one for each program variable, initially filled with 0, will store the drawn N independent samples. The control pointer vector V of algorithm $\mathcal{A}$ is replaced here by a pair of tensors mask,t. The tensor mask will keep track of which of the N instances is actually terminated (True), failed (False) or live (NaN) at the end of the execution. The tensor t is a counter, that will record the actual time at which each instance will reach nil or fail, if ever. Note that "stuttering" transitions from nil or fail do not increment t; but transitions from if-then-else's where one of the two branches is a nil or fail do increment t (cf. lines 19 and 31). The use of t is necessary because the N instances are not necessarily time-aligned at the end of the execution of f_S. The information contained in mask,t will be used at the end to discern which instances are actually terminated/failed/live at time t_0, and the corresponding samples from x (like in lines 42,43). Taken together, these form the wanted N i.i.d. samples from μ^{t_0} drawn according[6] to (13).

The clauses of the translation function T($\cdot$) should be self-explanatory, but possibly for the last one. In the case of while, a complication is that, within the given time limit of t_0, the N independently executed instances may exit the loop at different times, or even not exit the loop at all. In the corresponding code (lines 23–34), the execution of an instance is suspended as soon as it reaches either an exit condition or time t_0 (second branch of where in the definition of fbody, line 26); the while_loop stops when all instances have reached this situation (condition in fcond, lines 28,29). Next, the execution of all instances is resumed with S_2 (lines 30,31), but this will have an effect only for instances that had exited the loop normally, that is by invalidating ϕ before time t_0 without failing: the other instances are deemed either live or failed at time t_0 (lines 32, 33), and S_2 will have no effect for them. Further details are provided in the caption of Fig. 2. A simplification of this translation is possible for statements $S \in \mathcal{P}_0$ that rule out either while or if-then-else (this in fact covers all the examples we will see); see [6].

Technically, the TF execution model is based on *data flow graphs*: under the hood [19], the code is expressed as a computational graph, a data structure where nodes represent operations, and edges represent data, arranged as tensors, that flow through the graph between operations. A graph execution model is defined that leverages fine-grained, massive SIMD-parallelism. Better performances are obtained on GPU-based hardware.

We will refer to the framework resulting from our translation of $\mathcal{P}_0$ into TF as *TSI*, for *Tensorized Sampling and Inference*.

6.2   Evaluation of TSI

In the first subsection, we present experimental results obtained by putting TSI at work on four loop-free probabilistic programs taken from the literature: TrueSkill, Pearl's Burglar Alarm, ClickGraph and MontyHall. At the same time, we compare TSI with two state-of-the-art probabilistic languages (and inference tools), webPPL [11] and R2 [22]: we have chosen them because the corresponding languages are easy to compare with ours, in particular both feature observe or equivalent statements for conditioning. webPPL incorporates the principles of Church [10], a LISP based PPL with an elegant semantics. For the time being,

---

[6] Like for algorithm $\mathcal{A}$, since f is assumed to be the lifting of some h defined on Ω, we only keep the last element of each sequence, ω_{i,t_0} in the notation of (13).

we have left out of our comparison languages and tools that are more tailored to fitting parameters to data and do not feature observe statements, like Stan [7], or that lack a formally defined semantics.

In the second subsection, we apply TSI to probabilistic programs involving unbounded loops, and discuss the obtained results mainly from the point of view of accuracy. Code and examples for these experiments[7] are available online [6].

Loop Free Programs and Comparison with WebPPL and R2. We consider TrueSkill, Pearl's Burglar Alarm, ClickGraph and Monty Hall, four nontrivial probabilistic programs. On these models, we compare TSI with both WebPPL, a programming language based on JavaScript, and R2, a probabilistic programming system based on C#. In Table 1, we report the execution time required by the different approaches on the four programs, as the number of samples increases, together with the estimated expected value for their output variables. In each case, the value of t for TSI has been chosen large enough so guarantee termination or failure of all instances (no live terms). N is the number of samples generated by TSI, including non-accepted (failed) ones. Out of these, a number $N_a \leq N$ in each case is accepted: this N_a is the number of samples that the other tools are required to generate in each case. In all cases TSI is significantly faster—up to three orders of magnitude—than the other tools. In the table, we only report the central value of TSI's confidence interval of width 2ϵ, computed by (15) with $N_t = N_a$ and the rightmost term elided. We see that TSI also performs very well in terms of accuracy: e.g. for $N = 10^6$, the reported value coincides with the exact expectation, in all the analyzed models. The confidence is also good: for $N \geq 5 \times 10^6$ and $\epsilon = 0.005$, it always exceeds 0.99. Note that not being MCMC an exact sampling techniques, in the case of webPPL-MCMC and R2 the generated samples cannot be used to give formal guarantees. In conclusion, over the considered examples, TSI shows a significant advantage in terms of both execution time and accuracy.

Table 1. Vectorized Comparison of TSI, webPPL and R2 on TrueSkill (TS), Burglar Alarm (BA), ClickGraph (CG) and Monty Hall (MH) programs. *runtime*: execution time; *exp.*: estimated expected value; N: number of i.i.d. samples generated by TSI. For N_a = number of i.i.d. samples accepted by TSI, the estimated acceptance rates N_a/N for the four programs are: 0.14, 0.20, 0.025 and 1.

		TS(skillA = 1.057)				BA(burg = 0.029)				CG(simAll = 0.570)				MH(win = 0.666)			
		TSI	webppl rej	webppl mcmc	R2	TSI	webppl rej.	webppl mcmc	R2	TSI	webppl rej	webppl mcmc	R2	TSI	webppl rej.	webppl mcmc	R2
$N = 10^4$	runtime	0.001	0.051	0.041	0.063	0.003	0.036	0.016	0.046	**0.006**	0.032	0.011	0.031	**0.011**	0.104	0.097	0.015
	exp.	**1.057**	1.056	1.052	1.063	**0.029**	0.035	0.037	0.035	**0.560**	0.541	0.630	0.513	0.671	0.663	0.675	**0.668**
$N = 10^5$	runtime	0.001	0.305	0.147	0.640	0.005	0.191	0.065	0.296	**0.008**	0.119	0.052	0.141	**0.015**	0.692	0.523	0.141
	exp.	1.056	1.056	**1.057**	1.063	**0.029**	0.030	0.027	0.026	**0.561**	0.554	0.551	0.530	**0.666**	0.666	0.664	0.664
$N = 10^6$	runtime	0.002	2.765	1.061	6.370	**0.040**	1.154	0.344	2.893	**0.031**	0.954	0.223	1.309	**0.071**	6.426	4.925	1.454
	exp.	**1.057**	1.057	1.056	**1.057**	**0.029**	0.029	0.027	**0.029**	**0.570**	0.547	0.542	0.567	**0.666**	0.666	0.665	**0.666**
$N = 5 \times 10^6$	runtime	0.002	14.345	5.215	32.567	**0.042**	8.198	1.636	14.610	**0.125**	4.791	0.824	6.283	**0.344**	33.562	25.200	7.332
	exp.	**1.057**	1.057	1.058	**1.057**	**0.029**	0.029	0.028	**0.029**	**0.570**	0.547	0.551	**0.570**	**0.666**	0.666	0.666	**0.666**

---

[7] Our PC configuration is as follows. OS: Windows 10; CPU: 2.8 GHz Intel Core i7; GPU: Nvidia T500, driver v. 522.06; TF: v. 2.10.1; CUDA Toolkit v. 11.8; cuDNN SDK v. 8.6.0.

Translation function

```
1    T(nil) = return x,TRUE,t
2
3    T(fail) = return x,FALSE,t
4
5    T(xi = g.S) =
6        t=t+1
7        xi=g(x)
8        T(S)
9
10   T(xi ~ G.S) =
11       t=t+1
12       xi=G(x).sample()
13       T(S)
14
15   T( if φ then S1 else S2) =
16       T(S1)
17       T(S2)
18       x,mask,t'=tf.where(phi(x), {fS1(x,mask,t)}, {fS2(x,mask,t)})
19       t=tf.where(t==t',t+1,t')
20       return x,mask,t
21
22   T(while φ S1, S2) =
23       T(S1)
24       T(S2)
25       def fbody(x,mask,t):
26           x,mask,t=tf.where(phi(x)&(mask!=False)&(t<t0),{fS1(x,mask,t)},{x,mask,t})
27           return x,mask,t
28       fcond=lambda x,mask,t: Any(phi(x)&(mask!=False)&(t<t0))
29       x, mask, t =tf.while_loop(fcond, fbody, (x,mask,t))
30       x',mask',t'=fS2(x,mask,t)
31       t'=tf.where(t==t',t+1,t')
32       x,mask',t=tf.where(not(phi(x))&(t<t0)&(mask!=False),{x',mask',t'},{x,NAN,t})
33       mask=tf.where(mask!=False,mask',False)
34       return x,mask,t
35
36   T(S) =   def fS(x,mask,t):
37                T(S)
```

Workflow

Example assuming $m = 1$, hence x is x1. Compute est, the MC estimation (12) of $[S]^{t_0} f_{t_0}$, assuming f is the lifting of h.

```
38   x1=tf.fill(0,shape=(1,N))
39   mask=tf.fill(NaN,shape=(1,N))
40   t=tf.fill(1,shape=(1,N))
41   T(S)
42   x1,mask,t=fS(x1,mask,t)
43   term = (mask==True) & (t<=t0) # mask selecting terminated instances at time t0
44   est = sum(h(x1[term]))/N
```

Fig. 2. Outline of the translation from $\mathcal{P}_0$ into Python with TensorFlow. **Top, translation function.** Definition of functions $\mathbb{T}(S)$ and auxiliary $T(S)$, by structural induction on S. **Bottom, workflow.** The function f_S defined by $\mathbb{T}(S)$ is called with input arguments x,mask,t where $x = x1, \ldots, xm$: here xi,mask,t are tensors of shape $(1, N)$, filled with 0, NaN and 1, respectively. x,mask,t=f_S(x,mask,t) computes N independent samplings stored in x, and masks mask and t, to be used for selecting instances that are actually live/terminated/failed at the specified time t_0, and calculate estimations of $[S]^{t_0} f_{t_0}$ and $[\![S]\!]f$. **Notation.** TRUE (resp. FALSE, NAN) denote the $(1, N)$ tensor filled with True (resp. False, NaN). We use $\{...\}$ to denote concatenation of tensors x,mask,t $= x1, \ldots, xm,mask,t$ along the axis 0: assuming each of xi,mask,t is a tensor of shape $(1, N)$, then $\{x,mask,t\}$ returns a tensor of shape $(m + 2, N)$. Likewise, the left-hand side of any assignment, x,m,t = tf.where(...) actually denotes slicing of the tensor resulting from tf.where(...) into $m + 2$ $(1, N)$-tensors, which are then assigned to the variables x1, \ldots, xm,mask,t. In actual TF syntax, $\{...\}$ corresponds to using tf.concat(), and x,m,t = tf.where(...) to suitable tensor slicing operations. Any(b), for b a tensor of booleans, is the scalar boolean value obtained by OR-ing all elements in b. The usual Python's notation and rules to broadcast scalars to whole tensors apply, as well as those on boolean masking of tensors.

Unbounded Loops and Random Walks. Consider the programs S and S' defined respectively in (5) and (6) in Example 2, describing two random walks. Both programs use unbounded loops in an essential way that neither webPPL nor R2 are able to express easily. Moreover, both programs fall in sublanguage $\mathcal{P}_0$ we have considered in this section. Indeed, using the abbreviation introduced in (17), S and S' can be equivalently rewritten as

$$S = r \sim \rho_U.\, \mathsf{while}\,(|y| < 1)\,(y \sim \rho_G(y, 2 \cdot r).i = i + 1.\mathsf{nil})\,,\mathsf{nil}$$
$$S' = r \sim \rho_U.\, \mathsf{while}\,(|y| < 1)\,(y \sim \rho_G(y, 2 \cdot r).i = i + 1.\mathsf{nil})\,,\mathsf{obs}(i \geq 3)\,.$$

Continuing now with Example 2, consider $[\![S]\!]f$, that is the probability that S terminates correctly in at least 3 iterations. In the case of S', we consider $[\![S']\!]f'$, where f' (is the lifting of the function that) looks at the final value of variable r: that is, r conditioned on the observation that $i \geq 3$. We can compute upper and a lower bounds of $[\![S]\!]f$ and $[\![S']\!]f'$ by applying (8), for any fixed value of $t \geq 1$, the length of execution paths. As the exact expected values involved in (8) are not available, we proceed to their estimation via TSI, with confidence intervals provided by Hoeffding inequality (14), as discussed in Sect. 5. In detail, we fix an error threshold of $\epsilon = 0.005$ and $N = 10^6$ samples, and apply the bounds (15). We report the obtained results in Table 2, for different values of t. As t gets larger, TSI provides tighter and tighter confidence intervals for $[\![\cdot]\!]$. For example, at $t = 202$, for $[\![S]\!]f$ an interval of width 0.04 is computed in less than 0.5 seconds.

Table 2. Vectorized Confidence intervals for $[\![S]\!]f$ (left) and $[\![S']\!]f'$ (right), termination probability on paths of length $\leq t$ and corresponding execution time, as a function of t. The chosen path lengths t correspond to the execution of K iterations of the while loop, for $K = 5, 10, 50, 100$.

t	$[\![S]\!]f$	term. prob.	runtime	t	$[\![S']\!]f'$	term. prob.	runtime
12	0.623 ± 0.112	0.714	0.016	12	0.663 ± 0.337	0.445	0.016
22	0.576 ± 0.065	0.817	0.021	22	0.663 ± 0.336	0.647	0.022
102	0.538 ± 0.027	0.927	0.042	102	0.442 ± 0.115	0.861	0.062
202	0.530 ± 0.020	0.950	0.343	202	0.406 ± 0.079	0.904	0.268

7 Conclusion

We have presented a simple yet rigorous measure-theoretic small-step semantics for a probabilistic programming language, that lends itself to a direct, SIMD-parallel implementation. An approximation theorem reduces the effective estimation, with guarantees, of the program semantics (expectation) to the analysis of finite execution paths of a chosen length. TSI, a prototype implementation based on TensorFlow, has shown encouraging results.

References

1. Abadi, M., et al.: Tensorflow: a system for large-scale machine learning. In: Proceedings of the 12th USENIX Symposium on Operating Systems Design and Implementation (OSDI 2016) (2016). arXiv:1605.08695
2. Ash, R.: Real Analysis and Probability. Academic Press Inc., New York (1972)
3. Baier, C., Katoen, J.P.: Principles of Model Checking. MIT Press, Cambridge (2008)
4. Barthe, G., Katoen, J.P., Silva, A.: Foundations of Probabilistic Programming. Cambridge University Press, Cambridge (2020)
5. Bhat, S., Borgström, J., Gordon, A.D., Russo, C.: Deriving probability density functions from probabilistic functional programs. In: Proceedings of the International Conference on Tools and Algorithms for the Construction and Analysis of Systems (TACAS 2013) (2013)
6. Boreale, M., Collodi, L.: Python code for the experiments described in the present paper. https://github.com/Luisa-unifi/probabilistic_programming
7. Carpenter, B., et al.: A probabilistic programming language. J. Stat. Softw. **76**(1) (2017). https://doi.org/10.18637/jss.v076.i01
8. Chaganty, A., Nori, A.: Efficiently sampling probabilistic programs via program analysis. In: Artificial Intelligence and Statistics (AISTATS) (2013)
9. Gehr, T., Misailovic, S., Vechev, M.T.: PSI: exact symbolic inference for probabilistic programs. In: Proceedings of the 28th International Conference in Computer Aided Verification (CAV 2016), Toronto, pp. 62–83 (2016)
10. Goodman, N.D., Mansinghka, V., Roy, D., Bonawitz, K., Tenenbaum, J.: Church: a language for generative models. In: Proceedings of Uncertainty in Artificial Intelligence (2008)
11. Goodman, N.D., Stuhlmüller, A.: The design and implementation of probabilistic programming languages. http://dippl.org. Accessed 31 Aug 2023
12. Gordon, A.D., Henzinger, T.A., Nori, A.V., Rajamani, S.K.: Probabilistic programming. In: Proceedings of Future of Software Engineering Proceedings (FOSE 2014), pp. 167–181 (2014)
13. Gretz, F., Katoen, J.P., McIver, A.: Operational versus weakest pre-expectation semantics for the probabilistic guarded command language. Perform. Eval. **73**, 110–132 (2014)
14. Hoeffding, W.: Probability inequalities for sums of bounded random variables. J. Am. Stat. Assoc. **58**(301), 13–30 (1963)
15. Jansen, N., Dehnert, C., Kaminski, B.L., Katoen, J.P., Westhofen, L.: Bounded model checking for probabilistic programs. In: Proceedings of the International Symposium on Automated Technology for Verification and Analysis (ATVA 2016), vol. 9938, pp. 68–85 (2016)
16. Kozen, D.: Semantics of probabilistic programs. J. Comput. Syst. Sci. **22**(3), 328–350 (1981)
17. Maurer, A., Pontil, M.: Empirical bernstein bounds and sample-variance penalization. In: Proceedings of the 22nd Conference on Learning Theory (COLT 2009) (2009)
18. Milner, R.: Communication and Concurrency. Prentice Hall, International Series in Computer Science (1989)
19. Moldovan, D.: Autograph documentation. https://github.com/tensorflow/tensorflow/blob/master/tensorflow/python/autograph/g3doc/reference/index.md

20. Munos, R., Audibert, J.Y., Szepesvári, C.: Exploration-exploitation tradeoff using variance estimates in multi-armed bandits. Theoret. Comput. Sci. **410**(19), 1876–1902 (2009)
21. Narayanan, P., Carette, J., Romano, C.S.W., Zinkov, R.: Probabilistic inference by program transformation in hakaru (system description). In: Proceedings of the 13th International Symposium on Functional and Logic Programming (FLOPS 2016), pp. 62–79 (2016)
22. Nori, A.V., Hur, C.K., Rajamani, S.K., Samuel, S.: An efficient MCMC sampler for probabilistic programs. In: Proceedings of the Twenty-Eighth AAAI Conference on Artificial Intelligence, pp. 2476–2482 (2014)
23. Sankaranarayanan, S., Chakarov, A., Gulwani, S.: Static analysis for probabilistic programs: inferring whole program properties from finitely many paths. In: Proceedings of the 34th ACM SIGPLAN Conference on Programming Language Design and Implementation (2013)
24. Staton, S.: Commutative semantics for probabilistic programming. In: Proceedings of the 26th European Symposium on Programming (ESOP 2017), Uppsala, Sweden (2017)
25. Staton, S., Wood, F., Yang, H., Heunen, C., Kammar, O.: Semantics for probabilistic programming: higher-order functions. In: Proceedings of the 31st Annual ACM/IEEE Symposium on Logic in Computer Science (LICS) (2016)

Local Reasoning About Probabilistic Behaviour for Classical–Quantum Programs

Yuxin Deng[(✉)], Huiling Wu, and Ming Xu

Shanghai Key Laboratory of Trustworthy Computing, East China Normal University,
Shanghai, China
yxdeng@sei.ecnu.edu.cn

Abstract. Verifying the functional correctness of programs with both classical and quantum constructs is a challenging task. The presence of probabilistic behaviour entailed by quantum measurements and unbounded while loops complicate the verification task greatly. We propose a new quantum Hoare logic for local reasoning about probabilistic behaviour by introducing distribution formulas to specify probabilistic properties. We show that the proof rules in the logic are sound with respect to a denotational semantics. To demonstrate the effectiveness of the logic, we formally verify the correctness of non-trivial quantum algorithms including the HHL and Shor's algorithms.

Keywords: Quantum computing · Program verification · Hoare logic · Separation logic · Local reasoning

1 Introduction

Programming is an error-prone activity, and the situation is even worse for quantum programming, which is far less intuitive than classical computing. Therefore, developing verification and analysis techniques to ensure the correctness of quantum programs is an even more important task than that for classical programs.

Hoare logic [12] is probably the most widely used program logic to verify the correctness of programs. It is useful for reasoning about deterministic and probabilistic programs. A lot of efforts have been made to reuse the nice idea to verify quantum programs. Ying [28,29] was the first to establish a sound and relatively complete quantum Hoare logic to reason about pure quantum programs, i.e., quantum programs without classical variables. This work triggered a series of research in this direction. For example, Zhou et al. [34] proposed an applied quantum Hoare logic by only using projections as preconditions and

This work was supported by the National Natural Science Foundation of China under Grant Nos. 61832015, 62072176, 12271172 and 11871221, Shanghai Trusted Industry Internet Software Collaborative Innovation Center, and the "Digital Silk Road" Shanghai International Joint Lab of Trustworthy Intelligent Software under Grant No. 22510750100.

R. Dimitrova et al. (Eds.): VMCAI 2024, LNCS 14500, pp. 163–184, 2024.
https://doi.org/10.1007/978-3-031-50521-8_8

postconditions, which makes the practical use of quantum Hoare logic much easier. Barthe et al. [2] extended quantum Hoare logic to relational verification by introducing a quantum analogue of probabilistic couplings. Li and Unruh [16] defined a quantum relational Hoare logic with expectations in pre- and postconditions. Formalization of quantum Hoare logic in proof assistants such as Isabelle/HOL [20] and Coq [4] was accomplished in [17,33], respectively. Ying et al. [31] defined a class of disjoint parallel quantum programs and generalised the logic in [28] to this setting. As an extension of Hoare logic, separation logic turns out to be useful for verifying deterministic, concurrent, and probabilistic programs [1,3,5,15,21,22,25]. Zhou et al. [32] developed a quantum separation logic for local reasoning about quantum programs. However, all the work mentioned above cannot deal with programs that have both classical and quantum data.

In order to verify quantum programs found in almost all practical quantum programming frameworks such as Qiskit[1], $Q^{\#}$[2], Cirq[3], etc., we have to explicitly consider programs with both classical and quantum constructs. Verification techniques for this kind of hybrid programs have been put forward in the literature [6,7,9,10,13,14,26,27]. For example, Chadha et al. [6] proposed an ensemble exogenous quantum propositional logic for a simple quantum language with bounded iteration. The expressiveness of the language is very limited, and algorithms involving unbounded while loops such as the HHL and Shor's algorithms [11,24] cannot be described. Kakutani [13] presented a quantum Hoare logic for an imperative language with while loops, but the rule for them has no invariance condition. Instead, an infinite sequence of assertions has to be used. Unruh introduced a quantum Hoare logic with ghost variables to express properties such as that a quantum variable is disentangled with others [26] and a relational Hoare logic [27] for security analysis of post-quantum cryptography and quantum protocols. Deng and Feng [7] provided an abstract and a concrete proof system for classical–quantum programs, with the former being sound and relatively complete, while the latter being sound. Feng and Ying [10] introduced classical–quantum assertions, which are a class of mappings from classical states to quantum predicates, to analyse both classical and quantum properties. The approach was extended to verify distributed quantum programs in [9]. However, except for [14], all the work above offers no support for local reasoning, which is an obvious drawback. In the case that we have a large quantum register but only a few qubits are modified, it is awkward to always reason about the global states of the quantum register. Based on this observation, Le et al. [14] provided an interesting quantum interpretation of the separating conjunction, so to infuse separation logic into a Hoare-style framework and thus support local reasoning. However, a weakness of their approach is that it cannot handle probabilistic behaviour, which exists inherently in quantum programs, in a satisfactory way. Let us illustrate this with a simple example.

---

[1] https://qiskit.org.

[2] https://github.com/microsoft/qsharp-language.

[3] https://github.com/quantumlib/Cirq.

Example 1. The program **addM** defined below first initialises two qubits q_0 and q_1, and then applies the Hadamard gate H to each of them. By measuring them with the measurement operators $|0\rangle\langle 0|$ and $|1\rangle\langle 1|$, we add the measurement results and assign the sum to the variable v.

$$\mathbf{addM} \triangleq \quad \begin{aligned} &q_0 := |0\rangle; \ H[q_0]; \\ &q_1 := |0\rangle; \ H[q_1]; \\ &v_0 := M[q_0]; \\ &v_1 := M[q_1]; \\ &v := v_0 + v_1 \end{aligned}$$

Since the classical variable v_0 takes either 0 or 1 with equal chance, and similarly for v_1, the probability that variable v is assigned to 1 should be exactly $\frac{1}{2}$. However, in the program logic in [14], this property cannot be specified. □

We propose a novel quantum Hoare logic for a classical–quantum language. Two distribution formulas $\oplus_{i \in I} p_i \cdot F_i$ and $\oplus_{i \in I} F_i$ are introduced. A program state μ, which is a partial density operator valued distribution (POVD) [7], satisfies the formula $\oplus_{i \in I} p_i \cdot F_i$ if μ can be split into the weighted sum of i parts called μ_i and each μ_i satisfies the formula F_i. A state μ satisfies the formula $\oplus_{i \in I} F_i$ if there exists a collection of probabilities $\{p_i\}_{i \in I}$ with $\sum_{i \in I} p_i = 1$ such that $\oplus_{i \in I} p_i \cdot F_i$ can be satisfied. In other words, the splitting of μ does not necessarily follow a fixed set of weights. With distribution formulas, we can conveniently reason about the probabilistic behaviour mentioned in Example 1 (more details will be discussed in Example 2), and give an invariance condition in the proof rule for while loops. In addition, we adopt the labelled Dirac notation emphasised in [33] to facilitate local reasoning. Our program logic is shown to be sound and can be used to prove the correctness of non-trivial quantum algorithms including the HHL and Shor's algorithms. Therefore, the main contributions of the current work include the following aspects:

- We propose to use distribution formulas in a new quantum Hoare logic to specify the probabilistic behaviour of classical–quantum programs. Distribution formulas are useful to give an invariance condition in the proof rule for while loops, so to avoid an infinite sequence of assertions in the rule.
- We prove the soundness of our logic that allows for local reasoning in the spirit of separation logic.
- We demonstrate the effectiveness of the logic by proving the correctness of the HHL and Shor's algorithms.

The rest of the paper is structured as follows. In Sect. 2 we recall some basic notations about quantum computing. In Sect. 3 we review the syntax and denotational semantics of a classical–quantum imperative language considered in [7]. In Sect. 4 we define an assertion language and propose a proof system for local reasoning about quantum programs. We also prove the soundness of the system. In Sect. 5 we apply our framework to verify the HHL and Shor's algorithms. Finally, we conclude in Sect. 6 and discuss possible future work. Missing proofs are given in [8].

2 Preliminaries

We briefly recall some basic notations from linear algebra and quantum mechanics which are needed in this paper. For more details, we refer to [19].

A *Hilbert space* $\mathcal{H}$ is a complete vector space with an inner product $\langle\cdot|\cdot\rangle$: $\mathcal{H} \times \mathcal{H} \to \mathbb{C}$ such that

1. $\langle\psi|\psi\rangle \geq 0$ for any $|\psi\rangle \in \mathcal{H}$, with equality if and only if $|\psi\rangle = 0$,
2. $\langle\phi|\psi\rangle = \langle\psi|\phi\rangle^*$,
3. $\langle\phi| \sum_i c_i|\psi_i\rangle = \sum_i c_i\langle\phi|\psi_i\rangle$,

where $\mathbb{C}$ is the set of complex numbers, and for each $c \in \mathbb{C}$, c^* stands for the complex conjugate of c. For any vector $|\psi\rangle \in \mathcal{H}$, its length $\||\psi\rangle\|$ is defined to be $\sqrt{\langle\psi|\psi\rangle}$, and it is said to be *normalised* if $\||\psi\rangle\| = 1$. Two vectors $|\psi\rangle$ and $|\phi\rangle$ are *orthogonal* if $\langle\psi|\phi\rangle = 0$. An *orthonormal basis* of a Hilbert space $\mathcal{H}$ is a basis $\{|i\rangle\}$ where each $|i\rangle$ is normalised and any pair of them are orthogonal.

Let $\mathcal{L}(\mathcal{H})$ be the set of linear operators on $\mathcal{H}$. For any $A \in \mathcal{L}(\mathcal{H})$, A is *Hermitian* if $A^\dagger = A$ where $A^\dagger$ is the adjoint operator of A such that $\langle\psi|A^\dagger|\phi\rangle = \langle\phi|A|\psi\rangle^*$ for any $|\psi\rangle, |\phi\rangle \in \mathcal{H}$. A linear operator $A \in \mathcal{L}(\mathcal{H})$ is *unitary* if $A^\dagger A = AA^\dagger = I_\mathcal{H}$ where $I_\mathcal{H}$ is the identity operator on $\mathcal{H}$. The *trace* of A is defined as $\mathrm{tr}(A) = \sum_i \langle i|A|i\rangle$ for some given orthonormal basis $\{|i\rangle\}$ of $\mathcal{H}$. A linear operator $A \in \mathcal{L}(\mathcal{H})$ is *positive* if $\langle\phi|A|\phi\rangle \geq 0$ for any vector $|\phi\rangle \in \mathcal{H}$. The *Löwner order* $\sqsubseteq$ on the set of Hermitian operators on $\mathcal{H}$ is defined by letting $A \sqsubseteq B$ if and only if $B - A$ is positive.

Let $\mathcal{H}_1$ and $\mathcal{H}_2$ be two Hilbert spaces. Their *tensor product* $\mathcal{H}_1 \otimes \mathcal{H}_2$ is defined as a vector space consisting of linear combinations of the vectors $|\psi_1\psi_2\rangle = |\psi_1\rangle|\psi_2\rangle = |\psi_1\rangle \otimes |\psi_2\rangle$ with $|\psi_1\rangle \in \mathcal{H}_1$ and $|\psi_2\rangle \in \mathcal{H}_2$. Here the tensor product of two vectors is defined by a new vector such that

$$\left(\sum_i \lambda_i|\psi_i\rangle\right) \otimes \left(\sum_j \mu_j|\phi_j\rangle\right) = \sum_{i,j} \lambda_i\mu_j|\psi_i\rangle \otimes |\phi_j\rangle.$$

Then $\mathcal{H}_1 \otimes \mathcal{H}_2$ is also a Hilbert space where the inner product is defined as the following: for any $|\psi_1\rangle, |\phi_1\rangle \in \mathcal{H}_1$ and $|\psi_2\rangle, |\phi_2\rangle \in \mathcal{H}_2$,

$$\langle\psi_1 \otimes \psi_2|\phi_1 \otimes \phi_2\rangle = \langle\psi_1|\phi_1\rangle_{\mathcal{H}_1}\langle\psi_2|\phi_2\rangle_{\mathcal{H}_2}$$

where $\langle\cdot|\cdot\rangle_{\mathcal{H}_i}$ is the inner product of $\mathcal{H}_i$. Given $\mathcal{H}_1$ and $\mathcal{H}_2$, the *partial trace* with respect to $\mathcal{H}_2$, written $\mathrm{tr}_{\mathcal{H}_2}$, is a linear mapping from $\mathcal{L}(\mathcal{H}_1 \otimes \mathcal{H}_2)$ to $\mathcal{L}(\mathcal{H}_1)$ such that for any $|\psi_1\rangle, |\phi_1\rangle \in \mathcal{H}_1$ and $|\psi_2\rangle, |\phi_2\rangle \in \mathcal{H}_2$,

$$\mathrm{tr}_{\mathcal{H}_2}(|\psi_1\rangle\langle\phi_1| \otimes |\psi_2\rangle\langle\phi_2|) = \langle\psi_2|\phi_2\rangle|\psi_1\rangle\langle\phi_1|.$$

By applying quantum gates to qubits, we can change their states. For example, the Hadamard gate (H gate) can be applied on a single qubit, while the controlled-NOT gate (*CNOT* gate) can be applied on two qubits. Their representations in terms of matrices are given as

$$H = \tfrac{1}{\sqrt{2}} \begin{pmatrix} 1 & 1 \\ 1 & -1 \end{pmatrix} \quad \text{and} \quad CNOT = \begin{pmatrix} 1 & 0 & 0 & 0 \\ 0 & 1 & 0 & 0 \\ 0 & 0 & 0 & 1 \\ 0 & 0 & 1 & 0 \end{pmatrix}.$$

According to von Neumann's formalism of quantum mechanics [18], an isolated physical system is associated with a Hilbert space which is called the *state space* of the system. A *pure state* of a quantum system is a normalised vector in its state space, and a *mixed state* is represented by a density operator on the state space. Here a *density operator* ρ on Hilbert space $\mathcal{H}$ is a positive linear operator such that $\text{tr}(\rho) = 1$. A *partial density operator* ρ is a positive linear operator with $\text{tr}(\rho) \leq 1$.

The evolution of a closed quantum system is described by a unitary operator on its state space: if the states of the system at times t_1 and t_2 are ρ_1 and ρ_2, respectively, then $\rho_2 = U\rho_1 U^\dagger$ for some unitary operator U which depends only on t_1 and t_2.

A quantum *measurement* is described by a collection $\{M_m\}$ of positive operators, called measurement operators, where the indices m refer to the measurement outcomes. It is required that the measurement operators satisfy the completeness equation $\sum_m M_m^\dagger M_m = I_{\mathcal{H}}$. If the system is in state ρ, then the probability that measurement result m occurs is given by

$$p(m) = \text{tr}(M_m^\dagger M_m \rho),$$

and the state of the post-measurement system is $M_m \rho M_m^\dagger / p(m)$.

3   A Classical–Quantum Language

We recall the simple classical–quantum imperative language **QIMP** as defined in [7]. It is essentially similar to a few imperative languages considered in the literature [10,23,27,30]. We introduce its syntax and denotational semantics.

3.1   Syntax

We assume three types of data in our language: `Bool` for booleans, `Int` for integers, and `Qbt` for quantum bits (qubits). Let $\mathbb{Z}$ be the set of integer constants, ranged over by n. Let **Cvar**, ranged over by $x, y, \dots$, be the set of classical variables, and **Qvar**, ranged over by $q, q', \dots$, the set of quantum variables. It is assumed that both **Cvar** and **Qvar** are countable. We assume a set **Aexp** of arithmetic expressions over `Int`, which includes **Cvar** as a subset and is ranged over by $a, a', \dots$, and a set of boolean-valued expressions **Bexp**, ranged over by $b, b', \dots$, with the usual boolean constants **true**, **false** and boolean connectives such as $\neg, \wedge$ and $\vee$. We assume a set of arithmetic functions (e.g. $+, -, *$, etc.)

Table 1. Syntax of quantum programs.

(**Aexp**)	a	::=	$n \mid x, y, \ldots \mid f_m(a, \ldots, a)$
(**Bexp**)	b	::=	**true** $\mid$ **false** $\mid P_m(a, \ldots, a) \mid b \wedge b \mid \neg b \mid \forall x.b$
(**Com**)	c	::=	**skip** $\mid$ **abort** $\mid x := a \mid c; c$
			$\mid$ **if** b **then** c **else** c **fi** $\mid$ **while** b **do** c **od**
			$\mid q := \lvert 0 \rangle \mid U[\bar{q}] \mid x := M[\bar{q}]$

ranged over by the symbol f_m, and a set of boolean predicates (e.g. $=$, $\leq$, $\geq$, etc.) ranged over by P_m, where m indicates the number of variables involved. We further assume that only classical variables can occur freely in both arithmetic and boolean expressions.

We let U range over unitary operators, which can be user-defined matrices or built in if the language is implemented. For example, a concrete U could be the 1-qubit Hadamard operator H, or the 2-qubit controlled-NOT operator $CNOT$, etc. Similarly, we write M for the measurement described by a collection $\{M_i\}$ of measurement operators, with each index i representing a measurement outcome. For example, to describe the measurement of the qubit referred to by variable q in the computational basis, we can write $M := \{M_0, M_1\}$, where $M_0 = \lvert 0 \rangle_q \langle 0 \rvert$ and $M_1 = \lvert 1 \rangle_q \langle 1 \rvert$.

Sometimes, we use metavariables which are primed or subscripted, e.g. x', x_0 for classical variables. We abbreviate a tuple of quantum variables $\langle q_1, \ldots, q_n \rangle$ as $\bar{q}$ if the length n of the tuple is not important. If two tuples of quantum variables $\bar{q}$ and $\bar{q}'$ are disjoint, where $\bar{q} = \langle q_1, \ldots, q_n \rangle$ and $\bar{q}' = \langle q_{n+1}, q_{n+2}, \ldots, q_{n+m} \rangle$, then their concatenation is a larger tuple $\bar{q}\bar{q}' = \langle q_1, \ldots, q_n, q_{n+1}, \ldots, q_{n+m} \rangle$. If no confusion arises, we occasionally use a tuple to stand for a set.

The formation rules for arithmetic and boolean expressions as well as commands are defined in Table 1. An arithmetic expression can be an integer, a variable, or built from other arithmetic expressions by some arithmetic functions. A boolean expression can be a boolean constant, built from arithmetic expressions by some boolean predicates or formed by using the usual boolean operations. A command can be a skip statement, an abort statement, a classical assignment, a conditional statement, or a while-loop, as in many classical imperative languages. The command **abort** represents the unsuccessful termination of programs. In addition, there are three commands that involve quantum data. The command $q := \lvert 0 \rangle$ initialises the qubit referred to by variable q to be the basis state $\lvert 0 \rangle$. The command $U[\bar{q}]$ applies the unitary operator U to the quantum system referred to by $\bar{q}$. The command $x := M[\bar{q}]$ performs a measurement M on $\bar{q}$ and assigns the measurement outcome to x. It differs from a classical assignment because the measurement M may change the quantum state of $\bar{q}$, besides the fact that the value of x is updated.

For convenience, we further define the following syntactic sugar for the initialization of a sequence of quantum variables: $\bar{q} = \lvert 0 \rangle^{\otimes n}$, where $\bar{q} = \langle q_1, \ldots, q_n \rangle$, is an abbreviation of the commands:

$$q_1 = |0\rangle; q_2 = |0\rangle; \ldots; q_n := |0\rangle.$$

3.2 Denotational Semantics

In the presence of classical and quantum variables, the execution of a **QIMP** program involves two types of states: classical states and quantum states.

As usual, a classical state is a function $\sigma : \mathbf{Cvar} \to \mathbb{Z}$ from classical variables to integers, where $\sigma(x)$ represents the value of classical variable x. For each quantum variable $q \in \mathbf{Qvar}$, we assume a 2-dimensional Hilbert space $\mathcal{H}_q$ to be the state space of the q-system. For any finite subset V of $\mathbf{Qvar}$, we denote

$$\mathcal{H}_V = \bigotimes_{q \in V} \mathcal{H}_q.$$

That is, $\mathcal{H}_V$ is the Hilbert space spanned by tensor products of the individual state spaces of the quantum variables in V. Throughout the paper, when we refer to a subset of $\mathbf{Qvar}$, it is assumed to be finite. Given $V \subseteq \mathbf{Qvar}$, the set of *quantum states* consists of all partial density operators in the space $\mathcal{H}_V$, denoted by $\mathcal{D}^-(\mathcal{H}_V)$. A *machine state* is a pair $\langle \sigma, \rho \rangle$ where σ is a classical state and ρ a quantum state. In the presence of measurements, we often need to consider an ensemble of states. For that purpose, we introduce a notion of distribution.

Definition 1. [7] Suppose $V \subseteq \mathbf{Qvar}$ and Σ is the set of classical states, i.e., the set of functions of type $\mathbf{Cvar} \to \mathbb{Z}$. A *partial density operator valued distribution (POVD)* is a function $\mu : \Sigma \to \mathcal{D}^-(\mathcal{H}_V)$ with $\sum_{\sigma \in \Sigma} \mathrm{tr}(\mu(\sigma)) \leq 1$.

Intuitively, a POVD μ represents a collection of machine states where each classical state σ is associated with a quantum state $\mu(\sigma)$. The notation of POVD is called classical–quantum state in [10]. If the collection has only one element σ, we explicitly write $(\sigma, \mu(\sigma))$ for μ. The support of μ, written $\lceil \mu \rceil$, is the set $\{\sigma \in \Sigma \mid \mu(\sigma) \neq 0\}$. We can also define the addition of two distributions by letting $(\mu_1 + \mu_2)(\sigma) = \mu_1(\sigma) + \mu_2(\sigma)$.

We interpret programs as POVD transformers. We write **POVD** for the set of POVDs called distribution states. Given an expression e, we denote its interpretation with respect to machine state (σ, ρ) by $[\![e]\!]_{(\sigma,\rho)}$. The denotational semantics of commands is displayed in Table 2, where we omit the denotational semantics of arithmetic and boolean expressions such as $[\![a]\!]_\sigma$ and $[\![b]\!]_\sigma$, which is almost the same as in the classical setting because the quantum part plays no role for those expressions. A state evolves into a POVD after some quantum qubits are measured, with the measurement outcomes assigned to a classical variable. Two other quantum commands, initialisation of qubits and unitary operations, are deterministic and only affect the quantum part of a state. As usual, we define the semantics of a loop (**while** b **do** c **od**) as the limit of its lower approximations, where the n-th lower approximation of $[\![\mathbf{while}\ b\ \mathbf{do}\ c\ \mathbf{od}]\!]_{(\sigma,\rho)}$ is $[\![(\mathbf{if}\ b\ \mathbf{then}\ c\ \mathbf{fi})^n; \mathbf{if}\ b\ \mathbf{then}\ \mathbf{abort}\ \mathbf{fi}]\!]_{(\sigma,\rho)}$, where

Table 2. Denotational semantics of commands.

$$[\![\textbf{skip}]\!]_{(\sigma,\rho)} = (\sigma,\rho)$$

$$[\![\textbf{abort}]\!]_{(\sigma,\rho)} = \varepsilon$$

$$[\![x := a]\!]_{(\sigma,\rho)} = (\sigma[[\![a]\!]_\sigma/x],\rho)$$

$$[\![c_0; c_1]\!]_{(\sigma,\rho)} = [\![c_1]\!]_{[\![c_0]\!]_{(\sigma,\rho)}}$$

$$[\![\textbf{if } b \textbf{ then } c_0 \textbf{ else } c_1 \textbf{ fi}]\!]_{(\sigma,\rho)} = \begin{cases} [\![c_0]\!]_{(\sigma,\rho)} & \text{if } [\![b]\!]_\sigma = \textbf{true} \\ [\![c_1]\!]_{(\sigma,\rho)} & \text{if } [\![b]\!]_\sigma = \textbf{false} \end{cases}$$

$$[\![\textbf{while } b \textbf{ do } c \textbf{ od}]\!]_{(\sigma,\rho)} = \lim_{n\to\infty} [\![(\textbf{if } b \textbf{ then } c \textbf{ fi})^n; \textbf{if } b \textbf{ then abort fi}]\!]_{(\sigma,\rho)}$$

$$[\![q := |0\rangle]\!]_{(\sigma,\rho)} = (\sigma,\rho')$$
$$\text{where } \rho' := |0\rangle_q\langle0|\rho|0\rangle_q\langle0| + |0\rangle_q\langle1|\rho|1\rangle_q\langle0|$$

$$[\![U[\bar{q}]]\!]_{(\sigma,\rho)} = (\sigma, U\rho U^\dagger)$$

$$[\![x := M[\bar{q}]]\!]_{(\sigma,\rho)} = \mu$$
$$\text{where } M = \{M_i\}_{i\in I} \text{ and}$$
$$\mu(\sigma') = \sum_i \{M_i\rho M_i^\dagger \mid \sigma[i/x] = \sigma'\}$$

$$[\![c]\!]_\mu = \sum_{\sigma\in\lceil\mu\rceil} [\![c]\!]_{(\sigma,\mu(\sigma))}.$$

(**if** b **then** c **fi**) is shorthand for (**if** b **then** c **else skip fi**) and c^n is the command c iterated n times with $c^0 \equiv \textbf{skip}$. The limit exists because the sequence $([\![(\textbf{if } b \textbf{ then } c \textbf{ fi})^n; \textbf{if } b \textbf{ then abort fi}]\!]_{(\sigma,\rho)})_{n\in\mathbb{N}}$ is increasing and bounded with respect to the Löwner order [7, Lemma 3.2]. We write ε for the special POVD whose support is the empty set.

We remark that the semantics $[\![c]\!]_{(\sigma,\rho)}$ of a command c in initial state (σ,ρ) is a POVD. The lifted semantics $[\![c]\!]_\mu$ of a command c in initial POVD μ is also a POVD. Furthermore, the function $[\![c]\!]$ is linear in the sense that

$$[\![c]\!]_{p_0\mu_0+p_1\mu_1} = p_0[\![c]\!]_{\mu_0} + p_1[\![c]\!]_{\mu_1}$$

where we write $p\mu$ for the POVD defined by $(p\mu)(\sigma) = p \cdot \mu(\sigma)$.

Similarly as in [14], we take advantage of the labelled Dirac notation throughout the paper with subscripts identifying the subsystem where a ket/bra/operator lies or operates. For example, the subscript in $|a\rangle_{\bar{q}}$ indicates the Hilbert space $\mathcal{H}_{\bar{q}}$ where the state $|a\rangle$ lies. The notation $|a\rangle|a\rangle$ and $|a\rangle_{\bar{q}}|a\rangle_{\bar{q}}$ are the abbreviations of $|a\rangle \otimes |a\rangle$ and $|a\rangle_{\bar{q}} \otimes |a\rangle_{\bar{q}}$ respectively. We also use operators with subscripts like $A_{\bar{q}}$ to identify the Hilbert space $\mathcal{H}_{\bar{q}}$ where the operator A is applied.

Table 3. Syntax of assertion languages.

(Classical expression)	e	::=	$n \mid x, y, ... \mid f_m(e, ..., e)$
(Classical formula)	P	::=	$\textbf{true} \mid \textbf{false} \mid P_m(e, ..., e) \mid P \wedge P \mid \neg P \mid \forall x.P(x)$
(Quantum expression)	$\lvert s \rangle$	::=	$\lvert a \rangle_{\overline{q}} \mid \lvert s \rangle \otimes \lvert s \rangle$
(State formula)	F	::=	$P \mid \lvert s \rangle \mid F \odot F \mid F \wedge F \mid \neg F$
(Distribution formula)	D	::=	$\oplus_{i \in I} \, p_i \cdot F_i \mid \oplus_{i \in I} F_i$

4   Proof System

In this section we present a proof system for local reasoning about probabilistic behaviour of quantum programs. We first define an assertion language, then propose a Hoare-style proof system, and finally prove the soundness of the system.

4.1   Assertion Language

We now introduce an assertion language for our programs, whose syntax is given in Table 3. Classical expressions are arithmetic expressions with integer constants, ranged over by e. Classical formulas, ranged over by P, include the boolean constants $\textbf{true}$ and $\textbf{false}$, boolean predicates in the form P_m and any P connected by boolean operators such as negation, conjunction, and universal quantification. They are intended to capture properties of classical states. Quantum expressions, ranged over by $\lvert s \rangle$, include quantum states of the form $\lvert a \rangle_{\overline{p}}$ and the tensor product $\lvert s \rangle \otimes \lvert s \rangle$ which we abbreviate as $\lvert s \rangle \lvert s \rangle$. Here $\lvert a \rangle_{\overline{p}}$ can be any computational basis or their linear combinations in the Hilbert space $\mathcal{H}_{\overline{p}}$. State formulas, ranged over by F, are used to express properties on both classical and quantum states, which include the classical formula P, the quantum expression $\lvert s \rangle$ and any expression connected by boolean operators such as negation and conjunction. In addition, we introduce a new connective $\odot$ to express an assertion of two separable systems. Following [14], we use free(F) to denote the set of all free classical and quantum variables in F. For example, free($\lvert a_1 \rangle_{\overline{q_1}} \otimes \lvert a_2 \rangle_{\overline{q_2}}$) = $\overline{q_1 q_2}$. Moreover, we use qfree(F) to denote the set of all quantum variables in F. For the formula $F_1 \odot F_2$ to be well defined, we impose the syntactical restriction that free(F_1) $\cap$ free(F_2) = $\emptyset$. Intuitively, a quantum state satisfies $F_1 \odot F_2$ if the state mentions two disjoint subsystems whose states satisfy F_1 and F_2 respectively. Distribution formulas consist of some state formulas F_i connected by the connective $\oplus$ with the weights given by p_i satisfying $\sum_{i \in I} p_i = 1$ as well as the non-probabilistic formula $\oplus_{i \in I} F_i$. If there is a collection of distribution formulas $D_i = \oplus_j p_{ij} \cdot F_{ij}$ and a collection of probabilities p_i with $\sum_{i \in I} p_i = 1$, we sometimes write $\oplus_i p_i \cdot D_i$ to mean the formula $\oplus_{ij} p_i p_{ij} \cdot F_{ij}$.

We use the notation $\mu \models F$ to indicate that the state μ satisfies the assertion F. The satisfaction relation $\models$ is defined in Table 4. When writing $(\sigma, \rho) \models F$, we mean that (σ, ρ) is a machine state and ρ is its quantum part representing the status of the whole quantum system in a program under consideration. We

Table 4. Semantics of assertions.

$(\sigma, \rho) \models P$	if	$[\![P]\!]_\sigma = \mathbf{true}$
$(\sigma, \rho) \models \lvert s \rangle$	if	$\frac{\rho}{\mathrm{tr}(\rho)}\vert_{\bar{q}} = \lvert s \rangle\langle s \rvert \quad \text{where } \bar{q} = \mathrm{free}(\lvert s \rangle)$
$(\sigma, \rho) \models F_1 \wedge F_2$	if	$(\sigma, \rho) \models F_1 \wedge (\sigma, \rho) \models F_2$
$(\sigma, \rho) \models \neg F$	if	$(\sigma, \rho) \not\models F$
$(\sigma, \rho) \models F_1 \odot F_2$	if	$(\sigma, \rho\vert_{\mathrm{qfree}(F_1)}) \models F_1 \wedge (\sigma, \rho\vert_{\mathrm{qfree}(F_2)}) \models F_2$
$\mu \models F$	if	$\forall \sigma \in \lceil \mu \rceil.\ (\sigma, \mu(\sigma)) \models F$
$\mu \models \oplus_{i \in I} p_i \cdot F_i$	if	$\exists \mu_1 \cdots \exists \mu_m.\ [(\bigwedge_{i \in I} \mu_i \models F_i) \wedge \mu = \sum_{i \in I} p_i \cdot \mu_i]$ for $I = \{1, \ldots, m\}$
$\mu \models \oplus_{i \in I} F_i$	if	$\exists p_1 \cdots \exists p_m.\ [(\bigwedge_{i \in I} p_i \geq 0) \wedge \mu \models \oplus_{i \in I} p_i \cdot F_i]$ for $I = \{1, \ldots, m\}$

use $[\![P]\!]_\sigma$ to denote the evaluation of the classical predicate P with respect to the classical state σ. If V is set of quantum variables and $V' \subseteq V$, we write $\rho\vert_{V'}$ for the reduced density operator $\mathrm{tr}_{V \setminus V'}(\rho)$ obtained by restricting ρ to V'. A machine state (σ, ρ) satisfies the formula $\lvert s \rangle$ if the reduced density operator obtained by first normalising ρ and then restricting it to $\mathrm{free}(\lvert s \rangle)$ becomes $\lvert s \rangle\langle s \rvert$. The state (σ, ρ) satisfies the formula $F_1 \odot F_2$ if $\mathrm{free}(F_1)$ and $\mathrm{free}(F_2)$ are disjoint and the restrictions of ρ to $\mathrm{qfree}(F_1)$ and $\mathrm{qfree}(F_2)$ satisfy the two sub-formulas F_1 and F_2. The assertion F holds on a distribution μ when F holds on each pure state in the support of μ. A distribution state μ satisfies the distribution formula $\oplus_{i \in I} p_i \cdot F_i$ if μ is a linear combination of some μ_i with weights p_i and each μ_i satisfies F_i. In the special case that μ is a pure state, we have that $(\sigma, \rho) \models \oplus_{i \in I} p_i \cdot F_i$ means $(\sigma, \rho) \models F_i$ for every $i \in I$. The formula $\oplus_{i \in I} F_i$ is a nondeterministic version of the distribution formulas in the form $\oplus_{i \in I} p_i \cdot F_i$ without fixing the weights p_i, so the weights can be arbitrarily chosen as long as their sum is 1. For other assertions, the semantics should be self-explanatory.

From the relation $\models$, we can derive a quantitative definition of satisfaction, where a state satisfies a predicate only to certain degree.

Definition 2. Let F be an assertion and $p \in [0, 1]$ a real number. We say that the probability of a state μ satisfying F is p, written by

$$\mathbb{P}_\mu(F) = p,$$

if there exist two states μ_1 and μ_2 such that $\mu = p\mu_1 + (1 - p)\mu_2$, $\mu_1 \models F$ and $\mu_2 \models \neg F$, and moreover, p is the maximum probability for this kind of decomposition of μ.

In [14], assertions with disjunctions are used as the postconditions of measurement statements. However, this approach is awkward when reasoning about probabilities. Let us take a close look at the problem in Example 2.

Example 2. We revisit the program **addM** discussed in Example 1, which measures the variables q_0 and q_1, both in the state $|+\rangle\langle+|$, and subsequently adds their results of measurements. Here M is a projective measurement $\{|0\rangle\langle0|, |1\rangle\langle1|\}$.

$$\mathbf{addM} \triangleq \quad q_0:=|0\rangle; H[q_0];$$
$$q_1:=|0\rangle; H[q_1];$$
$$v_0:=M[q_0];$$
$$v_1:=M[q_1];$$
$$v:=v_0 + v_1$$

After the measurements on q_0 and q_1, the classical variables v_0 and v_1 are assigned to either 0 or 1 with equal probability. By executing the last command, we assign the sum of v_0 and v_1 to v, and obtain the following POVD μ.

$$
\begin{array}{rcl}
v_0 v_1 v & & q_0 q_1 \\
000 & \mapsto & \frac{1}{4}|00\rangle\langle00| \\
\mu: \quad 011 & \mapsto & \frac{1}{4}|01\rangle\langle01| \\
101 & \mapsto & \frac{1}{4}|10\rangle\langle10| \\
112 & \mapsto & \frac{1}{4}|11\rangle\langle11|
\end{array}
$$

The second column represents the four classical states while the last column shows the four quantum states. For example, we have $\sigma_1(v_0 v_1 v) = 000$ and $\rho_1 = |00\rangle\langle00|$. Let $\mu_i = (\sigma_i, \rho_i)$ for $1 \le i \le 4$. Then $\mu = \frac{1}{2}\mu_{14} + \frac{1}{2}\mu_{23}$, where $\mu_{14} = \frac{1}{2}\mu_1 + \frac{1}{2}\mu_4$ and $\mu_{23} = \frac{1}{2}\mu_2 + \frac{1}{2}\mu_3$. Since $\mu_i \models (v = 1)$ for $i = 2, 3$, it follows that $\mu_{23} \models (v = 1)$. On the other hand, we have $\mu_{14} \models (v \ne 1)$. Therefore, it follows that $\mathbb{P}_\mu(v = 1) = \frac{1}{2}$. That is, the probability for μ to satisfy the assertion $v = 1$ is $\frac{1}{2}$.

Alternatively, we can express the above property as a distribution formula. Let $D = \frac{1}{2} \cdot (v = 1) \oplus \frac{1}{2} \cdot (v \ne 1)$. We see that $\mu \models D$ due to the fact that $\mu = \frac{1}{2}\mu_{23} + \frac{1}{2}\mu_{14}$, $\mu_{23} \models (v = 1)$ and $\mu_{14} \models (v \ne 1)$.

In [14], there is no distribution formula. The best we can do is to use a disjunctive assertion to describe the postcondition of the above program.

$$F \triangleq \&\frac{1}{4} \cdot (v = 0 \wedge |00\rangle_{q_0 q_1}) \vee \frac{1}{4} \cdot (v = 1 \wedge |01\rangle_{q_0 q_1}) \vee$$
$$\frac{1}{4} \cdot (v = 1 \wedge |10\rangle_{q_0 q_1}) \vee \frac{1}{4} \cdot (v = 2 \wedge |11\rangle_{q_0 q_1}).$$

This assertion does not take the mutually exclusive correlations between different branches into account. For example, it is too weak for us to prove that $\mathbb{P}(v = 1) = \frac{1}{4} + \frac{1}{4} = \frac{1}{2}$, as discussed in more details in [14]. From this example, we see that distribution formulas give us a more accurate way of describing the behaviour of measurement statements than disjunctive assertions. □

4.2   Proof System

In this subsection, we present a series of inference rules for classical–quantum programs, which will be proved to be sound in the next section. As usual, we

Table 5. Inference rules for classical statements.

$$\frac{}{\{D\}\ \textbf{skip}\ \{D\}}[\text{Skip}] \qquad \frac{}{\{D\}\ \textbf{abort}\ \{\textbf{false}\}}[\text{Abort}]$$

$$\frac{}{\{D[a/x]\}\ x := a\ \{D\}}[\text{Assgn}] \qquad \frac{\{D_0\}\ c_0\ \{D_1\}\quad \{D_1\}\ c_1\ \{D_2\}}{\{D_0\}\ c_0; c_1\{D_2\}}[\text{Seq}]$$

$$\frac{\{F_1 \wedge b\}\ c_1\ \{F_1'\}\quad \{F_2 \wedge \neg b\}\ c_2\ \{F_2'\}}{\{p(F_1 \wedge b) \oplus (1-p)(F_2 \wedge \neg b))\}\ \textbf{if}\ b\ \textbf{then}\ c_1\ \textbf{else}\ c_2\ \textbf{fi}\ \{pF_1' \oplus (1-p)F_2'\}}[\text{Cond}]$$

$$\frac{}{\{\textbf{false}\}\ c\ \{D\}}[\text{Absurd}] \qquad \frac{D_0 \Rightarrow D_1 \quad \{D_1\}\ c\ \{D_2\}\quad D_2 \Rightarrow D_3}{\{D_0\}\ c\ \{D_3\}}[\text{Conseq}]$$

$$\frac{D = (F_0 \wedge b) \oplus (F_1 \wedge \neg b)\quad \{F_0 \wedge b\}\ c\ \{D\}}{\{D\}\ \textbf{while}\ b\ \textbf{do}\ c\ \textbf{od}\ \{F_1 \wedge \neg b\}}[\text{While}]$$

$$\frac{\{F_1\}\ c\ \{F_1'\}\quad \{F_2\}\ c\ \{F_2'\}}{\{F_1 \wedge F_2\}\ c\ \{F_1' \wedge F_2'\}}[\text{Conj}]$$

$$\frac{\{F_1\}\ c\ \{F_2\}\quad \text{free}(F_3) \cap \text{mod}(c) = \emptyset}{\{F_1 \odot F_3\}\ c\ \{F_2 \odot F_3\}}[\text{QFrame}]$$

$$\frac{\forall i \in I.\ \{D_i\}\ c\ \{D_i'\}\quad \sum_{i \in I} p_i = 1}{\{\oplus_{i \in I} p_i \cdot D_i\}\ c\ \{\oplus_{i \in I} p_i \cdot D_i'\}}[\text{Sum}]$$

Table 6. Inference rules for quantum statements.

$$\frac{}{\{\textbf{true}\}\ \bar{q} := |0\rangle\ \{|0\rangle_{\bar{q}}\}}[\text{QInit}] \qquad \frac{}{\{U_{\bar{q}}^\dagger F\}\ U[\bar{q}]\ \{F\}}[\text{QUnit}]$$

$$\frac{M = \{M_i\}_{i \in I}\quad p_i = \|M_{i\bar{q}}|v\rangle\|^2}{\{\wedge_i(P_i[i/x] \wedge |v\rangle_{\overline{qq'}})\}\ x := M[\bar{q}]\ \{\oplus_i p_i \cdot ((P_i \wedge M_{i\bar{q}}|v\rangle_{\overline{qq'}} / \sqrt{p_i}))\}}[\text{QMeas}]$$

use the Hoare triple $\{D_1\}\ S\ \{D_2\}$ to express the correctness of our programs, where S is a program and D_1, D_2 are the assertions specified in Table 3.

Table 5 lists the rules for classical statements, with most of them being standard and thus self-explanatory. The assertion $D[a/x]$ is the same as D except that all the free occurrences of variable x in D are replaced by a. The assertion $D_1 \Rightarrow D_2$ indicates that D_1 logically implies D_2. The quantum frame rule [QFrame] in Table 5 is introduced for local reasoning, which allows us to add assertion F_3 to the pre/post-conditions of the local proof $\{F_1\}\ S\ \{F_2\}$. We use $\text{mod}(c)$ to denote the set of all classical and quantum variables modified by c. Then $\text{free}(F_3) \cap \text{mod}(c) = \emptyset$ indicates that all the free classical and quantum variables in F_3 are not modified by program c. Note that when F_3 is a classical assertion, $\odot$ can be replaced by $\wedge$. The rule [Sum] allows us to reason about a probability distribution by considering each pure state individually.

Table 6 displays the inference rules for quantum statements. In rule [QInit] we see that the execution of the command $\bar{q} := |0\rangle_{\bar{q}}$ sets the quantum system $\bar{q}$ to $|0\rangle$, no matter what is the initial state. In rule [QUnit], for the postcondition F we have the precondition $U_{\bar{q}}^\dagger F$. Here $U_{\bar{q}}^\dagger$ distributes over those connectives

Table 7. Inference rules for entailment reasoning.

$$\frac{}{F \vdash \mathbf{true}}[\text{PT}] \qquad \frac{}{F \odot \mathbf{true} \dashv\vdash F}[\text{OdotE}]$$

$$\frac{}{F_1 \odot F_2 \dashv\vdash F_2 \odot F_1}[\text{OcotC}] \qquad \frac{}{F_1 \odot (F_2 \odot F_3) \dashv\vdash (F_1 \odot F_2) \odot F_3}[\text{OdotA}]$$

$$\frac{}{P_1 \odot P_2 \dashv\vdash P_1 \wedge P_2}[\text{OdotO}] \qquad \frac{}{P \odot F \dashv\vdash P \wedge F}[\text{OdotOP}]$$

$$\frac{}{P \wedge (F_1 \odot F_2) \dashv\vdash (P \wedge F_1) \odot F_2}[\text{OdotOA}]$$

$$\frac{}{F_1 \odot (F_2 \wedge F_3) \dashv\vdash (F_1 \odot F_2) \wedge (F_1 \odot F_3)}[\text{OdotOC}]$$

$$\frac{}{|u\rangle_{\overline{p}}|v\rangle_{\overline{q}} \dashv\vdash |v\rangle_{\overline{q}}|u\rangle_{\overline{p}}}[\text{ReArr}] \qquad \frac{}{|u\rangle_{\overline{p}}|v\rangle_{\overline{q}} \dashv\vdash |u \otimes v\rangle_{\overline{pq}}}[\text{Separ}]$$

$$\frac{}{|u\rangle_{\overline{p}}|v\rangle_{\overline{q}} \dashv\vdash |u\rangle_{\overline{p}} \odot |v\rangle_{\overline{q}}}[\text{OdotT}]$$

$$\frac{}{p_0 \cdot F \oplus p_1 \cdot F \oplus p_2 \cdot F' \dashv\vdash (p_0 + p_1) \cdot F \oplus p_2 \cdot F'}[\text{OMerg}]$$

$$\frac{}{\oplus_{i \in I} p_i \cdot F_i \vdash \oplus_{i \in I} F_i}[\text{Oplus}] \qquad \frac{\forall i \in I,\ F_i \vdash F_i'}{\oplus_{i \in I} p_i \cdot F_i \vdash \oplus_{i \in I} p_i \cdot F_i'}[\text{OCon}]$$

of state formulas and eventually applies to quantum expressions. For example, if $F = |v\rangle_{\overline{qq'}} \wedge P$, then $U_{\overline{q}}^{\dagger} F = U_{\overline{q}}^{\dagger}(|v\rangle_{\overline{qq'}}) \wedge P$ and $U_{\overline{q}}^{\dagger}(\neg F) = \neg(U_{\overline{q}}^{\dagger} F)$. In rule [QMeas], the combined state of the variables $\overline{qq'}$ is specified because there may be an entanglement between the subsystems for $\overline{q}$ and $\overline{q'}$. In that rule, we write $P_i[i/x]$ for the assertion obtained from P_i by replacing the variable x with value i. The postcondition is a distribution formula, with each assertion $P_i \wedge (M_{i[\overline{q}]}|v\rangle/\sqrt{p_i})$ assigned probability p_i, i.e., the probability of obtaining outcome i after the measurement M.

Table 7 presents several rules for entailment reasoning about quantum predicates. The notation $D_1 \vdash D_2$ says that D_1 proves D_2. Intuitively, it means that any state satisfying D_1 also satisfies D_2. We write $D_1 \dashv\vdash D_2$ if the other direction also holds.

The connective $\odot$ is commutative and associative, according to the rules [OdotC] and [OdotA]. If one or two assertions are classical, the rules [OdotO] and [OdotOP] replace $\odot$ with $\wedge$. The rule [OdotOA] replaces $P \wedge (F_1 \odot F_2)$ with $(P \wedge F_1) \odot F_2$ and vice versa. The rule [OdotOC] assists us to distribute $\odot$ into conjunctive assertions. The rule [ReArr] allows us to rearrange quantum expressions while [Separ] allows us to split/join the quantum expressions, given that $\overline{p}$ and $\overline{q}$ are not entangled with each other. The rules [ReArr] and [Separ] can be obtained naturally via the properties of tensor products. The rule [OdotT] replaces the $\odot$ connective with $\otimes$ when both assertions are state expressions. The rule [OMerg] allows us to merge the probabilities of two branches in a distribution formula if the two branches are the same. The rule [Oplus] is easy to be understood as $\oplus_{i \in I} F_i$ is essentially a relaxed form of $\oplus_{i \in I} p_i \cdot F_i$. The rule [OCon] says that if each F_i entails F_i', then the entailment relation is preserved between the combinations $\oplus_{i \in I} p_i \cdot F_i$ and $\oplus_{i \in I} p_i \cdot F_i'$.

We use the notation $\vdash \{D_1\}\ c\ \{D_2\}$ to mean that the Hoare triple $\{D_1\}\ c\ \{D_2\}$ is provable by applying the rules in Tables refFig.6–refFig.8.

Example 3. Suppose there are two separable systems $\bar{q}$ and $\bar{q}'$. They satisfy the precondition $\frac{1}{2}|u_1\rangle_{\bar{q}}|v_1\rangle_{\bar{q}'} \oplus \frac{1}{2}|u_2\rangle_{\bar{q}}|v_2\rangle_{\bar{q}'}$. After applying the operator $U[\bar{q}]$, they satisfy the postcondition $\frac{1}{2}(U_{\bar{q}}|u_1\rangle_{\bar{q}})|v_1\rangle_{\bar{q}'} \oplus \frac{1}{2}(U_{\bar{q}}|u_2\rangle_{\bar{q}})|v_2\rangle_{\bar{q}'}$. In other words, the following Hoare triple holds.

$$\vdash \{\tfrac{1}{2}|u_1\rangle_{\bar{q}}|v_1\rangle_{\bar{q}'} \oplus \tfrac{1}{2}|u_2\rangle_{\bar{q}}|v_2\rangle_{\bar{q}'}\}$$
$$U[\bar{q}] \tag{1}$$
$$\{\tfrac{1}{2}(U_{\bar{q}}|u_1\rangle_{\bar{q}})|v_1\rangle_{\bar{q}'} \oplus \tfrac{1}{2}(U_{\bar{q}}|u_2\rangle_{\bar{q}})|v_2\rangle_{\bar{q}'}\}$$

This Hoare triple can be proved as follows. Firstly, we apply the rule [QUnit] to obtain

$$\vdash \{|u_1\rangle_{\bar{q}}\}\ U[\bar{q}]\ \{|U_{\bar{q}}|u_1\rangle_{\bar{q}}\}. \tag{2}$$

Then we use the rules [QFrame] and [OdotT] to get

$$\vdash \{|u_1\rangle_{\bar{q}}|v_1\rangle_{\overline{q'}}\}\ U[\bar{q}]\ \{|U_{\bar{q}}|u_1\rangle_{\bar{q}}|v_1\rangle_{\overline{q'}}\}. \tag{3}$$

Similarly, we have

$$\vdash \{|u_2\rangle_{\bar{q}}|v_2\rangle_{\overline{q'}}\}\ U[\bar{q}]\ \{|U_{\bar{q}}|u_2\rangle_{\bar{q}}|v_2\rangle_{\overline{q'}}\}. \tag{4}$$

Combining (3) with (4) by rule [Sum], we obtain the Hoare triple in (1). □

Example 4. Let us consider the program **addM** and the distribution formula $D = \frac{1}{2} \cdot (v = 1) \oplus \frac{1}{2} \cdot (v \neq 1)$ discussed in Example 2. It can be formally proved that

$$\{\textbf{true}\}\ \textbf{addM}\ \{D\}\ .$$

A proof outline is given in [8]. Following [14], we use the following notations to highlight the application of the frame rule [QFrame]:

$$\Longleftrightarrow\ \{\ F_1\ \}\ c\ \{\ F_2\ \}\qquad \text{or equivalently}\qquad \begin{array}{l}\Longrightarrow \{\ F_1\ \}\\ c\\ \Longleftarrow \{\ F_2\ \}\end{array}$$

Both notations indicate that $\{F_1\}c\{F_2\}$ is a local proof for c and is useful for long proofs. The frame assertion F_3 can be deduced from the assertions before F_1 or after F_2. □

Our inference system is sound in the following sense: any Hoare triple in the form $\{D_1\}\ c\ \{D_2\}$ derived from the inference system is valid, denoted by $\models \{D_1\}c\{D_2\}$, meaning that for any state μ we have $\mu \models D_1$ implies $[\![c]\!]_\mu \models D_2$.

Theorem 1. (Soundness). *If $\vdash \{D_1\}\ c\ \{D_2\}$ then $\models \{D_1\}\ c\ \{D_2\}$.*

5   Examples

We apply the proof system to verify the functional correctness of two non-trivial algorithms: the HHL and Shor's algorithms in Subsects. 5.1 and 5.2, respectively.

Notice that the correctness of the HHL algorithm was verified in [34] where projections are used as assertions. The Shor's algorithm was verified in [10] with a quantitative interpretation of assertions. As we will see, our verification for both algorithms employs a qualitative reasoning, which is more in the style of the classical Hoare logic.

5.1   HHL Algorithm

The HHL algorithm [11] aims to obtain a vector x such that $Ax = b$, where A is a given Hermitian operator and b is a given vector. Suppose that A has the spectral decomposition $A = \Sigma_j \lambda_j |j\rangle\langle j|$, where each λ_j is an eigenvalue and $|j\rangle$ is the corresponding eigenvector of A. On the basis $\{|j\rangle\}_{j \in J}$, we have $A^{-1} = \Sigma_j \lambda_j^{-1} |j\rangle\langle j|$ and $|b\rangle = \Sigma_j b_j |j\rangle$. Then the vector x can be expressed as $|x\rangle = A^{-1}|b\rangle = \Sigma_j \lambda_j^{-1} b_j |j\rangle$. Here we require that $|j\rangle, |b\rangle$ and $|x\rangle$ are all normalized vectors. Hence, we have

$$\sum_j |\lambda_j^{-1} b_j|^2 = 1. \tag{5}$$

A quantum program implementing the HHL algorithm is presented in Table 8. The n-qubit subsystem $\overline{p}$ is used as a control system in the phase estimation step with $N = 2^n$. The m-qubit subsystem $\overline{q}$ stores the vector $|b\rangle = \sum_i b_i |i\rangle$. The one-qubit subsystem r is used to control the while loop. The measurement $M = \{M_0, M_1\}$ in the loop is the simplest two-value measurement: $M_0 = |0\rangle_r\langle 0|$ and $M_1 = |1\rangle_r\langle 1|$. The results of measurements will be assigned to the classical variable v, which is initialized to be 0. If the value of v is 0, then the while loop is repeated until it is 1. The unitary operator U_b is assumed to map $|0\rangle^{\otimes m}$ to $|b\rangle$. The controlled unitary operator U_f has a control system $\overline{p}$ and a target system $\overline{q}$, that is,

$$U_f = \sum_{\tau=0}^{N-1} |\tau\rangle_p\langle\tau| \otimes U^\tau,$$

where $U = e^{iAt}$. Equivalently, we have $U|j\rangle = e^{i\lambda_j t}|j\rangle$. We denote $\phi_j = \frac{\lambda_j t}{2\pi}$ and $\widetilde{\phi}_j = \phi_j \cdot N$, then we have $U|j\rangle = e^{2\pi i \phi_j}|j\rangle$. A given controlled unitary operator U_c has control system $\overline{p}$ and target system r, more precisely,

$$U_c|0\rangle_{\overline{p}}|0\rangle_r = |0\rangle_{\overline{p}}|0\rangle_r, \qquad U_c|j\rangle_{\overline{p}}|0\rangle_r = |j\rangle_{\overline{p}}(\sqrt{1 - \frac{C^2}{j^2}}|0\rangle + \frac{C}{j}|1\rangle)_r,$$

where $1 \leq j \leq N - 1$ and C is a given parameter.

The symbol $H^{\otimes n}$ represents n Hadamard gates applied to the variables $\overline{q}$; QFT and QFT^{-1} are the quantum Fourier transform and the inverse quantum Fourier transform acting on the variables in $\overline{p}$.

Table 8. A quantum program for the HHL algorithm.

$$\textbf{HHL} \triangleq$$

$1 : v := 0$

$2 : \textbf{while} \quad v = 0 \;\; \textbf{do}$

$3 : \qquad \bar{p} := |0\rangle^{\otimes n} \, ;$

$4 : \qquad \bar{q} := |0\rangle^{\otimes m} \, ;$

$5 : \qquad r := |0\rangle \, ;$

$6 : \qquad U_b[\bar{q}];$

$7 : \qquad H^{\otimes n}[\bar{p}];$

$8 : \qquad U_f[\overline{pq}];$

$9 : \qquad \text{QFT}^{-1}[\bar{p}];$

$10 : \qquad U_c[\overline{pr}];$

$11 : \qquad \text{QFT}[\bar{p}];$

$12 : \qquad U_f^\dagger[\overline{pq}];$

$13 : \qquad H^{\otimes n}[\bar{p}];$

$14 : \qquad v := M[r] \;\; \textbf{od}$

The correctness of the HHL algorithm can be specified by the Hoare triple:

$$\{ \textbf{ true } \} \; \textbf{HHL} \; \{ \, |x\rangle_{\bar{q}} \, \} \, .$$

Now let $D \triangleq (v = 0) \oplus ((|0\rangle_{\bar{p}}^{\otimes n}|x\rangle_{\bar{q}}|1\rangle_r) \wedge (v = 1))$, and S the body of the while loop of the HHL algorithm. The following Hoare triple can be proved.

$$\{ \, v = 0 \, \} \; S \; \{ \, D \, \} \, .$$

So D is an invariant of the while loop of the HHL algorithm. Then by rule [While] we obtain that

$$\{ \, D \, \} \; \textbf{while} \; (v = 0) \; \textbf{do} \; S \, \textbf{od} \; \{ \, |0\rangle_{\bar{p}}^{\otimes n}|x\rangle_{\bar{q}}|1\rangle_r \wedge (v = 1) \, \} \, .$$

Finally, we can establish the correctness of the HHL algorithm as given in Table 9, where we highlight the invariant of the while loop in red.

5.2 Shor's Algorithm

Shor's algorithm relies on the order-finding algorithm [24]. So we first verify the correctness of the latter. Given two co-prime positive integers x and N, the smallest positive integer r that satisfies the equation $x^r = 1 (\text{mod}) N$ is called the order of x modulo N, denoted by $\text{ord}(x, N)$. The problem of order-finding is to find the order r defined above, which is solved by the program presented in Table 10. Let $L \triangleq \lceil \log(N) \rceil$, $\epsilon \in (0, 1)$ and $t \triangleq 2L + 1 + \lceil \log(2 + 1/2\epsilon) \rceil$. The

Table 9. Proof outline of the HHL algorithm.

{ **true** }
{ $(v = 0)[0/v]$ }   by rule [Conseq]
$v := 0;$
{ $v = 0$ }   by rule [Assgn]
{ $(v = 0) \oplus (|0\rangle_{\overline{p}}^{\otimes n} |x\rangle_{\overline{q}} |1\rangle_r \wedge (v = 1))$ }   by rule [Oplus]
while $v = 0$ **do**
        { $v = 0$ }
        S
        { $(v = 0) \oplus (|0\rangle_{\overline{p}}^{\otimes n} |x\rangle_{\overline{q}} |1\rangle_r \wedge (v = 1))$ }
od
{ $|0\rangle_{\overline{p}}^{\otimes n} |x\rangle_{\overline{q}} |1\rangle_r \wedge (v = 1)$ }   by rule [While]
{ $|0\rangle_{\overline{p}}^{\otimes n} |x\rangle_{\overline{q}} |1\rangle_r$ }   by rule [Conseq]
{ $|0\rangle_{\overline{p}}^{\otimes n} \odot |x\rangle_{\overline{q}} \odot |1\rangle_r$ }   by rule [OdotT]
{ **true** $\odot |x\rangle_{\overline{q}} \odot$ **true** }   by rule [PT]
{ $|x\rangle_{\overline{q}}$ }   by rule [OdotE]

order-finding algorithm can successfully obtain the order of x with probability at least $(1 - \epsilon)/2 \log(N)$, by using $\mathrm{O}(L^3)$ operations as discussed in [19].

The variables in $\overline{q}$ correspond to a t-qubit subsystem while $\overline{p}$ represents an L-qubit subsystem. We introduce the variable z to store the order computed by the program **OF**, and initialize it to 1. The unitary operator U_+ maps $|0\rangle$ to $|1\rangle$, that is $U_+|0\rangle = |1\rangle$. The notation $H^{\otimes t}$ means t Hadamard gates applied to the system $\overline{p}$ and QFT^{-1} is the inverse quantum Fourier transform. The function $f(x)$ stands for the continued fraction algorithm which returns the minimal denomination n of all convergents m/n of the continued fraction for x with $|m/n - x| < 1/(2n^2)$ [10]. The unitary operator CU is the controlled-U, with $\overline{q}$ being the control system and $\overline{p}$ the target system, that is CU $|i\rangle_{\overline{q}}|j\rangle_{\overline{p}} = |i\rangle_{\overline{q}} U^i |j\rangle_{\overline{p}}$, where for each $0 \le y \le 2^L$,

$$U|y\rangle = \begin{cases} |xy \bmod N\rangle & \text{if } y \le N \\ |y\rangle & \text{otherwise.} \end{cases}$$

Note that the states defined by

$$|u_s\rangle \triangleq \frac{1}{\sqrt{r}} \sum_{k=0}^{r-1} e^{-2\pi i s k/r} |x^k \bmod N\rangle$$

for integer $0 \le s \le r - 1$ are eigenstates of U and $\frac{1}{\sqrt{r}} \sum_{s=0}^{r-1} |u_s\rangle = 1$.

Table 10. A quantum program for the order-finding algorithm.

$$\mathbf{OF}(x, N) :\equiv$$

1 : $z := 1$;

2 : $b := x^z \pmod{N}$;

3 : **while** $(b \neq 1)$ **do**

    4 : $\bar{q} := |0\rangle^{\otimes t}$; .

    5 : $\bar{p} = |0\rangle^{\otimes L}$;

    6 : $H^{\otimes t}[\bar{q}]$;

    7 : $U_+[\bar{p}]$;

    8 : $CU[\overline{qp}]$;

    9 : $QFT^{-1}[\bar{q}]$;

  10 : $z' := M[\bar{q}]$;

  11 : $z := f(\dfrac{z'}{2^t})$;

  12 : $b := x^z \pmod{N}$ **od**

The variable b stores the value of $(x^z \bmod N)$, and the operator $(a \bmod b)$ computes the modulo of a divided by b. If the value of b is not equal to 1, which means that the value of z computed by **OF** is not equal to the actual order of x modulo N, then the program will repeat the body of the while loop until $b = 1$. The while loop in the **OF** program exhibits probabilistic behaviour due to a measurement in the loop body.

The correctness of the order-finding algorithm can be specified as

$$\{\, 2 \leq x \leq N - 1 \wedge \gcd(x, N) = 1 \wedge N \bmod 2 \neq 0 \,\} \ \mathbf{OF} \ \{\, z = r \,\} \,.$$

Now let $D' \triangleq (z = r \wedge b = 1) \oplus (z \neq r \wedge b \neq 1)$, and S' the body of the while loop **OF**. We can establish the correctness of S' as follows:

$$\{\, z \neq r \wedge b \neq 1 \,\} \ S' \ \{\, D' \,\} \,.$$

So the invariant of the while loop of **OF** can be D', and by rule [While] we have

$$\{\, D' \,\} \ \mathbf{while} \ (b \neq 1) \ \mathbf{do} \ S' \ \mathbf{od} \ \{\, z = r \wedge b = 1 \,\} \,.$$

Finally, a proof outline of **OF** is given in Table 11.

Table 11. Proof outline of the **OF** program.

$\{2 \leq x \leq N - 1\}$
$\Longrightarrow \{(z = 1)[1/z]\}$
$z := 1;$
$\Longleftarrow \{z = 1\}$ by rule [Assgn]
$\Longrightarrow \{(z = 1 \wedge b = x^1 \bmod N)[(x^z \bmod N)/b]\}$
$b := x^z \bmod N;$
$\Longleftarrow \{z = 1 \wedge b = x^1 \bmod N\}$ by rule [Assgn]
$\Longrightarrow \{z \neq \mathrm{ord}(x, N) \wedge b \neq 1\}$ by rule [Conseq]
$\{(z = \mathrm{ord}(x, N) \wedge b = 1) \oplus (z \neq \mathrm{ord}(x, N) \wedge b \neq 1)\}$ by rule [Conseq]
while $(b \neq 1)$ **do**
$\qquad \{z \neq \mathrm{ord}(x, N) \wedge b \neq 1\}$
$\qquad S'$
$\qquad \{(z = \mathrm{ord}(x, N) \wedge b = 1) \oplus (z \neq \mathrm{ord}(x, N) \wedge b \neq 1)\}$
od
$\Longleftarrow \{z = \mathrm{ord}(x, N) \wedge b = 1\}$ by rule [While]
$\{2 \leq x \leq N - 1 \wedge z = \mathrm{ord}(x, N)\}$ by rule [Conseq]

Then we introduce Shor's algorithm in Table 12. The function $\mathrm{random}(a,b)$ is used to randomly generate a number between a and b. The function $\gcd(a,b)$ returns the greatest common divisor of a and b. The operator $\equiv_N$ represents identity modulo N. $\mathbf{OF}(x,N)$ is the order-finding algorithm given before, which will return the order of x modulo N and assign the order to the classical variable z. The classical variable y stores one of the divisors of N and we use $y|N$ to represent that N is divisible by y.

The correctness of Shor's algorithm can be specified by the Hoare triple:

$$\{\ \mathrm{cmp}(N)\ \}\ \mathbf{Shor}\ \{\ y|N \wedge y \neq 1 \wedge y \neq N\ \}$$

where $\mathrm{cmp}(N)$ is a predicate stating that N is a composite number greater than 0. The invariant of the while loop in Shor's algorithm can be

$$y|N \wedge y \neq N .$$

A proof outline for the correctness of the algorithm is given in [8].

Table 12. A program for Shor's algorithm.

```
1:  if 2 | N then
2:      y := 2;
3:  else
4:      x := random(2, N − 1);
5:      y := gcd(x, N);
6:      while y = 1 do
7:          z := OF(x, N);
8:          if 2 | z and x^{z/2} ≢_N −1 then
9:              y' := gcd(x^{z/2} − 1, N);
10:             if 1 < y' < N then
11:                 y := y';
12:             else
13:                 y := gcd(x^{z/2} + 1, N);
14:             fi
15:         else
16:             x := random(2, N − 1);
17:             y := gcd(x, N);
18:         fi
19:     od
20: fi
```

6 Conclusion and Future Work

We have presented a new quantum Hoare logic for classical–quantum programs. It includes distribution formulas for specifying probabilistic properties of classical assertions naturally, and at the same time allows for local reasoning. We have proved the soundness of the logic with respect to a denotational semantics and exhibited its usefulness in reasoning about the functional correctness of the HHL and Shor's algorithms, which are non-trivial algorithms involving probabilistic behaviour due to quantum measurements and unbounded while loops.

We have not yet precisely delimited the expressiveness of our logic. It is unclear whether the logic is relatively complete, which is an interesting future work to consider. We would also like to embed the logic into a proof assistant so to alleviate the burden of manually reasoning about quantum programs as done in Sect. 5.

Usually there are two categories of program logics when dealing with probabilistic behaviour: satisfaction-based or expectation-based [7]. Our logic belongs to the first category. In an expectation-based logic, e.g. the logic in [28,29], the Hore triple $\{P\}c\{Q\}$ is valid in the sense that the expectation of an initial state satisfying P is a lower bound of the expectation of the final state satisfying Q. It would be interesting to explore local reasoning in expectation-based logics for classical–quantum programs such as those proposed in [9,10].

References

1. Barthe, G., Hsu, J., Liao, K.: A probabilistic separation logic. Proc. ACM Program. Lang. **4**(POPL), 55:1–55:30 (2020)
2. Barthe, G., Hsu, J., Ying, M., Yu, N., Zhou, L.: Relational proofs for quantum programs. Proc. ACM Program. Lang. **4**(POPL), 1–29 (2019)
3. Batz, K., Kaminski, B.L., Katoen, J., Matheja, C., Noll, T.: Quantitative separation logic: a logic for reasoning about probabilistic pointer programs. Proc. ACM Program. Lang. **3**(POPL), 34:1–34:29 (2019)
4. Bertot, Y., Castéran, P.: Interactive Theorem Proving and Program Development - Coq'Art: The Calculus of Inductive Constructions. Springer (2004). https://doi.org/10.1007/978-3-662-07964-5
5. Brookes, S.: A semantics for concurrent separation logic. Theoret. Comput. Sci. **375**(1–3), 227–270 (2007)
6. Chadha, R., Mateus, P., Sernadas, A.: Reasoning about imperative quantum programs. Electron. Notes Theor. Comput. Sci. **158**, 19–39 (2006)
7. Deng, Y., Feng, Y.: Formal semantics of a classical-quantum language. Theoret. Comput. Sci. **913**, 73–93 (2022)
8. Deng, Y., Wu, H., Xu, M.: Local reasoning about probabilistic behaviour for classical-quantum programs (full version). https://arxiv.org/abs/2308.04741
9. Feng, Y., Li, S., Ying, M.: Verification of distributed quantum programs. ACM Trans. Comput. Log. **23**(3), 19:1–19:40 (2022)
10. Feng, Y., Ying, M.: Quantum Hoare logic with classical variables. ACM Trans. Quantum Comput. **2**(4), 16:1–16:43 (2021)
11. Harrow, A.W., Hassidim, A., Lloyd, S.: Quantum algorithm for linear systems of equations. Phys. Rev. Lett. **103**(15), 150502 (2009)
12. Hoare, C.A.R.: An axiomatic basis for computer programming. Commun. ACM **12**(10), 576–580 (1969)
13. Kakutani, Y.: A logic for formal verification of quantum programs. In: Datta, A. (ed.) ASIAN 2009. LNCS, vol. 5913, pp. 79–93. Springer, Heidelberg (2009). https://doi.org/10.1007/978-3-642-10622-4_7
14. Le, X.B., Lin, S.W., Sun, J., Sanan, D.: A quantum interpretation of separating conjunction for local reasoning of quantum programs based on separation logic. Proc. ACM Program. Lang. **6**(POPL), 1–27 (2022)
15. Li, J.M., Ahmed, A., Holtzen, S.: Lilac: a modal separation logic for conditional probability. Proc. ACM Program. Lang. **7**(PLDI), 148–171 (2023)
16. Li, Y., Unruh, D.: Quantum relational Hoare logic with expectations. In: 48th International Colloquium on Automata, Languages, and Programming. Leibniz International Proceedings in Informatics (LIPIcs), vol. 198, pp. 136:1–136:20 (2021)
17. Liu, J., et al.: Formal verification of quantum algorithms using quantum Hoare logic. In: Dillig, I., Tasiran, S. (eds.) CAV 2019. LNCS, vol. 11562, pp. 187–207. Springer, Cham (2019). https://doi.org/10.1007/978-3-030-25543-5_12
18. von Neumann, J.: States, Effects and Operations: Fundamental Notions of Quantum Theory. Princeton University Press (1955)
19. Nielsen, M., Chuang, I.: Quantum Computation and Quantum Information. Cambridge University Press, Cambridge (2000)
20. Nipkow, T., Wenzel, M., Paulson, L.C. (eds.): Isabelle/HOL. LNCS, vol. 2283. Springer, Heidelberg (2002). https://doi.org/10.1007/3-540-45949-9
21. O'Hearn, P.W.: Resources, concurrency, and local reasoning. Theoret. Comput. Sci. **375**(1–3), 271–307 (2007)

22. Reynolds, J.C.: Separation logic: a logic for shared mutable data structures. In: Proceedings of the 17th IEEE Symposium on Logic in Computer Science, pp. 55–74. IEEE Computer Society (2002)

23. Selinger, P.: Towards a quantum programming language. Math. Struct. Comput. Sci. **14**(4), 527–586 (2004)

24. Shor, P.W.: Algorithms for quantum computation: discrete logarithms and factoring. In: Proceedings of the 35th Annual Symposium on Foundations of Computer Science, pp. 124–134. IEEE Computer Society (1994)

25. Tassarotti, J., Harper, R.: A separation logic for concurrent randomized programs. Proc. ACM Program. Lang. **3**(POPL), 64:1–64:30 (2019)

26. Unruh, D.: Quantum Hoare logic with ghost variables. In: Proceedings of the 34th Annual ACM/IEEE Symposium on Logic in Computer Science, pp. 1–13. IEEE (2019)

27. Unruh, D.: Quantum relational Hoare logic. Proc. ACM Program. Lang. **3**(POPL), 1–31 (2019)

28. Ying, M.: Floyd-Hoare logic for quantum programs. ACM Trans. Program. Lang. Syst. **33**(6), 1–49 (2012)

29. Ying, M.: Foundations of Quantum Programming. Morgan Kaufmann (2016)

30. Ying, M., Feng, Y.: A flowchart language for quantum programming. IEEE Trans. Software Eng. **37**(4), 466–485 (2011)

31. Ying, M., Zhou, L., Li, Y., Feng, Y.: A proof system for disjoint parallel quantum programs. Theor. Comput. Sci. **897**, 164–184 (2022)

32. Zhou, L., Barthe, G., Hsu, J., Ying, M., Yu, N.: A quantum interpretation of bunched logic & quantum separation logic. In: Proceedings of the 36th Annual ACM/IEEE Symposium on Logic in Computer Science, pp. 1–14. IEEE (2021)

33. Zhou, L., Barthe, G., Strub, P.Y., Liu, J., Ying, M.: CoqQ: foundational verification of quantum programs. Proc. ACM Program. Lang. **7**(POPL), 833–865 (2023)

34. Zhou, L., Yu, N., Ying, M.: An applied quantum Hoare logic. In: Proceedings of the 40th ACM SIGPLAN Conference on Programming Language Design and Implementation, pp. 1149–1162 (2019)

Program and System Verification

Deductive Verification of Parameterized Embedded Systems Modeled in SystemC

Philip Tasche[1]([✉]) [ID], Raúl E. Monti[1] [ID], Stefanie Eva Drerup[2] [ID],
Pauline Blohm[2] [ID], Paula Herber[2] [ID], and Marieke Huisman[1] [ID]

[1] University of Twente, Enschede, The Netherlands
{p.b.h.tasche,r.e.monti,m.huisman}@utwente.nl
[2] University of Münster, Münster, Germany
{s.dr,pauline.blohm,paula.herber}@uni-muenster.de

Abstract. Major strengths of deductive verification include modular verification and support for functional properties and unbounded parameters. However, in embedded systems, crucial safety properties often depend on concurrent process interactions, events, and time. Such properties are global in nature and thus difficult to verify in a modular fashion. Furthermore, the execution and scheduling semantics of industrially used embedded system design languages such as SystemC are typically only informally defined. In this paper, we propose a deductive verification approach for embedded systems that are modeled with SystemC. Our main contribution is twofold: 1) We provide a formal encoding and an automated transformation of SystemC designs for verification with the VerCors deductive verifier. 2) We present a novel approach for invariant construction to abstractly capture global dependencies. Our encoding enables an automated formalization and deductive verification of parameterized SystemC designs, and the invariant construction enables local reasoning about global properties with comparatively low manual effort. We demonstrate the applicability of our approach on three parameterized case studies, including an automotive control system.

1 Introduction

Embedded systems have become ubiquitous in our daily life, including in safety-critical systems such as cars, airplanes or medical instruments. This makes systematically ensuring their correctness crucial and formal verification highly desirable. However, there are two major challenges when it comes to the formal verification of embedded systems: First, the languages that are used for embedded system design in industry are typically only informally defined, and their semantics is hard to capture formally due to their concurrency and timed behavior. Second, important properties of embedded systems, such as deadlock freedom or a timely reaction to events, typically depend on the scheduling, concurrent process interactions, events and time, and thus are global in nature. This makes it difficult to tackle them with modular verification approaches.

© The Author(s), under exclusive license to Springer Nature Switzerland AG 2024
R. Dimitrova et al. (Eds.): VMCAI 2024, LNCS 14500, pp. 187–209, 2024.
https://doi.org/10.1007/978-3-031-50521-8_9

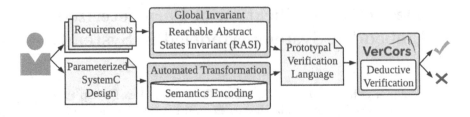

Fig. 1. Transformation-based Deductive Verification.

A language for embedded system design that is widely used in industry is SystemC [26]. In SystemC, concurrent processes are executed by a non-preemptive scheduler and are triggered by events. The execution semantics of SystemC is informally defined in [26]. To verify SystemC designs, several approaches have been proposed, e.g. [10,15,22,24,27,28]. However, these are based on model checking and are thus limited, e.g. when it comes to unbounded parameters. This is especially relevant for SystemC, since it is often used at early development stages before system parameters, like buffer sizes or threshold values, are fixed.

This paper presents a deductive verification approach for parameterized embedded systems that are modeled in SystemC. Unlike model checking, deductive verification can cope with unbounded parameters as well as functional properties by using inference rules and mathematical deduction. Furthermore, deductive verification supports modular verification using local reasoning based on method contracts. This enables compositional verification, even for partially implemented systems, and significantly increases reusability. Our goal is to use deductive verification to obtain formal and exhaustive guarantees while mitigating the state space explosion incurred by parameterized systems. Our main contribution is twofold: First, we provide a formal encoding of key aspects of the SystemC execution semantics and an automated transformation from SystemC into PVL, an input language of the VerCors deductive verifier [7]. Second, we present an approach for invariant generation that allows us to use modular verification for global properties.

Our approach is based on five key ideas. To encode the execution semantics of SystemC, we 1) provide a predefined abstract representation of the scheduler, 2) capture the event mechanism by defining shared global variables that describe event notifications and process states, and 3) define a global mutex that ensures non-preemptive execution. To use local reasoning for global properties, we 4) associate the global mutex with a global invariant that relates the states of events and processes to the desired properties, and 5) automatically generate a *Reachable Abstract States Invariant* (RASI), which enumerates reachable abstract process and event states and is included in the global invariant.

Our encoding enables us to automatically transform a given parameterized SystemC design into PVL, as shown in Fig. 1. For the automated transformation, we use our encoding of the SystemC semantics and additionally translate user-defined processes, methods, and data structures into PVL. The designer can then formally analyze crucial safety properties with VerCors by adding specifications to the transformed program, while the RASI restricts the analysis to the relevant part of the state space. Note that, while our approach is specialized for SystemC, the core ideas can equally be applied to similar systems, such as RTOS.

We have implemented our approach and demonstrate its applicability on three parameterized case studies, including an automotive control system that could not be verified with existing approaches using, for example, the UPPAAL model checker [5,21] or the CPAchecker [6,20].

This paper is structured as follows. Section 2 introduces SystemC and VerCors. Section 3 presents our transformation from SystemC into PVL, and Sect. 4 shows how to use our encoding for verification. Section 5 shows experimental results and discusses the applicability and scalability of our approach. Section 6 summarizes related work and Sect. 7 concludes the paper.

2   Background

In this section, we introduce the necessary background for the remainder of this paper, namely SystemC and VerCors.

2.1   SystemC

systemc is a modeling language and a simulation framework for hardwaresoftware co-design. It is implemented as a C++ library and its semantics is informally defined in an IEEE standard [26]. SystemC enables modeling and simulation at different levels of abstraction, from functional over transaction level design down to register transfer level. The SystemC library implements primitives for the design of interacting processes as well as an event-driven simulation kernel. A SystemC design consists of *modules* that are connected by *channels*. A module defines *processes* that run concurrently and interact through *events*. SystemC uses non-preemptive scheduling, i.e. a process will not be interrupted until it finishes or gives up control with a `wait` statement. Processes can wait for a given amount of time or for an event. Like typical hardware simulators, SystemC uses the concept of *delta cycles* to impose a partial order on parallel processes, i.e. concurrent processes are executed sequentially but in zero time.

The SystemC simulation kernel controls the simulation time and the execution of processes, handles event notifications and updates channels for communication. The execution semantics can be summarized as follows [26]: 1. Initialization: execute each process once; 2. Evaluation: while processes are ready to run, execute them in arbitrary order[1]; 3. Update: update channels; 4. If there are delta-delayed notifications, wake up the corresponding processes and go to step 2; 5. If there are timed notifications, advance simulation time to the earliest pending timed notification and go to step 2; 6. If there are no timed notifications remaining, simulation is finished.

---

[1] While implementations of the simulation semantics often use a deterministic execution order, the IEEE standard does not define any such order [26].

```
1    SC_MODULE(Robot) {              13      void controller() {
2      sc_event od;                  14        bool flag = false;
3      void sensor() {               15        while (true) {
4        int dist = -1;              16          wait(od);
5        while(true) {               17          flag = true;
6          wait(2, SC_MS);           18        }
7          dist = read_sensor();     19      }
8          if(dist < MIN_DIST) {     20      SC_CTOR(Robot) {
9            od.notify(SC_ZERO_TIME); 21        SC_THREAD(sensor);
10         }                         22        SC_THREAD(controller);
11       }                           23      }
12     }                             24    };
```

Fig. 2. Running example of a simple robot in SystemC.

Figure 2 shows a SystemC design of a simple robot, which we use as a running example in this paper. It consists of just one module, declared in line 1, with two processes `sensor` and `controller`, defined as methods and declared as thread processes in lines 21 and 22. The `sensor` process periodically reads sensor values by waiting for 2 ms (line 6) and then reading a new `dist` value (line 7). If `dist` falls below a threshold, which is given by system parameter `MIN_DIST`, the sensor notifies an obstacle detection event `od`. The event notification is delayed by one delta cycle, indicated by the `SC_ZERO_TIME` keyword. The `controller` process waits for `od` (line 16) and sets a flag (line 17) if it receives this event.

2.2 VerCors

vercors is a deductive program verifier that specializes in data race freedom, memory safety and functional correctness of concurrent programs [7]. It supports different concurrency models for programs written in Java, OpenCL, OpenMP for C and PVL, its own Prototypal Verification Language. We use PVL for our encoding because of its flexibility and support for different language features. PVL is an object-oriented language with a syntax and semantics similar to Java.

VerCors uses *contract-based reasoning*. Methods are annotated with pre- and postconditions specified in permission-based separation logic (PBSL) [3]. VerCors *automatically* verifies whether the annotated code complies with its contract in a modular way: It verifies in isolation that each method fulfils its contract and replaces the method by its contract at the call site. Pre- and postconditions are indicated by the keywords `requires` and `ensures`, respectively.

Besides first order logic formulas, contracts in VerCors also include heap *permission* obligations. To access a heap variable, a thread must have permission to do so. An annotation $Perm(x, \pi)$ with $\pi \in (0, 1]$ specifies read access permission to the heap location x, and write access permission if π is 1 [2]. Permissions can be split and merged as long as the total sum of permissions to a field stays constant, e.g. $Perm(x, 1)$ can be split into two terms $Perm(x, 1\backslash3)$ ** $Perm(x, 2\backslash3)$, e.g. to distribute it over two threads, and later recombined into $Perm(x, 1)$. Here, the operator ** represents the separating conjunction from PBSL and indicates that resources on both sides are disjoint. VerCors uses these annotations to reason about absence of data races and memory safety in concurrent programs.

```
1   class BankAccount{
2     int balance;
3     boolean active;
4     resource lock_invariant() = Perm(balance, 1) ** balance >= 0;
5
6     ensures Perm(active, 1) ** active;
7     BankAccount(){                {= Perm(active, 1) ** Perm(balance, 1) =}
8       active = true;              {= Perm(active, 1) ** active ** Perm(balance, 1) =}
9       balance = 0;                {= ... ** Perm(balance, 1) ** balance == 0 =}
10    }                            {= Perm(active, 1) ** active =}
11
12    requires amount > 0;
13    requires Perm(active, read) ** active;
14    void deposit(int amount){    {= ... =}
15      lock(this);                {= ... ** Perm(balance, 1) ** balance >= 0 =}
16      balance = balance + amount; {= ... ** Perm(balance, 1) ** balance > 0 =}
17      unlock(this);              {= ... =}
18    }
19  }
```

Fig. 3. Intrinsic locks and lock invariants in VerCors.

Locks are the main way to synchronize threads in PVL. They are not explicitly defined in code; rather, every object in PVL is associated with an intrinsic lock that can be acquired and released with the keywords lock and unlock. To reason about correct synchronization, a lock declares a *lock invariant*, which includes access permissions and functional properties about the shared state protected by the lock [34]. A lock invariant must first be established in the constructor. Then, when a thread acquires the lock, it also acquires all permissions and assumes all properties in the invariant. When the lock is released, the invariant must be reestablished and the permissions are returned to the lock [2].

Figure 3 illustrates the use of contracts and lock invariants in VerCors. The annotations between {= and =} are not part of the PVL program, but indicate the permissions and knowledge at a point. The BankAccount class implements a simple bank account that has two fields, balance and active, and allows concurrent deposits. To avoid data races, access to the balance field is protected by a lock by including its permission in the lock invariant (line 4). The lock invariant also specifies a functional property that ensures a nonnegative balance. Since the lock invariant must be established every time the lock is released (e.g. in line 17), this encodes a global property about balance. The constructor generates permissions for both fields. However, it passes the permission to balance to the lock (line 10) and only returns permission to active (line 6). The precondition of the method deposit (lines 12 and 13) uses this permission to require that the account is active (line 13). It requires read permission, which means an arbitrary amount of permission $\pi \in (0, 1]$. It also needs permission to update balance, which it gets from the lock (line 15). Additionally, it can assume the global invariant (balance >= 0) at this point. Since it requires a positive deposit (line 12) and the mathematical integers VerCors uses cannot overflow, the assignment in line 16 cannot violate the global invariant and the method is able to reestablish the lock invariant in line 17 when it releases the lock.

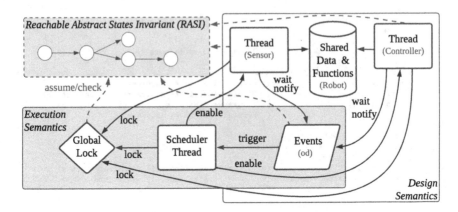

Fig. 4. Overview of the SystemC Encoding in PVL.

3   Encoding the SystemC Semantics in PVL

This section presents our formal encoding of the execution semantics of SystemC and our automated transformation into PVL. Our transformation formalizes the informal SystemC semantics and enables its deductive verification.

Figure 4 illustrates the overall structure of our encoding. Rectangular boxes indicate threads, the diamond a lock, the rhomboid a data structure for events, and the cylinder a class containing other data structures and functions. Our encoding consists of two parts: the general *Execution Semantics* of SystemC and the *Design Semantics* of a given SystemC design.

To encode the SystemC *Execution Semantics*, we define a model of the scheduler, which is run in its own *Scheduler Thread*, and we encode *Events* as globally accessible variables. The scheduler *enables* threads based on process states and event triggers. The *Global Lock* serves as a mutex to encode non-preemptive scheduling. Every thread, including the scheduler, contends on this lock. If the process that acquires the lock is *enabled*, it executes, otherwise it releases the lock again.

To encode the *Design Semantics* of a given SystemC design, we translate each process into a runnable class, instantiated in its own *Thread*, as illustrated with the *Sensor* and the *Controller* from our running example (see Fig. 2). We add the necessary variables for the events (*od* in our example), and shared data and functions (from the *Robot* module in our example). The threads can notify or wait on events and can access shared data and functions.

Figure 4 also shows the embedding of the *Reachable Abstract States Invariant* (RASI) into our encoding of the SystemC semantics. The RASI abstractly describes the reachable states by enumerating all reachable combinations of event and process states. When verifying our designs, we include the RASI in the global lock invariant to restrict the analysis to interesting states. We explain the generation of the RASI and its use in deductive verification in Sect. 4.

In the following, Sect. 3.1 defines the SystemC subset we currently support. Section 3.2 explains our encoding of the SystemC execution semantics and Sect. 3.3 describes our transformation of a given SystemC design.

3.1 Assumptions

The language subset that we support comprises key features for hardware/-software co-design and embedded systems modeling, most importantly the non-preemptive scheduling semantics, concurrent processes, dynamic event sensitivity, and time. This allows us to reason about crucial properties of embedded systems, such as deadlock freedom and a timely reaction to events. However, we still make some assumptions. As is standard among formal verification approaches for SystemC, we assume a static design without dynamic process creation, dynamic port binding or dynamic memory allocation. Without loss of generality[2], we also assume a flattened design without static sensitivity lists. Supported data types are restricted by VerCors to booleans, (mathematical) integers and arrays thereof. VerCors offers only limited support for pointers and floats, although there is ongoing research on extending this support. Currently unsupported features that we can conceptually support and that we plan to add to our automatic transformation in the future include primitive channels and multiple-event sensitivity.

3.2 Encoding of Events and Concurrent Processes

To verify a SystemC design, we need to formally encode the semantics of SystemC, in particular the event mechanism and the process scheduling. This section presents our approach to encode key aspects of the SystemC semantics.

Non-Preemptive Execution. We introduce a *global lock* to encode the non-preemptive execution of concurrent processes as established by SystemC. This lock guards the whole state of the design, ensuring that there is always a single process executing. When a process starts execution, its first statement is to lock on the *global lock*, as shown in line 2 of Fig. 5a. The process is then non-preemptively executed until it releases control with a wait statement. We encode this by continuously releasing the *global lock* and trying to reacquire it as long as the process has not been woken up due to the event, as shown in lines 8 and 9 of Fig. 5a. When the process releases the lock, other processes can execute. Whether a process is ready to run is determined by the current state of events and the scheduler. Conceptually, all processes compete for the lock, and one of them is nondeterministically chosen according to the semantics of locks in PVL. If it is ready to execute, then it resumes execution, otherwise it releases the lock and another process is chosen. This adheres to the SystemC simulation semantics [26], where processes are executed in arbitrary order, and it ensures that the requirements are verified for all possible execution orders.

---

[2] Static sensitivity is subsumed by dynamic sensitivity and a hierarchical design can be flattened by using, for example, prefixing.

194   P. Tasche et al.

```
1   void run {
2     lock(m);
3     while (true) {
4       m.event_state[0] = 2;
5       m.process_state[0] = 0;
6       while (m.process_state[0] != -1
7              || m.event_state[0] != -2) {
8         unlock(m);
9         lock(m);
10      }
11      dist = read_sensor();
12      if (dist < MIN_DIST) {
13        m.event_state[2] = -1;
14      }
15    }
16    unlock(m);
17  }
```

```
1   while (true) {
2     lock(this);
3     immediate_wakeup();
4     reset_events_no_delta();
5     if (no_process_ready()) {
6       reset_occurred_events();
7       int d = min_advance(event_state);
8       advance_time(d);
9       wakeup_after_wait();
10      reset_all_events();
11    }
12    unlock(this);
13  }
```

(a) Encoding of the sensor process.   (b) Encoding of the scheduler.

Fig. 5. Encoding of the scheduler and an example process.

Events. We encode events as globally accessible variables and associate each event with a global event ID. To encode triggers and event sensitivities, and to enable processes to continue their execution after a wait statement, we also associate each process with a distinguishing global ID. The state of all processes and events in the design is encoded in the integer arrays[3] process_state and event_state, to which we refer as *scheduling variables* in this paper. These arrays are indexed by the process and event ID, respectively. Position p in process_state holds the ID of the event that process p is waiting on, or -1 if the process is not waiting. Position e in event_state contains the remaining time until the occurrence of event e if it is nonnegative, -1 if the event is notified with delta delay, -2 if the event occurred in the current delta cycle and -3 if the event is not notified.

wait and notify. Figure 5a shows how we use the scheduling variables to encode event handling in our translation of the Sensor process from Fig. 2. The scheduling variables are attributes of a top level class Main, which is referenced with the variable m. We use this class both for globally shared variables and as the global lock. The example process starts by waiting for 2 ms. We encode timed waits by defining a dedicated timeout event (here with ID 0). The process notifies this event with the given delay and then waits on it. This is shown in lines 4 to 10 of Fig. 5a. For the event notification, the process sets the event_state of the timeout event to the required delay (2 in this case) in line 4. Lines 5 to 10 then encode the process waiting for the event. In line 5, it sets its own entry (ID 0) in the process_state array to the ID of the event it is waiting on. Then, in lines 6 to 10, it remains in a busy wait. It releases the global lock so that other processes can execute and then tries to reacquire it to continue execution. The loop condition in lines 6 and 7 makes sure the sensor can only continue

---
[3] For simplicity, we speak of arrays here. In our implementation, all collections have the PVL-intrinsic type seq, which reduces verification time.

```
1   // Context: The lock is held by the scheduler
2   requires held(this) ** scheduler_permission_invariant();
3   ensures held(this) ** scheduler_permission_invariant();
4   // The event state is not affected by this method
5   ensures event_state == \old(event_state);
6   // If any process is waiting on an event with 0 delay, it is woken up
7   ensures (\forall int i = 0 .. |process_state|;
8                (  \old(process_state[i]) >= 0
9                && \old(event_state[\old(process_state[i])]) == 0)
10               ==> (process_state[i] == -1));
11  // Any process not waiting on an event with delay 0 stays the same
12  ensures (\forall int i = 0 .. |process_state|;
13               (!(  \old(process_state[i]) >= 0
14               && \old(event_state[\old(process_state[i])]) == 0))
15               ==> (process_state[i] == \old(process_state[i])));
16  void immediate_wakeup();
```

Fig. 6. One of the abstract scheduler methods.

execution if it is woken up due to the occurrence of the event. Line 13 encodes a delta-delayed notification of the event with ID 2 (od in Fig. 2). To indicate the delta-delayed notification of the event, its event_state entry is set to −1.

The Scheduler. Figure 5b shows our model of the scheduler. The scheduler uses the scheduling variables to decide which event should occur next and which processes should be woken up. It also advances time by subtracting the due time to the next event from all event delays in the event_state array. Since the scheduler runs in its own thread, it needs to hold the global lock to execute, as seen in lines 2 and 12. Lines 3 and 4 wake up processes waiting on events with immediate notifications and reset these events. This encodes SystemC's evaluation phase. If there are still no processes ready in line 5, the scheduler advances to the next delta cycle. Line 6 resets the events that occurred in the previous delta cycle. Lines 7 and 8 advance time by finding the delay until the next event (0 if there is a pending delta-delayed notification) and subtracting it from all pending event notifications, before lines 9 and 10 again wake up processes waiting for all events that occur after the delay.

Note that the methods used in the scheduler model are not implemented but abstractly specified by their postconditions. This avoids additional verification effort for these methods. The structure of the abstract specifications is shown in Fig. 6 for the case of the immediate_wakeup method. Lines 2 and 3 ensure that the method has the required permissions to change the scheduling variables and that the scheduler is holding the global lock. Line 5 ensures that the events are not changed by this method. Lines 6 to 15 encode the method's effect on the process state. If a process is waiting for an event with an immediate notification, then it is woken up (Lines 7 to 10). If these conditions are not fulfilled, then that process's state should not be changed (Lines 12 to 15). This encoding comprehensively describes the method's behavior.

3.3 Encoding of SystemC Designs

Our encoding of the SystemC scheduling, non-preemptive execution and event mechanism captures the key concepts of the SystemC semantics. To transform a SystemC design into a PVL program, we add to this a translation of processes, modules and methods from the design. The result is a PVL program consistent with the semantics of the original SystemC design. Except for wait and notify, we map most SystemC statements simply onto their PVL equivalents. We treat ports and channels as external method calls. Our encoding of processes and modules is described in the following.

Processes. To encode a SystemC process in PVL, we create a runnable class for that process. Each such class is associated with a unique process ID and contains an attribute m as a reference to the global lock. It also defines a method run that encodes the SystemC process definition. This method starts by acquiring the global lock and ends by releasing it. Figure 5a illustrates the translation of the process body with the sensor process from our running example (lines 3 to 12 in Fig. 2). The SystemC code runs in an infinite loop that waits for 2 ms, reads new sensor data, and then notifies the od event if the given safety threshold is violated, i.e. if an obstacle is detected within the range of MIN_DIST. The PVL translation shown in Fig. 5a first acquires the lock in line 2 and then goes into the while(true) loop. It encodes waiting for 2 ms in lines 4 to 10, reads new sensor data in line 11, checks the distance in line 12 and notifies the *od* event in line 13. Finally, it releases the lock at termination.

Modules. If a module defines exactly one process, then any other functionality of the module, including methods and attributes, is also included in this runnable class. If the module defines more than one process, we generate a runnable class for each process and transform the remaining functionality of the module into another PVL class that holds the module's shared data and methods (see Fig. 4). Each class that encodes a process holds a reference to an instance of this shared module class. If there are multiple instances of a module, we duplicate all relevant classes for each instance to be able to consider dependencies between them. In our automated transformation, we use prefixing to distinguish between these classes.

4 Deductive Verification of SystemC Designs

With our encoding of the SystemC semantics, we can automatically transform a given SystemC design into a PVL representation. This enables us to use VerCors for deductive verification, i.e. we can verify that functional as well as safety properties hold for all possible input scenarios and all possible parameter values. This section explains how to specify and verify such properties. It also introduces the *Reachable Abstract States Invariant* (RASI), a technique to restrict the state space to consider and with that partially automate the verification process and lessen the burden on the user.

```
1  resource global_invariant() =          1  loop_invariant true
2    ...                                   2    ** Perm(m, 1\2)
3    // SENSOR                             3    ** m != null
4    ** Perm(sensor, read)                 4    ** held(m)
5    ** sensor != null                     5    ** m.global_invariant()
6    ** Perm(sensor.dist, write)           6    ** m.sensor == this
7    ** Perm(sensor.pc, 1\2)               7    ** Perm(pc, 1\2)
8    ...                                   8    ** pc == 1
```

(a) Part of the global invariant   (b) Generated loop invariant

Fig. 7. Generated permissions for the sensor process from Fig. 2.

4.1 Property Specification and Verification

Our transformation to PVL enables us to use the concepts supported by VerCors for the specification and verification of SystemC designs. These are *method contracts*, *loop invariants* and *assertions*. Each allow the user to specify different aspects of the code. Method contracts describe the behavior of a method with a pre- and a postcondition. To allow modular verification, VerCors proves that the method, given the precondition, fulfils the postcondition, and that the precondition is fulfilled at every call site. Loop invariants describe properties that should hold throughout the execution of a loop and are proven before and after each loop iteration. Assertions describe properties that should hold at one specific point in the code. By using the appropriate specifications, we can reason about a wide variety of properties, both functional and global, that the original SystemC design should fulfil.

Generated Permissions. For VerCors to be able to guarantee memory safety of concurrent programs, a program must be annotated with appropriate permission obligations. To access a heap variable, a method must have permission to this variable; otherwise, verification will fail. We have designed our transformation such that we can already automatically generate almost all of the annotations needed for verification. We store permissions to all variables in the global invariant, which is included in local contracts and invariants as well as the global lock invariant. Since we require the global lock to execute, processes are guaranteed to have sufficient permissions without user intervention. The user is then left with only the task of adding the desired properties.

Figure 7 illustrates these generated permissions on the example of the sensor process from Fig. 2. Figure 7a shows the relevant part of the global invariant. It contains permissions to the sensor object itself as well as its attribute dist. It also contains permission to the sensor's program counter pc to allow other processes to read its value, which is useful for some properties. However, since pc is only updated by the sensor itself, the global invariant only contains half permission; the other half is retained by the process. This is shown in Fig. 7b, which displays a loop invariant in the sensor process. It ensures that the global lock is held for execution and that the global invariant is preserved, but it also requires the remaining half permission to `sensor.pc` to allow the sensor to update its own program counter.

Declaring Properties. As an example of declaring a property about a SystemC design with our approach, take the simple robot from Fig. 2. Assume the user wants to verify that the controller process is only woken up if the sensor reads a distance below the threshold. In this case, after the transformation, the user can add the assertion `assert m.sensor.dist < m.MIN_DIST` to the controller process after the initial `wait` statement. If the verification succeeds, then the property is guaranteed to hold whenever the controller is woken up. Note, however, that this verification might require some auxiliary properties to succeed.

4.2 Reachable Abstract States Invariant

While functional properties can usually be verified locally, this is not the case for global properties. Typical safety properties for embedded systems, such as a timely reaction to external events, generally depend on the interplay between processes and on the scheduling. This requires additional information about the global state for local verification.

For deductive verification, this information comes in the form of user-defined invariants that capture the global state. Such invariants allow the verifier to exclude impossible states that might otherwise wrongly falsify the desired property, but they are often challenging to define by hand. To overcome this problem and reduce user effort, we lean on ideas from model checking. We enumerate the design's abstract state space with regard to its scheduling in the RASI to capture process and event dependencies. This provides a systematic and automatable way to capture the global system behavior for local verification.

To mitigate the state space explosion problem while capturing the state space, we construct an abstraction of the reachable state space that can be automatically generated. We abstract the program to a small subset A of program variables, replacing all other variables by arbitrary values. Section 4.3 explains the selection of these variables. The RASI then consists of the reachable states of this abstract program. Each state is represented by a conjunction of the values of the variables in A, and the RASI is represented by a disjunction of such states, as defined in the following.

Definition 1. Let p be a PVL program, V_p the set of all program variables of p, and $A \subseteq V_p$. An **abstract state** s with regard to A is a valuation that maps each $x \in A$ into $dom(x)$. The Boolean representation of an abstract state is defined as the condition formula $cond(s) := \bigwedge_{x \in A} x = s(x)$.

Definition 2. Let p be a PVL program and $A \subseteq V_p$. The **abstract program** $p|_A$ is the program that only takes assignments to the variables in A into consideration and performs a nondeterministic overapproximation for all other variables, i.e. all variables in $V_p \setminus A$ are treated as arbitrary values. Let now c be a program location in p. An abstract state s is **reachable with regard to A** at c in p if an execution of $p|_A$ exists for which $cond(s)$ holds at c.

```
1   resource global_invariant() =
2       /* >>> Permissions <<< */
3       ...
4       /* >>> Abstract state space <<< */
5       ** ( (process_state == [-1, -1] && event_state == [-3, -3, -3]
6               && sensor.pc == 0 && controller.pc == 0)
7          || (process_state == [0, -1] && event_state == [2, -3, -3]
8               && sensor.pc == 1 && controller.pc == 0)
9          || (process_state == [-1, 2] && event_state == [-3, -3, -3]
10              && sensor.pc == 0 && controller.pc == 1)
11         || ... )
12      /* >>> Establish the property <<< */
13      ** (event_state[2] > -2 ==> sensor.dist < MIN_DIST) // when sensor notifies od
14      ** (event_state[2] == -2 ==> sensor.dist < MIN_DIST); // when od occurs
```

Fig. 8. Excerpt of the RASI for the robot example.

Definition 3. Let p be a PVL program, c a program location in p and $A \subseteq V_p$. Let $S(A, c)$ be the set of all reachable states of p at c with regard to A. The **Reachable Abstract States Invariant (RASI)** at c with regard to A is $\mathcal{R}_A^c := \bigvee_{s \in S(A,c)} cond(s)$. The full RASI of p with regard to A is $\mathcal{R}_A := \bigvee_{c \in p} \mathcal{R}_A^c$.

Note that, while we usually speak of the RASI as the entire state space invariant $\mathcal{R}_A$ in this paper, it is often more useful to use an invariant $\mathcal{R}_A^c$ that only captures states that are reachable at program location c. This provides a smaller and stronger invariant than the full abstract reachable state space.

Figure 8 shows an example of the RASI for the robot example from Fig. 2 with A containing the scheduling variables `process_state` and `event_state` as well as program counters for both processes. Lines 5 to 11 contain 3 out of 8 reachable abstract states regarding A. Lines 5 and 6 represent the initial state, where both the sensor (process ID 0) and the controller (process ID 1) are enabled and no event has been notified or occurred yet. This state has two possible successors: either the sensor runs first and reaches the point where it waits for 2 ms (lines 7 and 8), or the controller runs first and waits for the obstacle detected event (event ID 2, lines 9 and 10).

4.3 Variables in the Abstract State

In principle, the RASI can be based on any set A of variables. To capture global inter-process dependencies, we always include the scheduling variables `process_state` and `event_state`. We also include program counters for each process in the design to capture the execution state of other processes. Data variables should generally be avoided to mitigate the state space explosion problem.

However, in some cases, the smaller abstract state can be outweighed by an increase in the potential behavior of $p|_A$. Whenever the program branches on a variable that is not in A, $p|_A$ instead performs a nondeterministic overapproximation. If the branches of the nondeterministic choice influence the variables in A, then this introduces spurious states in the RASI. For example, a buffer read operation might branch on the size of the buffer and wait if the buffer is empty. In this case, a data variable influences the process and event states. Abstracting from the buffer size causes an overapproximation of the state space, which negatively affects verification time. In such a situation, it can be beneficial to include the concerned data variable in A as well. On the other hand, this can also add new states to the RASI any time this variable changes. The decision whether to include a variable is nontrivial in the general case, but often easy to see for a domain expert. For now, we leave it to the user.

4.4 RASI Generation and Soundness

We have implemented two RASI generators, based on different abstraction methods, which are included in our artifact [1]. The first generator is based on simulation. An abstract design is repeatedly executed while recording all visited states. This approach is conceptually simple to use, but it may lead to an underapproximation of the state space and fail the verification with VerCors.

The second RASI generator is based on systematic state space exploration. The user provides transition systems for the processes in the given design, abstracted to A. The tool then systematically explores all reachable states of the composition of these transition systems. This approach is more flexible than simulation, since it explores the abstract $p|_A$. We currently require the user to provide the necessary abstractions for each process, but we plan to automate this with existing approaches such as abstract interpretation [11] or predicate abstraction [4] in the future.

The reason that we need to impose this additional effort of abstracting the program to the variables in A is that VerCors overapproximates variables about which it has no information. An invariant that does not use the same overapproximation would not be able to be verified. This means that the state space can only depend on variables that are also included in each state's condition.

While this reasoning of VerCors creates some challenges when it comes to generating a useful RASI, it also directly implies its soundness. We include the RASI in the global invariant, and local subsets of the RASI in loop invariants and contracts. Since these are included in proof obligations for VerCors, a successful verification automatically guarantees that the RASI contains all reachable program states. If the program would enter a state that is not contained in the RASI, the verification would fail.

4.5 Verifying Global Properties with the RASI

The RASI is useful for restricting the state space under consideration and avoiding spurious verification failures for global properties. It also makes verification significantly easier for multiple classes of global properties.

Cross-Process Properties. Properties that are established in one process and need to be proven in another are difficult to prove deductively. For example, in the robot design from Fig. 2, the sensor only notifies the controller process if the value sensor.dist that it observes is below the threshold MIN_DIST. However, proving `sensor.dist < MIN_DIST` in the controller once it is woken up requires complex global invariants that capture the dependencies between the two processes. These invariants are hard to come up with and are sensitive to changes in the system, limiting reusability. The RASI allows an easy and reusable verification technique by following the desired property from the point at which it is established to the point at which it is proven, as shown in lines 13 and 14 in Fig. 8. If the system changes, e.g. if the controller also waits before trying to prove the property (see *robot-1MS* example in Sect. 5), the user can add similar invariants for this new event with little effort.

Timing Properties. Timing constraints are usually very difficult to conceptualize for deductive verification. However, as the RASI contains timing information for events, it can help with the formalization and verification of such properties. By adding a new event, notifying this event with the desired delay to start measuring time, and checking that it has not yet occurred at the goal location, a timing property can be verified with little manual effort.

Unreachability Properties. Properties that exclude a certain abstract state, e.g. a deadlock state in which no process is ready to execute and no event is notified, are direct corollaries from the RASI. If the RASI does not contain such a state and is verified, then the program can never reach that state. Note that reachability cannot be guaranteed in this way, since the RASI might overapproximate the program behavior.

5   Evaluation

We have implemented our transformation from SystemC to PVL on top of the STATE tool [23]. To demonstrate the applicability of our approach, we conducted case studies on three example designs and some variants of them. Our tools and experiments, along with their reproduction instructions, can be found at [1]. For all three case studies, we used our transformation from SystemC to PVL to generate a formal representation. We automatically generated a RASI for an appropriate variable subset and manually added properties that capture the system requirements to the global invariant. Finally, we used VerCors to deductively verify the properties.

5.1   Case Studies

We evaluate our approach using three parameterized SystemC designs as case studies, for each of which we verify the RASI and some sample properties. Two of the case studies are smaller examples, showcasing different features of SystemC, namely our running example of a simple robot and a system of a producer and a consumer communicating via a FIFO channel. The third case study is a functional model of an automotive control system, containing an anti-slip regulation (ASR) and anti-lock braking system (ABS). It is a slightly modified version of a case study that was developed by a student according to the Bosch specification [37]. Existing verification using, for example, the UPPAAL model checker [5,19,21] or the CPAchecker [6,20] were not able to verify this case study, and UPPAAL runs out of 64 GB of main memory both for the original and for our modified version. This is caused by the data-dependence of the properties we prove together with the missing data abstractions in UPPAAL. Even if we combine UPPAAL with an abstract interpretation as in [21], data-dependent behavior has to be unfolded in UPPAAL to avoid spurious counterexamples.

Simple Robot. The first case study is our running example of a parameterized simple robot (see Fig. 2). This case study showcases event-driven communication between processes. We verify the property that, if the controller sets the flag, then the sensor must have sensed an obstacle closer than the symbolic system parameter MIN_DIST. We consider several variations of the *robot* case study.

In *robot-1MS*, the controller waits 1 ms before setting the flag. This complicates the proof of the property, as it adds the requirement that the value of dist is unchanged during the controller's waiting period. However, since the controller waits for a shorter time than the sensor, the property still holds.

For the *robot-dummy* variant, we introduce an extra dummy process. As this dummy process does not interfere with the sensor or the controller, the property should still be verifiable.

Producer-Consumer. The second case study is a system where a producer and a consumer communicate via a FIFO channel, showcasing communication through a user-defined channel. This example has also been used in several existing works on the formal verification of SystemC, e.g. [10,14,25,43]. However, these works could only analyze the example with a fixed buffer size. In [18], the verification is performed for a variety of different fixed buffer sizes, but verification time exponentially increases for larger buffer sizes. We use a parameterized version of the producer-consumer example where the buffer size is set arbitrarily by a parameter. On this parametric design, we verify two properties. We prove that the FIFO channel, which is implemented as a circular buffer, works as expected. We further prove the global property that all items read by the consumer were sent by the producer.

ABS/ASR. Our third case study is an anti-lock braking system and anti-slip regulation (ABS/ASR) design. It contains four *tick counter* modules that measure the speed of each wheel and send it to a central *electronic control unit (ECU)* that gathers the inputs, estimates wheel and vehicle speed and acceleration, and executes the ABS or ASR algorithm. The communication takes place via FIFO channels that a dedicated process in the ECU reads from. Another ECU process reacts to the input by periodically adjusting braking pressure at the wheels. Figure 9 shows the architecture of the design with 6 processes communicating through 4 channels.

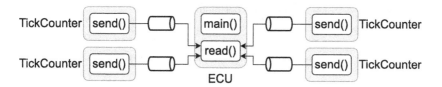

Fig. 9. Architecture of the ABS/ASR design.

We prove several properties that are crucial for this kind of automotive control system. First, we leverage the strengths of deductive verification by proving several functional properties about the correctness of the ABS and ASR algorithms. We prove that the estimation of the vehicle speed works as expected, that ABS and ASR calculate the correct slack on each wheel, that the ASR behaves correctly depending on the slack and that the ABS follows the state machine representation defined in the Bosch specification [37]. Previous work that uses model checking struggles with such properties [19,21]. Second, we prove some global safety properties, namely that the system is deadlock-free and that the ECU receives the correct data from the sensors in less than 1 ms.

To evaluate the scalability of our approach for verifying global properties, we additionally consider a variant of the ABS/ASR example in which we verify only local properties. This variant is equal to the base ABS/ASR case study except that it does not contain the global property specifications and the manual and generated invariants needed to verify them.

5.2   Experimental Results

Our experimental results are summarized in Table 1. We characterize the complexity of each of our case studies by the number of processes and events they contain, the number of methods that were generated in the PVL encoding and the number of properties we verify about them. The number of manual invariants necessary for verification and the number of states in the RASI give an indication of the verification effort. The RASI is generally based on scheduling variables and program counters; only the ABS/ASR case study also includes FIFO queue sizes to limit the overapproximation of the state space. Each case study was successfully verified in the given time, running on an i7-8700 6-core 3.2 GHz CPU with 64 GB of main memory available.

5.3   Scalability

Being able to deal with parameterized systems at all is a significant advantage of our approach over alternatives based on model checking. In addition, deductive verification is a promising technique to improve the scalability of SystemC verification. Model checking suffers from the state space explosion problem, limiting its applicability to complex designs. Due to its modular nature and local

P. Tasche et al.

Table 1. Experimental results.

Design	# procs	# events	# methods (in PVL)	# props	# man. invariants	RASI size $|\mathcal{R}_A|$	Verif. time (sec.)
robot	2	3	6	1	2	8	13
robot-1MS	2	3	6	1	4	12	15
robot-dummy	3	4	8	1	4	24	34
producer-consumer	2	4	11	2	8	72	52
ABS/ASR (local)	6	17	22	4	n.a	n.a	167
ABS/ASR	6	17	22	13	9	247*	10400

* Manually reduced with data variables, see Sect. 4.3

reasoning, deductive verification does not generally share this problem, making it a good candidate for working towards a more scalable solution.

The problem with the scalability of a deductive verification approach is the inclusion of the global state for local reasoning. In general, relating the local state to the global state is associated with high manual effort for the user. While an efficient encoding of such dependencies can lead to comparatively low verification times, the time and expertise required on the user's side often outweighs this advantage. In this work, we therefore opted for an assisted approach that would minimize user effort while maintaining some of the potential advantage in scalability that deductive verification offers. The RASI's state space enumeration allows verification of several common property classes with minimal user effort. The downside of this approach is that the RASI construction is closely related to model checking and may, therefore, suffer from the state space explosion problem.

This issue is illustrated in Table 1. While all the smaller case studies were successfully verified in less than a minute, the larger ABS/ASR case study took almost three hours to verify. A portion of this time discrepancy can be explained by the additional methods, processes and events that need to be considered and the higher number of properties that we verify about this design. However, the RASI that, even after a simple manual reduction, is still significantly larger than in the other cases[4], has the largest effect on this verification time by far. This can be seen in the comparison with the local variant of the ABS/ASR example. A potential explosion of the RASI size is a major problem for our approach.

Even so, our approach still improves the scalability of SystemC verification over model checking approaches, as is evidenced by our ability to verify the ABS/ASR example and the inability of UPPAAL and CPAChecker to do the same. The RASI, as we use it, does suffer from the state space explosion prob-

---

[4] Note also that, despite not necessarily increasing the state space, more processes and events also mean that the representation of each state in the RASI will have more atomic conditions.

lem, but its abstract representation mitigates this effect. Furthermore, the level of abstraction of the RASI can be controlled by the user. This makes this approach promising for a more scalable verification technique that might not be affected by state space explosion. If a property can be proven with knowledge about only a fraction of the global state, this would allow the RASI to abstract from other factors that could induce a state space explosion. While finding the right abstraction level for this is nontrivial, our approach enables such potential advances. We plan to investigate this in future work.

5.4 Discussion

Our experimental results demonstrate that our approach is applicable to the verification of parameterized SystemC designs. We were able to verify both functional and global safety properties of our case studies, including the relatively complex ABS/ASR design. In contrast to existing work based on model checking, we can verify the designs for all possible parameter values with reasonable effort with our deductive approach. This is a major advantage, in particular as SystemC is often used for design space exploration and as a golden reference model, such that concrete implementation details are fixed later in the development process. Our deductive approach also allows us to reason about unbounded data, which model checkers cannot deal with. This is vital to proving functional properties about reactive systems.

The experiments also show how the RASI can reduce user effort. While functional properties did not pose a problem, many global properties proved challenging to verify by hand. With the RASI and the verification techniques described in Sect. 4.5, it took relatively little manual effort to verify them. We could also reuse some invariants, e.g. in the *robot* case study. In our artifact documentation [1], we elaborate on the manual steps that are needed for our approach.

The main challenge proved to be the engineering of the RASI to avoid an explosion of the abstract state space. As the ABS/ASR shows, although we only take a small part of the full state space into consideration, verification time can grow quickly if the abstract state space gets too large. However, we expect it to stay manageable with careful selection of the abstract state variables as long as process interleavings are limited. Since embedded system designers try to maintain predictability in their designs, this is typically the case.

Overall, our results demonstrate the ability of our approach to verify crucial properties of parameterized embedded systems. By leveraging the strengths of deductive verification in combination with ideas from model checking, we can verify a wide range of properties with reduced user effort.

6   Related Work

There exist several approaches to enable formal verification for SystemC. Some of them only cope with a synchronous subset (e.g. [13,38,39]), others rely on a transformation of SystemC designs into some sort of state machine, as done in [16–19, 23, 27, 33, 41, 43], or they use process algebras, petri-nets or a C representation for the verification of SystemC designs [8–10, 12, 20, 28, 32]. In [24, 25, 30], the authors present an approach to verify SystemC designs using symbolic simulation and partial order reduction. In [14, 15, 27, 40], the authors present a combination of induction and bounded model checking to formally verify SystemC designs. However, all of these approaches rely on model checking at some point. They cannot deal with parameterized systems, and they suffer from the state space explosion problem. In [29], the authors present an approach for automatic hardware/software partitioning, which enables the underlying verification method to analyze hardware parts more efficiently. However, they still suffer from state space explosion for the software part. In [31, 42], the authors present an approach for component-based hardware/software co-verification. However, the approach requires that the designer specifies subsystems and properties that can be separately verified. In contrast to this, we present an encoding that precisely captures the SystemC execution semantics and enables us to map SystemC designs into a formal representation fully automatically. Furthermore, with our deductive approach, we gain access to the mature and sophisticated verification capabilities of VerCors. We also extend the applicability of VerCors to embedded systems that are modeled in SystemC, and gain new insights on how to encode reactivity and discrete-event mechanisms in VerCors.

7   Conclusion

In this paper, we have presented a novel approach for the deductive verification of embedded systems that are modeled in SystemC. We have proposed a formal encoding of SystemC designs in PVL that is consistent with the SystemC semantics informally defined in [26]. With that, we can use the VerCors deductive verifier to verify these designs. Our main contributions are twofold: First, we have presented an encoding of the SystemC semantics that enables an automated transformation of a given SystemC design into PVL. This spares designers the tedious task of formalizing a system themselves. Second, we have presented an approach to generate an abstract enumeration of the reachable state space with respect to process scheduling (the RASI). Both contributions are supported by the tools in our artifact [1]. Our results show that this approach allows us to deductively verify parameterized SystemC designs with comparatively low manual effort, as the RASI provides a simple and automatable technique to connect global properties with the local state.

In future work, we plan to expand the subset of SystemC we support with our automated transformation, e.g. with dynamic process creation and dynamic memory allocation. We want to automate many of the manual steps that are currently necessary, if possible, and increase the scalability of the RASI, which is currently the greatest bottleneck. We also plan to investigate the use of the VerCors-built-in support for hybrid verification [35, 36] to verify global properties by model checking and then use them locally for deductive verification.

References

1. [Artifact] Deductive Verification of Parameterized Embedded Systems modeled in SystemC. https://doi.org/10.4121/a7e780c9-87fa-486c-b484-a76a459a9d53
2. Amighi, A., Blom, S., Darabi, S., Huisman, M., Mostowski, W., Zaharieva-Stojanovski, M.: Verification of concurrent systems with VerCors. In: Bernardo, M., Damiani, F., Hähnle, R., Johnsen, E.B., Schaefer, I. (eds.) SFM 2014. LNCS, vol. 8483, pp. 172–216. Springer, Cham (2014). https://doi.org/10.1007/978-3-319-07317-0_5
3. Amighi, A., Haack, C., Huisman, M., Hurlin, C.: Permission-based separation logic for multithreaded Java programs. Log. Methods Comput. Sci. **11**(1) (2015). https://doi.org/10.2168/LMCS-11(1:2)2015
4. Ball, T., Majumdar, R., Millstein, T., Rajamani, S.K.: Automatic predicate abstraction of c programs. In: Proceedings of the ACM SIGPLAN 2001 Conference on Programming Language Design and Implementation, pp. 203–213 (2001)
5. Behrmann, G., David, A., Larsen, K.G.: A tutorial on UPPAAL. In: Bernardo, M., Corradini, F. (eds.) SFM-RT 2004. LNCS, vol. 3185, pp. 200–236. Springer, Heidelberg (2004). https://doi.org/10.1007/978-3-540-30080-9_7
6. Beyer, D., Keremoglu, M.E.: CPACHECKER: a tool for configurable software verification. In: Gopalakrishnan, G., Qadeer, S. (eds.) CAV 2011. LNCS, vol. 6806, pp. 184–190. Springer, Heidelberg (2011). https://doi.org/10.1007/978-3-642-22110-1_16
7. Blom, S., Darabi, S., Huisman, M., Oortwijn, W.: The VerCors tool set: verification of parallel and concurrent software. In: Polikarpova, N., Schneider, S. (eds.) IFM 2017. LNCS, vol. 10510, pp. 102–110. Springer, Cham (2017). https://doi.org/10.1007/978-3-319-66845-1_7
8. Cimatti, A., Micheli, A., Narasamdya, I., Roveri, M.: Verifying SystemC: a software model checking approach. In: Formal Methods in Computer-Aided Design (FMCAD), pp. 51–59. IEEE (2010). https://dl.acm.org/doi/10.5555/1998496.1998510
9. Cimatti, A., Griggio, A., Micheli, A., Narasamdya, I., Roveri, M.: KRATOS – a software model checker for SystemC. In: Gopalakrishnan, G., Qadeer, S. (eds.) CAV 2011. LNCS, vol. 6806, pp. 310–316. Springer, Heidelberg (2011). https://doi.org/10.1007/978-3-642-22110-1_24
10. Cimatti, A., Narasamdya, I., Roveri, M.: Software model checking SystemC. IEEE Trans. Comput. Aided Des. Integr. Circuits Syst. **32**(5), 774–787 (2013). https://doi.org/10.1109/TCAD.2012.2232351
11. Cousot, P.: Abstract interpretation. ACM Comput. Surv. (CSUR) **28**(2), 324–328 (1996)
12. Garavel, H., Helmstetter, C., Ponsini, O., Serwe, W.: Verification of an industrial SystemC/TLM model using LOTOS and CADP. In: IEEE/ACM International Conference on Formal Methods and Models for Co-design (MEMOCODE '09), pp. 46–55 (2009). https://doi.org/10.1109/MEMCOD.2009.5185377
13. Große, D., Kühne, U., Drechsler, R.: HW/SW co-verification of embedded systems using bounded model checking. In: Great Lakes Symposium on VLSI, pp. 43–48. ACM Press (2006). https://doi.org/10.1145/1127908.1127920
14. Große, D., Le, H.M., Drechsler, R.: Proving transaction and system-level properties of untimed SystemC TLM designs. In: MEMOCODE, pp. 113–122. IEEE (2010). https://doi.org/10.1109/MEMCOD.2010.5558643

15. Große, D., Le, H.M., Drechsler, R.: Formal verification of SystemC-based cyber components. In: Jeschke, S., Brecher, C., Song, H., Rawat, D.B. (eds.) Industrial Internet of Things. SSWT, pp. 137–167. Springer, Cham (2017). https://doi.org/10.1007/978-3-319-42559-7_6

16. Habibi, A., Moinudeen, H., Tahar, S.: Generating finite state machines from SystemC. In: Design, Automation and Test in Europe, pp. 76–81. IEEE (2006). https://doi.org/10.1109/DATE.2006.243777

17. Habibi, A., Tahar, S.: An approach for the verification of SystemC designs using AsmL. In: Peled, D.A., Tsay, Y.-K. (eds.) ATVA 2005. LNCS, vol. 3707, pp. 69–83. Springer, Heidelberg (2005). https://doi.org/10.1007/11562948_8

18. Herber, P., Fellmuth, J., Glesner, S.: Model checking SystemC designs using timed automata. In: International Conference on Hardware/Software Codesign and System Synthesis (CODES+ISSS), pp. 131–136. ACM Press (2008). https://doi.org/10.1145/1450135.1450166

19. Herber, P., Glesner, S.: A HW/SW co-verification framework for SystemC. ACM Trans. Embed. Comput. Syst. (TECS) **12**(1s), 1–23 (2013). https://doi.org/10.1145/2435227.2435257

20. Herber, P., Hünnemeyer, B.: Formal verification of SystemC designs using the BLAST software model checker. In: ACESMB@ MoDELS, pp. 44–53 (2014). https://dblp.org/rec/conf/models/HerberH14

21. Herber, P., Liebrenz, T.: Dependence analysis and automated partitioning for scalable formal analysis of SystemC designs. In: 18th ACM-IEEE International Conference on Formal Methods and Models for System Design (MEMOCODE), pp. 1–6. IEEE (2020). https://doi.org/10.1109/MEMOCODE51338.2020.9314998

22. Herber, P., Liebrenz, T., Adelt, J.: Combining forces: how to formally verify informally defined embedded systems. In: Huisman, M., Păsăreanu, C., Zhan, N. (eds.) FM 2021. LNCS, vol. 13047, pp. 3–22. Springer, Cham (2021). https://doi.org/10.1007/978-3-030-90870-6_1

23. Herber, P., Pockrandt, M., Glesner, S.: STATE - a SystemC to timed automata transformation engine. In: ICESS. IEEE (2015). https://doi.org/10.1109/HPCC-CSS-ICESS.2015.188

24. Herdt, V., Große, D., Drechsler, R.: Formal verification of SystemC-based designs using symbolic simulation. In: Enhanced Virtual Prototyping, pp. 59–117. Springer, Cham (2021). https://doi.org/10.1007/978-3-030-54828-5_4

25. Herdt, V., Le, H.M., Große, D., Drechsler, R.: Verifying SystemC using intermediate verification language and stateful symbolic simulation. IEEE Trans. Comput.-Aided Des. Integr. Circuits Syst. **38**(7), 1359–1372 (2018). https://doi.org/10.1109/TCAD.2018.2846638

26. IEEE Standards Association: IEEE Std. 1666-2011, Open SystemC Language Reference Manual. IEEE Press (2011). https://doi.org/10.1109/IEEESTD.2012.6134619

27. Jaß, L., Herber, P.: Bit-precise formal verification for SystemC using satisfiability modulo theories solving. In: Götz, M., Schirner, G., Wehrmeister, M.A., Al Faruque, M.A., Rettberg, A. (eds.) IESS 2015. IAICT, vol. 523, pp. 51–63. Springer, Cham (2017). https://doi.org/10.1007/978-3-319-90023-0_5

28. Karlsson, D., Eles, P., Peng, Z.: Formal verification of SystemC designs using a Petri-Net based Representation. In: Design, Automation and Test in Europe (DATE), pp. 1228–1233. IEEE Press (2006). https://doi.org/10.1109/DATE.2006.244076

29. Kroening, D., Sharygina, N.: Formal verification of SystemC by automatic hardware/software partitioning. In: Proceedings of MEMOCODE 2005, pp. 101–110. IEEE (2005). https://doi.org/10.1109/MEMCOD.2005.1487900

30. Le, H.M., Große, D., Herdt, V., Drechsler, R.: Verifying SystemC using an intermediate verification language and symbolic simulation. In: 2013 50th ACM/EDAC/IEEE Design Automation Conference (DAC), pp. 1–6. IEEE (2013). https://doi.org/10.1145/2463209.2488877

31. Li, J., Sun, X., Xie, F., Song, X.: Component-based abstraction and refinement. In: Mei, H. (ed.) ICSR 2008. LNCS, vol. 5030, pp. 39–51. Springer, Heidelberg (2008). https://doi.org/10.1007/978-3-540-68073-4_4

32. Man, K.L., Fedeli, A., Mercaldi, M., Boubekeur, M., Schellekens, M.: SC2SCFL: automated SystemC to $SystemC^{FL}$ translation. In: Vassiliadis, S., Bereković, M., Hämäläinen, T.D. (eds.) SAMOS 2007. LNCS, vol. 4599, pp. 34–45. Springer, Heidelberg (2007). https://doi.org/10.1007/978-3-540-73625-7_6

33. Niemann, B., Haubelt, C.: Formalizing TLM with communicating state machines. Forum Specification Des. Lang. (2006). https://doi.org/10.1007/978-1-4020-6149-3_14

34. O'Hearn, P.W.: Resources, concurrency and local reasoning. Theor. Comput. Sci. **375**(1–3), 271–307 (2007). https://doi.org/10.1016/j.tcs.2006.12.035

35. Oortwijn, W.: Deductive techniques for model-based concurrency verification. Ph.D. thesis, University of Twente, Netherlands, December 2019. https://doi.org/10.3990/1.9789036548984

36. Oortwijn, W., Gurov, D., Huisman, M.: Practical abstractions for automated verification of shared-memory concurrency. In: Beyer, D., Zufferey, D. (eds.) VMCAI 2020. LNCS, vol. 11990, pp. 401–425. Springer, Cham (2020). https://doi.org/10.1007/978-3-030-39322-9_19

37. Reif, K.: Bremsen und Bremsregelsysteme. Bosch Fachinformation Automobil, Vieweg+Teubner Verlag Wiesbaden (2010). https://doi.org/10.1007/978-3-8348-9714-5

38. Ruf, J., Hoffmann, D.W., Gerlach, J., Kropf, T., Rosenstiel, W., Müller, W.: The simulation semantics of SystemC. In: Design, Automation and Test in Europe, pp. 64–70. IEEE Press (2001). https://doi.org/10.1109/DATE.2001.915002

39. Salem, A.: Formal semantics of synchronous SystemC. In: Design, Automation and Test in Europe (DATE), pp. 10376–10381. IEEE Computer Society (2003). https://doi.org/10.1109/DATE.2003.1253637

40. Schwan, S., Herber, P.: Optimized hardware/software co-verification using the UCLID satisfiability modulo theory solver. In: 29th IEEE International Conference on Enabling Technologies: Infrastructure for Collaborative Enterprises, WETICE 2020, Virtual Event, France, 10–13 September 2020, pp. 225–230. IEEE (2020). https://doi.org/10.1109/WETICE49692.2020.00051

41. Traulsen, C., Cornet, J., Moy, M., Maraninchi, F.: A SystemC/TLM semantics in PROMELA and its possible applications. In: Bošnački, D., Edelkamp, S. (eds.) SPIN 2007. LNCS, vol. 4595, pp. 204–222. Springer, Heidelberg (2007). https://doi.org/10.1007/978-3-540-73370-6_14

42. Xie, F., Yang, G., Song, X.: Component-based hardware/software co-verification for building trustworthy embedded systems. J. Syst. Softw. **80**(5), 643–654 (2007). https://doi.org/10.1016/j.jss.2006.08.015

43. Zhang, Y., Vedrine, F., Monsuez, B.: SystemC waiting-state automata. In: International Workshop on Verification and Evaluation of Computer and Communication Systems (2007). https://dl.acm.org/doi/abs/10.5555/2227445.2227453

Automatically Enforcing Rust Trait Properties

Twain Byrnes[✉][iD], Yoshiki Takashima[iD], and Limin Jia[iD]

Carnegie Mellon University, Pittsburgh, PA, USA
{binarynewts,ytakashima,liminjia}@cmu.edu

Abstract. As Rust's popularity increases, the need for ensuring correctness properties of software written in Rust also increases. In recent years, much work has been done to develop tools to analyze Rust programs, including Property-Based Testing (PBT), model checking, and verification tools. However, developers still need to specify the properties that need to be analyzed and write test harnesses to perform the analysis. We observe that one kind of correctness properties that has been overlooked is correctness invariants of Rust trait implementations; for instance, implementations of the equality trait need to be reflexive, symmetric, and transitive. In this paper, we develop a fully automated tool that allows developers to analyze their implementations of a set of built-in Rust traits. We encoded the test harnesses for the correctness properties of these traits and use Kani to verify them. We evaluated our tool over six open-source Rust libraries and identified three issues in *PROST!*, a protocol buffer library with nearly 40 million downloads.

Keywords: Rust · Traits · Software model checking

1 Introduction

The Rust programming language [17] has been steadily increasing in popularity for projects requiring a high level of precision and performance. Rust is used in the Linux Kernel [10,25], Dropbox's file-syncing engine [13], Discord [12], several Amazon Web Services (AWS) projects including S3, EC2, and Lambda [19], and many other high-profile projects. Rust has gained such wide adoption, in large part, due to its memory safety guarantees. However, Rust developers must still perform additional testing or verification to ensure functional correctness properties, specific to their code.

To help programmers analyze Rust programs, several tools are being developed [8,11,14,16,18]. For instance, proptest [23], a popular Property-Based Testing (PBT) [20] tool for Rust, is used in over 1500 crates according to crates.io [2]. There are several formal verification tools for Rust [1], including concurrency checkers and dynamic symbolic executors [3,5,6].

However, one road block preventing more Rust developers from using these tools is that they need to write test or verification harnesses for each of the

R. Dimitrova et al. (Eds.): VMCAI 2024, LNCS 14500, pp. 210–223, 2024.
https://doi.org/10.1007/978-3-031-50521-8_10

properties that they want to analyze. This is particularly problematic for Rust developers without a formal methods background, or for those who do not wish to spend time creating these tests. Therefore, exploring avenues for lowering developers' burden and automating the specification and analysis of correctness properties of Rust programs can be beneficial towards making Rust software more correct and secure. For instance, prior work has explored automatically converting existing test harnesses that developers wrote for proptest to harnesses that Kani, a model checker for Rust can use [22]. As a result, developers can verify their code using Kani without additional manual effort.

In this paper, we identify further areas where Rust developers can benefit from automated code verification. We observe that one kind of correctness properties that has been overlooked is correctness invariants of Rust trait implementations; for instance, a function implementing the equality trait for objects of a particular type must be transitive, symmetric, and reflexive. Rust traits, similar to interfaces in other languages, support shared functionality across types. Like equality, some Rust traits have invariants that are not checked by Rust's type-checker, and are thus left to the programmers to implement correctly.

We develop `TraitInv` to verify Rust trait invariants, as an addition to the Kani VS Code Extension [4]. `TraitInv`, like PBT, is based on the idea that certain properties of the output of specific methods must exist, regardless of the input. We identify invariants of commonly-used traits and create modular harnesses that can be inserted into a user's code with little intervention. Once inserted, these harnesses are verified by Kani. The harnesses themselves can be used by proptest by replacing symbolic value generators with random ones.

We evaluate `TraitInv` on 42 trait implementations from six libraries from `crates.io` [2]. The evaluation results show that `TraitInv` can create Kani testing harnesses on a wide variety of libraries over many traits. To answer the question of whether developers need their trait implementations verified for correctness invariants, we then use Kani to verify all the created harnesses. We discovered three issues in one trait implementation in *PROST!*, a popular and heavily tested Rust library with 40 million downloads. The tool is open source at https://github.com/binarynewts/kani-vscode-extension.

The rest of this paper is organized as follows. In Sect. 2, we review Rust traits and Kani and discuss related work. Then we describe specific traits and their invariants that `TraitInv` supports in Sect. 3. Next, we present the implementation and evaluation results in Sect. 4. Limitations of `TraitInv` and future work are discussed in Sect. 5.

2   Background and Related Work

In this section, we first discuss Rust traits, then provide the background of Kani, the Rust model checker that our tool is built on; we then discuss related tools.

2.1 Rust Traits

Rust traits define shared behavior abstractly, and are used like interfaces in other languages. Traits allow for the same methods to be called on many different types with a shared expected behavior.

For example, the `PartialOrd` trait allows for a partial order to be instantiated for a custom type, and the `Ord` trait can be used to implement a total order. `PartialOrd` requires the following partial comparison method:

$$partial\_cmp(\&self, other: \&Rhs) \to Option<Ordering>.$$

This method takes two objects and returns a `Some` constructor of the proper `Ordering` type ($>$, $<$, or $=$) if they can be compared and a `None` constructor otherwise. The total order trait, `Ord`, requires the following method, which takes two objects, compares them, and returns an object of the `Ordering` type:

$$cmp(\&self, other: \&Self) \to Ordering.$$

Some basic Rust trait implementations can be derived automatically by the compiler. However, traits that require more complex behavior need to be user-defined. Listed below is an example user-defined implementation of a `PartialOrd` and `Ord`, taken from `crypto-bigint/src/limb/cmp.rs`, lines 93–116.

```
1   impl Ord for Limb {
2       fn cmp(&self, other: &Self) -> Ordering {
3           let mut n = 0i8;
4           n -= self.ct_lt(other).unwrap_u8() as i8;
5           n += self.ct_gt(other).unwrap_u8() as i8;
6
7           match n {
8               -1 => Ordering::Less,
9               1 => Ordering::Greater,
10              _ => {
11                  debug_assert_eq!(n, 0);
12                  debug_assert!(bool::from(self.ct_eq(other)));
13                  Ordering::Equal
14              }
15          }
16      }
17  }
18
19  impl PartialOrd for Limb {
20      #[inline]
21      fn partial_cmp(&self, other: &Self) -> Option<Ordering> {
22          Some(self.cmp(other))
23      }
24  }
```

Rust traits have properties that the programmer must enforce on their implementation. For example, `partial_cmp` needs to be transitive.

2.2 Kani

The Kani Rust Verifier [24] uses bit-precise model checking and is built on top of CBMC [15]. Kani has been used to verify components of `s2n-quic` [21], an AWS-developed cryptographic library, and several popular Rust crates [22].

Kani harnesses analyze properties that should be maintained within code. For example, we may want to check that a function that takes in a `u64` and is supposed to return an even integer actually meets this return specification. To do this, we could write the following Kani harness, where the function `return_even` is called and the assert statement checks that the result is even.

```
1   #[kani::proof]
2   fn from_test() {
3       let num: u64 = kani::any();
4       assert!(return_even(num)
5   }
```

Kani is able to provide formal guarantees by symbolically executing the harness, allowing it to analyze properties checked by the harness for every possible input. In order for Kani to be able to generate symbolic inputs of a type, the type must implement the `kani::Arbitrary` trait. The Kani developers have implemented this trait for most primitive types and some types in the standard library. If the objects are enumeration or structure types whose fields implement `kani::Arbitrary`, it can be derived by placing `#[cfg_attr(kani, derive(kani::Arbitrary))]` above the type declaration. If not, the Kani user will need to implement `kani::Arbitrary`. This trait contains only the method `any() -> Self`, which takes no argument and returns a value of the type that `kani::Arbitrary` implements (`Self`). There are limitations on types that Kani can efficiently generate symbolic inputs for, which we detail in Sect. 5.

2.3 Related Work

`propproof` [22] is a tool that converts existing PBT harnesses into Kani harnesses so that they can be formally verified. `TraitInv` implements Kani harnesses and can be combined with `propproof` to verify trait invariants and application-specific invariants of Rust programs.

Erdin developed a tool for verifying user-defined correctness properties of Rust programs, including trait invariants like ours [9]. Unlike `TraitInv`, which automatically generates test harnesses based on trait names, their tool requires user-provided annotations to generate test harnesses. It uses Prusti [7], a Rust verifier based on the Viper verification infrastructure.

3   Verifying Rust Trait Invariants

We explain how programmers can use `TraitInv` to analyze their trait implementations and explain its capabilities.

3.1   Workflow for Verifying Trait Invariants

After launching the Kani VS Code Extension with `TraitInv`, Rust programmers can easily add Kani testing harnesses. To use this tool, one must navigate to the code file, highlight the line with their trait implementation declaration, bring up the Command Palette, and select "Add Trait Test". Upon selecting this option, the tool generates the applicable Kani testing harness and inserts it directly above the implementation line.

3.2   Harnesses for Trait Invariants

We implemented harnesses for invariants of frequently used traits within the Rust standard library. Next, we describe all the supported traits and their invariants. `From/Into`: Rust allows for conversion of values of a source type to values of a (different) target type using the `From` and `Into` traits. In the following code, we convert from a u8 array of length 12 to a custom type, `ObjectId`, a struct with an `id` field of type length 12 u8 array. We do this by simply returning an `ObjectId` instance, populating its `id` parameter with the array passed in.

```
1   struct ObjectId {
2       id: [u8; 12],
3   }
4
5   impl From<[u8; 12]> for ObjectId {
6       fn from(bytes: [u8; 12]) -> ObjectId {
7           Self { id: bytes }
8       }
9   }
```

The only requirement for such a conversion is that it does not panic (e.g. unwrapping `None` or dividing by 0). Therefore, the harness consists of code that first instantiates a symbolic input of the source type to be converted, and then calls the `from` method to create an instance of the target type from the value. Kani will report any panic behavior. Below, we provide an example of a Kani test harness for the above implementation of `From`.

```
1   #[kani::proof]
2   fn from_test() {
3       let t: [u8; 12] = kani::any();
4       let _ = ObjectId::from(t);
5   }
```

PartialEq/Eq: Rust types that implement the `PartialEq` trait allow a partial equality relation between objects of those types. `PartialEq` can be implemented for one type or across two types (e.g., comparing `i32` and `u32`).

To implement this trait, the user must implement the following method:

$$eq(\&self, other: \&Rhs) \rightarrow bool$$

This method takes two objects and returns a boolean stating whether or not these two objects are equal. This method is allowed to throw errors, and states that the two objects are not comparable in this instance. This distinction allows for instances like floating point numbers, where we do not want anything to equal `NaN`, including `NaN`.

`PartialEq` must obey transitivity, symmetry, and be consistent between not equal and the negation of equality. If this trait is implemented between two types, all `PartialEq` properties must hold across any combination of types.

The equality trait, `Eq`, must obey the rules of `PartialEq`, but additionally requires reflexivity and totality. It can only compare instances of the same type.

Below, we include an example of a `TraitInv`-generated Kani harness for properties of `PartialEq` implemented on the type `MyType`, a placeholder type for illustrative purposes. Comments point out what each assertion checks for.

```
1   #[kani::proof]
2   fn partialeq_test() {
3       let a: MyType = kani::any();
4       let b: MyType = kani::any();
5       let c: MyType = kani::any();
6       if a == b { assert!(b == a); } // symmetry
7       if (a == b) && (b == c) { assert!(a == c); } //
    ↪   transitivity
8       if a != b { assert!(!(a == b)); } // ne eq consistency
9       if !(a == b) { assert!(a != b); } // eq ne consistency
10  }
```

Note that we do not need to add a test for symmetry by swapping `a` and `b`, since they are of the same type, and this case will be tested regardless. **PartialOrd/Ord:** As shown in Sect. 2, Rust has a partial order trait called `PartialOrd`, which requires a method

$$partial\_cmp(\&self, other: \&Rhs) \rightarrow Option<Ordering>.$$

This method takes two objects and, if they can be compared, returns the Some constructor of the proper Ordering type, or else returns the None constructor. This method must be consistent with the following infix comparators: <, >, ==. By consistent, we mean that two objects can be compared with one of the operators if and only if invoking the partial_cmp method on those two objects yields the Some constructor of the same Ordering type. Additionally, >= and <= must expand in the natural way. All of these requirements, other than consistency with ==, are ensured by the default implementation of the four methods describing less than (or equal to) and greater than (or equal to). However, the programmer may choose to override these default implementations, in which case conformance is not guaranteed and must be checked.

PartialOrd requires transitivity between <, >, and ==, as well as consistency between not equal and the negation of equality. Since PartialOrd is required to be consistent with PartialEq, this last consistency is ensured. Finally, PartialOrd requires duality: that a < b if and only if b > a.

The total order trait, Ord, requires a method

```
cmp(&self, other: &Self) -> Ordering,
```

which must be consistent with PartialOrd's partial_cmp. This consistency can be broken when some traits are derived and others are implemented manually, or if they are implemented manually in a way that is not future proof.

AsMut/AsRef: These traits provide an interface to define types that represent static or mutable references of another type. For example, Vec<T> has an AsRef implementation for [T] because static vector references are also slices. Both AsRef and AsMut conversions have the same requirement as the From trait, in that they must not crash during conversion.

4   Implementation and Evaluation

We implemented harnesses for the traits described in Sect. 3 as an addition to the Kani VS Code Extension, automating the generation and insertion of these harnesses. We evaluated TraitInv on Rust libraries to answer the following questions: can TraitInv enable efficient verification of trait invariants across different traits, types, and libraries; and can TraitInv find bugs in trait implementations in libraries.

4.1   Implementation

TraitInv is implemented as a fork of the Kani VS Code Extension. Including this tool, the extension contains around 195K lines of typescript code, with around 500 lines of typescript code added to original, containing the harness encodings. Other than containing the encodings, this code reads the highlighted line, strips it down to figure out what trait and types it must create a harness for, creates the harness, and inserts it into the user's code immediately above the highlighted line. The tool is launched along with the rest of the Kani VS Code extension, enabling potential users to use the tool without installing any tooling from outside the VS Code GUI.

4.2   Evaluation

Experiment Setup We ran our benchmarks on an Intel(R) Core(TM) i7-7700HQ with 4 cores and 16GB of RAM running Ubuntu 20.04.5 LTS. We used Kani Rust Verifier version 0.33.0 on CBMC version 5.88.1 with CaDiCaL.

Dataset. We pulled highly-downloaded Rust crates from the Rust community's crate registry, `crates.io` [2]. We looked for crates that contained some of the implementations of the above traits and had at least several million downloads. We created 42 harnesses across six libraries, and ran these harnesses with Kani. The traits and libraries are summarized in Table 1. The Types column details the types that the trait was implemented on or between, and may be slightly edited from the original implementation line for ease of reading and comprehension.

For traits implemented on polymorphic types, we can only analyze concrete instantiations of them. This is because Kani is bit-precise and needs concrete types to know the memory layout of the values. We list the concrete instantiations of the polymorphic types below.

Test 1. Since `chrono::DateTime` is parametric, we were only able to test it on one possible timezone. We chose `chrono::Utc`, since the other options were `chrono::Local`, which would lead to inconsistent testing based on where the testing takes place geographically, and `chrono::FixedOffset`, which requires an input to offset from UTC time, and is thus similar to simply using `chrono::Utc`. These other options could potentially lead to bugs that choosing `chrono::Utc` might not, which we did not test.

Tests 11, 12. Both of these tests create symbolic values of `Checked<T>`, which is parametric, so we were only able to test it on one possible type. We chose `U64`, a custom type from the `crypto-bigint` library similar to the standard `u64`, since that was the only type that we found used with `Checked` in the entire library.

Tests 29, 34. These tests were for the `StackVec<usize>` type, so we were forced to choose a `usize` to implement these tests for. Without choosing, Rust would not know what size to allot for this, and would lead to an error.

Tests 39–42. `Uint<usize>` is generated in these tests, and takes in a `usize`. We had to fix a `usize` to implement `kani::Arbitrary` for and test on, otherwise the items would not have a fixed size and running Kani would lead to an error.

Table 1. Harnesses used to evaluate effectiveness of synthesis. **`chrono::DateTime<T:`
`chrono::TimeZone>` for Bson.

	Trait	Library	Types	Time(s)	Mem(GB)
1	From	bson-rust	*	11.55	0.120
2	From	bson-rust	u8 for BinarySubtype	2.80	0.046
3	From	bson-rust	BinarySubtype for u8	2.63	0.045
4	From	bson-rust	f32 for Bson	7.89	0.038
5	From	bson-rust	f64 for Bson	5.52	0.048
6	From	bson-rust	bool for Bson	4.54	0.048
7	From	bson-rust	i32 for Bson	3.87	0.048
8	From	bson-rust	i64 for Bson	4.76	0.048
9	From	bson-rust	Decimal128 for Bson	4.70	0.048
10	From	bson-rust	[u8; 12] for ObjectId	2.42	0.045
11	From	crypto-bigint	Checked<T> for CtOption<T>	0.74	0.029
12	From	crypto-bigint	Checked<T> for Option<T>	0.74	0.030
13	From	crypto-bigint	u8 for Limb	0.61	0.029
14	From	crypto-bigint	u16 for Limb	0.61	0.029
15	From	crypto-bigint	u32 for Limb	0.60	0.029
16	From	crypto-bigint	u64 for Limb	0.62	0.030
17	From	crypto-bigint	Limb for Word	0.70	0.029
18	From	crypto-bigint	Limb for WideWord	0.61	0.029
19	From	proptest	(usize, usize) for SizeRange	0.48	0.026
20	From	proptest	usize for SizeRange	0.46	0.026
21	From	proptest	f64 for Probability	0.50	0.026
22	From	proptest	Probability for f64	0.49	0.026
23	From	prost	Timestamp for DateTime	434.63	0.452
24	From	prost	DateTime for Timestamp	3.18	0.073
25	From	prost	EncodeError for std::io::Error	4.21	0.282
26	PartialEq	crypto-bigint	Limb	1.85	0.043
27	PartialEq	rust-lexical	f16	5.09	0.093
28	PartialEq	rust-lexical	bf16	0.67	0.022
29	PartialEq	rust-lexical	StackVec<usize>	5.13	0.112
30	PartialEq	sharded-slab	State	0.42	0.118
31	PartialEq	sharded-slab	DontDropMe	1.73	0.135
32	Eq	crypto-bigint	Limb	1.92	0.043
33	Eq	prost	Timestamp	1.58	0.207
34	Eq	rust-lexical	StackVec<usize>	8.14	0.158
35	PartialOrd	crypto-bigint	Limb	37.90	0.334
36	PartialOrd	rust-lexical	f16	56.84	0.446
37	PartialOrd	rust-lexical	bf16	5.69	0.064
38	Ord	crypto-bigint	Limb	28.75	0.226
39	AsRef	crypto-bigint	[Word; usize] for Uint<usize>	1.30	0.057
40	AsRef	crypto-bigint	[Limb] for Uint<usize>	1.23	0.056
41	AsMut	crypto-bigint	[Word; usize] for Uint<usize>	1.34	0.057
42	AsMut	crypto-bigint	[Limb] for Uint<usize>	1.32	0.057

4.3   Results

We ran our 42 Kani test harnesses and collected the results into Table 1.

Performance of TraitInv. The last two columns of Table 1 document the time spent verifying the given harness in seconds, and maximum memory usage measured in gigabytes, respectively. Most of our tests were able to complete in under ten seconds, many of them under one second. Though we had an outlier at 434.64 s, none of the other tests took over a minute to complete. Each test uses under 0.5 GB of memory, with only a little over a quarter of the tests using over 0.1GB of memory. Thus, our tool has been shown to be efficient.

There is a notable disparity in time taken and memory used between tests 23 and 24. Though these From implementations are made to convert between the same two types, they appear to take vastly different amounts of time to verify. Conversion from Timestamp to DateTime takes over 400 s, by far the longest amount of time on the table, while its counterpart, converting in the other direction, takes only three seconds. This is due to the fact that the former takes over 400 s to verify as correct, while its counterpart takes only three seconds to terminate due to finding bugs. The same factor is responsible for the difference in maximum memory usage between benchmarks 23 and 24.

Verification for tests 29 and 34 took a 5.13 and 8.14 s respectively, considerably longer than those taking under a second. This is because these types involved loops: Kani will indefinitely unroll loops unless a maximum loop bound is obvious. Finding a maximum bound is very rare, and it is much more common for Kani to get stuck attempting to unwind a loop. By telling Kani how far to unwind, it is able to set an upper bound and try to verify from there. The lower the bound, the more efficient Kani is, but when set too low Kani may not be able to formally verify every case and will mark a failed check. The necessary unwinding on these tests explain the relatively lengthy test time.

Bugs Identified. Using TraitInv, we identified three issues with benchmark 24, an implementation of the From trait that converts objects of *PROST!*'s DateTime type to objects of its Timestamp type. There is already a PBT harness for this trait implementation, yet the errors had not been found. We followed a responsible disclosure process in reporting the bug described in this paper, reporting the bug on the *PROST!* GitHub issues page.

The from method that coverts from DateTime to Timestamp was implemented by calling several different functions, including ones which convert the number of years into seconds and the number of months into seconds. All of the errors were found in the function converting the year into a number of seconds. The problematic code from prost/prost-types/src/datetime.rs is shown below. We leave out sections of the function with no problems, since the full function is around 80 lines. Lines 2–3 set up some variables for the rest of the function. Line 6 contains an early return for years 1900–2038, since these are assumed to be seen more often and should thus be able to be computed more quickly. Lines 10–11 set up more variables to be used in the code excluded on line 13. Finally, lines 15–19 do some ending computations and return the tuple of the number of seconds and a boolean for whether or not it is a leap year.

```
1   pub(crate) fn year_to_seconds(year: i64) -> (i128, bool) {
2       let is_leap;
3       let year = year - 1900;
4
5       // Fast path for years 1900 - 2038.
6       if year as u64 <= 138 { ... }
7
8       ...
9
10      let mut cycles: i64 = (year - 100) / 400;
11      let mut rem: i64 = (year - 100)
12
13      ...
14
15      (
16          i128::from((year - 100) * 31_536_000) +
    ↪   i128::from(leaps * 86400 + 946_684_800 + 86400),
17          is_leap,
18      )
19  }
```

Next, we discuss the three bugs found.

Subtraction Overflow on Line 3. At line 3 in the above function, we see a subtraction of 1900 from the year. There is no check to ensure that the year is within a valid range, but the conversion must work for any year. For example, when this function is called with the year `i64::MIN + 1800`, we will overflow.

Subtraction Overflow on Line 10. There is a similar error on line 10, where no check is done to ensure that 100 can actually be subtracted from the current value of the year. For example, we would overflow if we input the year `i64::MIN + 1950`, since we first reset this to 50 on line 3, then attempt to subtract 100 from it. Though line 11 also contains subtraction by 100 without a check, Kani does not point this out, as it is the exact same issue. If the issue on line 10 were fixed without also addressing the subtraction on line 11, Kani would point to line 11 as an issue.

Multiplication Overflow on Line 16. The return value computation has a multiplication overflow, which would yield an incorrect number of seconds. Setting the `year` to a large negative number causes the multiplication `(year - 100) * 31_536_000` on line 16 to overflow. Kani returned `year=-6192467633871255600` as a concrete example that causes such an overflow.

5   Limitations and Future Work

We have seen that we can implement some traits across two types. We may wish to implement some traits across three or more types. We can do this by implementing a trait across two types in one implementation, and two types in a separate implementation with overlap. There are some traits which have special invariants for this case. Both `PartialEq` and `PartialOrd` may be defined across an arbitrary number of types, in which case transitivity must hold across types. This invariant is not currently checked by `TraitInv`, as it looks at one implementation line at a time. There were no such instances of a trait implemented across more than two types in the test set, so this is likely a rare occurrence.

Since `TraitInv` is built on top of Kani, it inherits all of Kani's limitations. Kani can derive some implementations of its `Arbitrary` trait, though it cannot for many types, often leading to manual implementations. For instance, programmers cannot implement `kani::Arbitrary` for types with unbounded size, like strings and vectors. Kani can create inputs for the above types of any size less than or equal to some bound with large verification time, and is much more efficient when told to create objects of a specific size. Kani also struggles to generate `Arbitrary` implementations for polymorphic types like `struct InnerArray<T, const N: usize>([T; N]);`, which requires both a type and size input in order to implement `kani::Arbitrary`; we cannot test on all potential parameterizations. The standard workaround for a parametric type is to pick a specific parameterization and implement `kani::Arbitrary` for that type.

Improving generation in these cases would be extremely beneficial, increasing modularity and decreasing the amount of time spent by users on formal verification. Additionally, there are several performance issues with Kani, stopping it from working on larger crates. We used our tool to create tests for the AWS SDK for Rust, but Kani fails to verify these tests, despite their similarity to tests that Kani has no trouble with in smaller libraries.

6   Conclusion

We introduced a fully automated harness synthesis tool as an extension to the Kani VS Code Extension, allowing Rust programmers with no formal verification expertise to ensure correctness of their Rust trait implementations. We enable Rust programmers to generate Kani testing harnesses for their trait implementations at the click of a button, which can be checked by Kani and provide formal verification that their trait implementations follow the necessary invariants. Our evaluation shows that such harnesses can be used across a wide variety of Rust libraries, and can identify issues in heavily tested real-world code.

Acknowledgements. We would like to thank the anonymous reviewers for their feedback on our paper. This work was partially funded by the National Science Foundation via grants CNS2114148 and an Amazon Research Award, Fall 2022 CFP. Any opinions, findings, and conclusions or recommendations expressed in this material are those of the author(s) and do not reflect the views of Amazon.

References

1. Rust verification tools (2021). https://rust-formal-methods.github.io/tools.html
2. crates.io: Rust Package Registry (2023). https://crates.io/
3. haybale (2023). https://github.com/PLSysSec/haybale
4. Introducing the kani vs code extension (2023). https://model-checking.github.io/kani-verifier-blog/2023/06/30/introducing-the-kani-vscode-extension.html
5. Loom (2023). https://github.com/tokio-rs/loom
6. Shuttle (2023). https://www.shuttle.rs/
7. Astrauskas, V., et al.: The Prusti project: formal verification for rust (invited). In: Deshmukh, J.V., Havelund, K., Perez, I. (eds.) NFM 2022. LNCS, vol. 13260, pp. 88–108. Springer, Cham (2022). https://doi.org/10.1007/978-3-031-06773-0_5
8. Denis, X., Jourdan, J.H., Marché, C.: Creusot: a foundry for the deductive verification of rust programs. In: Riesco, A., Zhang, M. (eds.) ICFEM 2022. LNCS, vol. 13478, pp. 90–105. Springer, Cham (2022). https://doi.org/10.1007/978-3-031-17244-1_6, https://hal.inria.fr/hal-03737878
9. Erdin, M.: Verification of Rust Generics, Typestates, and Traits. Master's thesis, ETH Zürich (2019)
10. Filho, W.A.: Rust in the Linux kernel, April 2021. https://security.googleblog.com/2021/04/rust-in-linux-kernel.html
11. Ho, S., Protzenko, J.: Aeneas: rust verification by functional translation. Proc. ACM Program. Lang. 6(ICFP), 116:711–116:741 (2022). https://doi.org/10.1145/3547647
12. Howarth, J.: Why discord is switching from go to rust (2020). https://discord.com/blog/why-discord-is-switching-from-go-to-rust
13. Jayakar, S.: Rewriting the heart of our sync engine (2020). https://dropbox.tech/infrastructure/rewriting-the-heart-of-our-sync-engine
14. Jung, R., Jourdan, J.H., Krebbers, R., Dreyer, D.: RustBelt: securing the foundations of the Rust programming language. Proc. ACM Program. Lang. 2(POPL), 66:1–66:34 (2017). https://doi.org/10.1145/3158154
15. Kroening, D., Tautschnig, M.: CBMC – C bounded model checker. In: Ábrahám, E., Havelund, K. (eds.) TACAS 2014. LNCS, vol. 8413, pp. 389–391. Springer, Heidelberg (2014). https://doi.org/10.1007/978-3-642-54862-8_26
16. Lehmann, N., Geller, A., Vazou, N., Jhala, R.: Flux: Liquid Types for Rust, November 2022. https://doi.org/10.48550/arXiv.2207.04034, http://arxiv.org/abs/2207.04034
17. Matsakis, N.D., Klock, F.S.: The rust language. In: Proceedings of the 2014 ACM SIGAda Annual Conference on High Integrity Language Technology. HILT '14, pp. 103–104. Association for Computing Machinery, New York, NY, USA, October 2014. https://doi.org/10.1145/2663171.2663188
18. Matsushita, Y., Tsukada, T., Kobayashi, N.: RustHorn: CHC-based Verification for Rust Programs. ACM Trans. Program. Lang. Syst. 43, 15:1–15:54 (2021). https://doi.org/10.1145/3462205
19. Miller, S., Lerche, C.: Sustainability with Rust | AWS Open Source Blog, February 2022. https://aws.amazon.com/blogs/opensource/sustainability-with-rust/, section: Developer Tools

20. Paraskevopoulou, Z., Hrițcu, C., Dénès, M., Lampropoulos, L., Pierce, B.C.: Foundational property-based testing. In: Urban, C., Zhang, X. (eds.) ITP 2015. LNCS, vol. 9236, pp. 325–343. Springer, Cham (2015). https://doi.org/10.1007/978-3-319-22102-1_22
21. Schwartz-Narbonne, D.: Use Kani action in CI by danielsn · Pull Request #1556 · aws/s2n-quic, October 2022. https://github.com/aws/s2n-quic/pull/1556
22. Takashimá, Y.: Propproof: free model-checking harnesses from PBT. In: ESEC/FSE (2023)
23. The proptest developers: Proptest, May 2023. https://github.com/proptest-rs/proptest
24. VanHattum, A., Schwartz-Narbonne, D., Chong, N., Sampson, A.: Verifying dynamic trait objects in rust. In: Proceedings of the 44th International Conference on Software Engineering: Software Engineering in Practice. ICSE-SEIP '22, pp. 321–330. Association for Computing Machinery (2022). https://doi.org/10.1145/3510457.3513031
25. Vaughan-Nichols, S.J.: Linux kernel 6.1: rusty release could be a game-changer (2023). https://www.theregister.com/2022/12/09/linux_kernel_61_column/

Borrowable Fractional Ownership Types for Verification

Takashi Nakayama[✉] [iD], Yusuke Matsushita[iD], Ken Sakayori[iD],
Ryosuke Sato[iD], and Naoki Kobayashi[iD]

The University of Tokyo, Tokyo, Japan
{takashi-nakayama,yskm24t,sakayori,rsato,koba}@is.s.u-tokyo.ac.jp

Abstract. Automated verification of functional correctness of imperative programs with references (a.k.a. pointers) is challenging because of
reference aliasing. Ownership types have recently been applied to address
this issue, but the existing approaches were limited in that they are effective only for a class of programs whose reference usage follows a certain
style. To relax the limitation, we combine the approaches of CONSORT
(based on fractional ownership) and RUSTHORN (based on borrowable
ownership), two recent approaches to automated program verification
based on ownership types, and propose the notion of *borrowable fractional ownership types*. We formalize a new type system based on the
borrowable fractional ownership types and show how we can use it to
automatically reduce the program verification problem for imperative
programs with references to that for functional programs without references. We also show the soundness of our type system and the translation,
and conduct experiments to confirm the effectiveness of our approach.

Keywords: Automated program verification · Ownership types ·
Imperative programs

1 Introduction

Various notions of ownership types have been used to prevent race conditions in
concurrent programs and also to enable strong updates of knowledge in sequential programs by controlling references (or pointers) aliases [4,5,7,8,12,28].
Among others, Boyland [6] introduced fractional ownership (a.k.a. fractional
permissions), where a fractional value in $[0,1]$ is associated with each reference
and the full ownership (of 1) allows write access whereas a non-zero ownership
allows read access. The Rust programming language [1,21] incorporates ownership with the *borrow* mechanism (which we call *borrowable ownership*), where
ownership for a mutable reference can be *borrowed* during a specific lifetime, and
the borrowed reference can be used for write access during the lifetime.

Both notions of ownership (i.e., fractional ownership and borrowable ownership) have recently been used for fully automated verification of the functional correctness of imperative programs [3,9,14,20,22,31]. Among them, CON-
SORT [31] used fractional ownership to enable strong updates of refinement

R. Dimitrova et al. (Eds.): VMCAI 2024, LNCS 14500, pp. 224–246, 2024.
https://doi.org/10.1007/978-3-031-50521-8_11

types for automated verification of the absence of assertion failures, whereas RUSTHORN [22] used borrowable ownership to reduce assertion verification to CHC solving. Although both approaches have been shown effective, they also have unavoidable limitations: they are inadequate when a given program does not follow a specific pattern of reference usage, i.e., when the type system conservatively rejects the program. Therefore, relaxing this limitation by extending the expressive power of ownership type systems is essential for automated verification of imperative programs leveraging ownership types.

In this paper, we propose the notion of *borrowable fractional ownership*, which combines both approaches to extend the expressive power of ownership types and develop a new method to verify the functional correctness of programs. Hereafter, we first review the ideas of CONSORT and RUSTHORN, showing how their ownership types work for automated verification, and then explain our new approach.

In CONSORT, each reference is given a type of the form $\tau\,\mathbf{ref}^{\,r}$, where r, called (fractional) ownership, ranges over $[0, 1]$, and τ is a refinement type that describes the content of the reference. Additionally, a reference type should satisfy the well-formedness that no refinement information (that is, only $\mathbf{true}$) is allowed for zero-ownership references. Figure 1 shows an example program of CONSORT and how it is typed. The type of x on the first line shows that x is a reference with full ownership, pointing to a cell holding the value 0. On the second line, the aliasing reference y is created, and all the ownership has been transferred to y, which deprives x of refinement information in its type. On the third line, the type of y can be strongly updated, as y has the full ownership. The $\mathbf{alias}$ command on the fourth line tells the type system that x and y are aliases; based on that, the information on y is propagated to x, and the ownership of y and its refinement information is redistributed between x and y. The fifth line type-checks, as the types of x and y say that they both store 1. In this manner, CONSORT uses fractional ownership to allow strong updates of refinement types and share information among references based on $\mathbf{alias}$ annotations (which may automatically be inserted by a separate pointer analysis).

```
let x = mkref 0 in // x : {ν : int | ν = 0} ref¹
let y = x in       // x : {ν : int | true} ref⁰,  y : {ν : int | ν = 0} ref¹
y := 1;            // x : {ν : int | true} ref⁰,  y : {ν : int | ν = 1} ref¹
alias(x = y);      // x : {ν : int | ν = 1} ref⁰·⁵,  y : {ν : int | ν = 1} ref⁰·⁵
assert( *x + *y = 2 ) // ok
```

Fig. 1. A program example of CONSORT.

```
let x = mkref 0 in // x : int ref α
let y = x in        // y : int ref β;  x is invalidated during β
y := 1;
endlft β;           // dispose y;  x : int ref α
assert( *x = 1 )
```

Fig. 2. Typing in RustHorn.

On the other hand, the type system of RustHorn (which inherits that of Rust[1]) expresses the reference type as of the form τ **ref**$^\alpha$, where α is a *lifetime*, a symbol representing an abstract time period during which the reference is valid. In addition, the borrowed references are invalidated during the corresponding lifetime in Rust. Figure 2 shows a variation of the program in Fig. 1 and how it is typed in the type system of RustHorn. On the first line, a reference x with lifetime α is created. On the second line, y is created as an alias, and the (full) ownership of x is *borrowed* to y during the lifetime β, which invalidates x. On the third line, y can be safely updated as it has temporally borrowed ownership. On the fourth line, the lifetime β ends, and the borrowed ownership is returned to x, and thus x can be safely accessed on the fifth line. In reality, **endlft** commands are automatically inserted by the Rust compiler.

To verify the assertions, RustHorn uses the technique of *prophecy* [2,7,17] to reduce the problem of verification to that of CHC solving. Here, for the sake of clarity, we instead reduce them to the verification problem for functional programs. In the reduction, a mutably borrowing reference y is modeled as a pair of $\langle v, y_\circ \rangle$, where the first component v is the current value of the reference y and the second component $y_\circ$, called y's *prophecy*, expresses the future value at the end of y's lifetime. For verification, the prophecy is encoded by standard non-determinism. Figure 3 shows how the translation works for the program in Fig. 2. Notice that on the second line, the mutably borrowing reference y becomes a pair of the current value (x) and the prophecy value initialized non-deterministically (represented by _), then x takes the prophecy value of y on the third line. After the value of y is changed on the fourth line, the **assume** command on the fifth line, which is converted from **endlft** β in the original program, *resolves* the prophecy value of y; in other words, the formally non-deterministic second value of y is determined by the current value (1) so that x recover its correct value and the last assertion succeeds. Note that this program has no references and is easier to verify than the original one.

As mentioned, both approaches have their limitations: CoNSORT relies on **alias** annotations and thus works effectively only for programs where exact alias information is available statically (Fig. 4 shows one such program), and RustHorn relies on Rust's type system, hence is inapplicable to programs written in other imperative languages like C.

---

[1] For formalization, RustHorn inherits the type system of λ_{Rust}, a core calculus for Rust by Jung et al. [16].

```
let x = 0 in           // x = 0
let y = (x, _) in      // x = 0, y = ⟨0, _⟩
let x = snd y in       // x = _, y = ⟨0, _⟩
let y = (1, snd y) in  // x = _, y = ⟨1, _⟩
assume(fst y = snd y); // x = 1, y = ⟨1, 1⟩
assert(x = 1)  // ok
```

Fig. 3. A pure functional program equivalent to Fig. 2.

In this paper, to overcome the limitations and take the best of both approaches, we introduce *borrowable fractional ownership types*. The type of a reference is now of the form $\mathbf{int\,ref}\,_B^{\alpha,r}$, where α is the lifetime of the reference, r is the (fractional) ownership of the reference ranging over $[0,1]$, and B is the borrowing information of the reference. A borrowing B is either a pair (β, s) or $\emptyset$, where β is the lifetime for which the reference lends the ownership and s is the amount of (fractional) ownership others borrow from the reference. Our type system extends those of both CONSORT and Rust; for the sake of simplicity, however, we consider only integer references, excluding nested reference types like $\mathbf{int\,ref}\,_{B_1}^{\alpha,r}\,\mathbf{ref}\,_{B_2}^{\beta,s}$. (We briefly discuss this in Sect. 7). The reference types of CONSORT can be considered a particular case where B is always $\emptyset$, and all the lifetimes are identical.[2] And those of Rust can be considered a particular case where the ownership r is always 0 or 1.[3] Figure 4 shows an example program that can be handled by neither CONSORT nor RUSTHORN but by our new type system. (We will show how the program is typed in our type system in Sect. 3). Together with the type system, we develop a verification method to prove the absence of assertion failures by a type-directed translation from imperative programs with mutable references into stateless functional programs without references, in a similar way to RUSTHORN.

```
minmax(x, y) { if *x < *y then (x, y) else (y, x) }
rand_choose(x, y) { if _ then x else y }
let x = mkref _ in  let y = mkref _ in
let (p, q) = minmax(x, y) in
let z = rand_choose(x, y) in
assert( *p <= *z && *z <= *q );
z := 1
```

Fig. 4. Example program that can be handled by neither CONSORT nor RUSTHORN but by our method.

---

[2] Unlike CONSORT, we do not have refinement predicates because we follow the translation-based approach of RUSTHORN for verification.

[3] This is not the case for shared (a.k.a. immutable) references in Rust, but our type system also subsumes these references.

Our main contributions are:

1. Proposal of the new notion of borrowable fractional ownership.
2. Formalization of a type system with borrowable fractional ownership and a proof of the preservation of types and ownership invariants.
3. Type-directed translation from imperative programs to functional programs without references, and proof of its soundness: if the target program has no assertion failures, neither does the source program.

The rest of this paper is organized as follows. Section 2 introduces the source language. Section 3 formalizes the type system with borrowable fractional ownership. Section 4 shows the reduction to stateless functional programs. Section 5 reports preliminary experimental results. Section 6 discusses related work and Sect. 7 concludes the paper and discusses future work.

2   Source Language

This section introduces a simple imperative language with mutable references and recursive functions with lifetime polymorphism, which serves as the target of our type-based verification method.

2.1   Syntax

The syntax of the language is as follows:

$$o \text{ (arithmetic expressions)} ::= n \mid x \mid o_1 \, op \, o_2$$
$$e \text{ (expressions)} ::= x \mid \texttt{let } x = o \texttt{ in } e \mid \texttt{let } x = y \texttt{ in } e$$
$$\mid \texttt{let } x = \texttt{mkref } y \texttt{ in } e \mid \texttt{let } x = \star y \texttt{ in } e \mid x := y; e$$
$$\mid \texttt{ifz } x \texttt{ then } e_1 \texttt{ else } e_2 \mid \texttt{let } x = f\langle\overrightarrow{\alpha}\rangle(y_1,\ldots,y_n) \texttt{ in } e$$
$$\mid \texttt{alias}\,(x = y); e \mid \texttt{newlft } \alpha \texttt{ in } e \mid \texttt{endlft } \alpha; e \mid \texttt{fail}$$
$$d \text{ (function definitions)} ::= f \mapsto \langle\overrightarrow{\alpha}\rangle\,(x_1,\ldots,x_n)e$$
$$P \text{ (programs)} ::= \langle\{d_1,\ldots,d_n\}, e\rangle$$

The meta-variables $x, y, \ldots$ range over the set **Var** of program variables, f ranges over the set of function names, and $\alpha, \beta \in$ **LftVar** range over a set of symbols called *lifetime variables*.

Program variables are introduced by function parameters or let bindings, and the lifetime variables are bound by $\texttt{newlft}$. We assume that all program and lifetime variables are alpha-renamed as necessary, so that each variable is unique for each binder.

For simplicity, an expression e is restricted to a de-nested, simplified form. An arithmetic expression o consists of integer constants, integer variables, and arithmetic binary operations $(+, -, \text{etc.})$ denoted by op. The expression $\texttt{let } x = y \texttt{ in } e$ creates a new alias x of y and evaluates e. We have primitives $\texttt{mkref } y$ for creating a new reference, $\star y$ for reading from a reference, and $x := y$ for updating a reference. The conditional branch $\texttt{ifz } x \texttt{ then } e_1 \texttt{ else } e_2$ evaluates e_1 if the value of z is 0, and evaluates e_2 otherwise. A function call $f\langle\overrightarrow{\alpha}\rangle(y_1,\dots,y_n)$ takes lifetime arguments $\overrightarrow{\alpha}$ along with program arguments $y_1,\dots,y_n$. The expression `fail` aborts program execution with an error.

An *alias assumption* `alias` $(x = y)$ has been inherited from CONSORT [31]; it assumes that x and y are aliasing references. This information is exploited by our type system described in Sect. 3. Moreover, we inherit `newlft` α and `endlft` α, which respectively introduces and eliminates a lifetime α, from λ_{Rust} [16]. They play essential roles in our type system but do not affect program execution.

A function declaration d consists of the function name f, the parameter variables $x_1,\dots,x_n$, and the function body e. We allow mutual recursion between functions. A program P is a pair $\langle D, e\rangle$ where $D = \{d_1,\dots,d_n\}$ is a set of function definitions and e is the main expression.

Henceforth, we write programs in an abbreviated, sugared style to avoid unnecessary complications. For example, the program in Fig. 1 is a valid abbreviated program in our source language. Here, the expression `assert` $(e_1 = e_2)$ is syntactic sugar of $(\texttt{let } c = e_1 - e_2 \texttt{ in ifz } c \texttt{ then } 0 \texttt{ else fail})$.

2.2 Operational Semantics

We now introduce the operational semantics of our source language. Let **Addr** be a countable set of heap addresses. We define a set of runtime values $\mathbf{Val}_{src}$ as $\mathbb{Z} \cup \mathbf{Addr}$.

A configuration (runtime state) of this language is a quadruple $\langle H, R, \overrightarrow{F}, e\rangle$, consisting of a heap, a register, a call stack, and a currently reducing expression. A heap H is a partial function from **Addr** to $\mathbf{Val}_{src}$. A register R is a partial mapping from **Var** to $\mathbf{Val}_{src}$. A call stack $\overrightarrow{F}$ is a sequence of return contexts of the form $(\texttt{let } x = [] \texttt{ in } e)$. A program $\langle D, e\rangle$ is executed by repeatedly stepping from the initial configuration $\langle\emptyset,\emptyset,\cdot,e\rangle$ according to the one-step reduction relation $\longrightarrow_D$, which is defined by the rules in Fig. 5. We write $H\{a \mapsto v\}$ for the heap that maps the address a to v and behaves as H for the other addresses. We also use a similar notation $R\{x \mapsto v\}$ for registers. In (RS-ARITH), we write $[\![o]\!]_R$ for the computed integer value of o, where variables are interpreted according to the register R. By (RS-FAIL), we fall into the 'hard' failure state **Fail** out of `fail`. Also, by (RS-ALIASFAIL), in case the alias assumption `alias` $(x = y)$ is not satisfied, we fall into **AliasFail**, a 'soft' failure state distinct from **Fail**.

The goal of our verification method is to check that a given program does not reach **Fail**. To that end, we introduce a type system based on the new notion of borrowable fractional ownership types in the next section and use it to reduce the verification problem to that for functional programs in Sect. 4.

3 Type System

In this section, we introduce our borrowable fractional ownership type system for the source language. This type system will be used for the translation described later in Sect. 4 to verify programs.

$$\dfrac{}{\left\langle H, R, (\texttt{let } x' = [] \texttt{ in } e) : \vec{F}, x \right\rangle \longrightarrow_D \left\langle H, R, \vec{F}, [x/x']e \right\rangle} \text{ (Rs-Var)}$$

$$\dfrac{[\![o]\!]_R = n \qquad x' \notin \mathrm{dom}(R)}{\left\langle H, R, \vec{F}, \texttt{let } x = o \texttt{ in } e \right\rangle \longrightarrow_D \left\langle H, R\{x' \mapsto n\}, \vec{F}, [x'/x]e \right\rangle} \text{ (Rs-Arith)}$$

$$\dfrac{x' \notin \mathrm{dom}(R)}{\left\langle H, R, \vec{F}, \texttt{let } x = y \texttt{ in } e \right\rangle \longrightarrow_D \left\langle H, R\{x' \mapsto R(y)\}, \vec{F}, [x'/x]e \right\rangle} \text{ (Rs-Let)}$$

$$\dfrac{f \mapsto \langle \vec{\alpha} \rangle (x_1, \dots, x_n)e \in D \qquad x' \notin \mathrm{dom}(R)}{\begin{array}{l} \left\langle H, R, \vec{F}, \texttt{let } x = f\langle \vec{\alpha} \rangle (y_1, \dots, y_n) \texttt{ in } e' \right\rangle \\ \longrightarrow_D \left\langle H, R, (\texttt{let } x' = [] \texttt{ in } [x'/x]e') : \vec{F}, [y_1/x_1] \cdots [y_n/x_n]e \right\rangle \end{array}} \text{ (Rs-Call)}$$

$$\dfrac{a \notin \mathrm{dom}(H) \qquad x' \notin \mathrm{dom}(R)}{\begin{array}{l} \left\langle H, R, \vec{F}, \texttt{let } x = \texttt{mkref } y \texttt{ in } e \right\rangle \\ \longrightarrow_D \left\langle H\{a \mapsto R(y)\}, R\{x' \mapsto a\}, \vec{F}, [x'/x]e \right\rangle \end{array}} \text{ (Rs-MkRef)}$$

$$\dfrac{R(y) = a \qquad H(a) = v \qquad x' \notin \mathrm{dom}(R)}{\left\langle H, R, \vec{F}, \texttt{let } x = \star y \texttt{ in } e \right\rangle \longrightarrow_D \left\langle H, R\{x' \mapsto v\}, \vec{F}, [x'/x]e \right\rangle} \text{ (Rs-Deref)}$$

$$\dfrac{R(x) = a \qquad a \in \mathrm{dom}(H)}{\left\langle H, R, \vec{F}, x := y; e \right\rangle \longrightarrow_D \left\langle H\{a \mapsto R(y)\}, R, \vec{F}, e \right\rangle} \text{ (Rs-Assign)}$$

$$\dfrac{R(x) = 0}{\left\langle H, R, \vec{F}, \texttt{ifz } x \texttt{ then } e_1 \texttt{ else } e_2 \right\rangle \longrightarrow_D \left\langle H, R, \vec{F}, e_1 \right\rangle} \text{ (Rs-IfTrue)}$$

$$\dfrac{R(x) \neq 0}{\left\langle H, R, \vec{F}, \texttt{ifz } x \texttt{ then } e_1 \texttt{ else } e_2 \right\rangle \longrightarrow_D \left\langle H, R, \vec{F}, e_2 \right\rangle} \text{ (Rs-IfFalse)}$$

$$\dfrac{R(x) = R(y)}{\left\langle H, R, \vec{F}, \texttt{alias } (x = y); e \right\rangle \longrightarrow_D \left\langle H, R, \vec{F}, e \right\rangle} \text{ (Rs-Alias)}$$

$$\dfrac{R(x) \neq R(y)}{\left\langle H, R, \vec{F}, \texttt{alias } (x = y); e \right\rangle \longrightarrow_D \textbf{AliasFail}} \text{ (Rs-AliasFail)}$$

$$\dfrac{}{\left\langle H, R, \vec{F}, \texttt{newlft } \alpha \texttt{ in } e \right\rangle \longrightarrow_D \left\langle H, R, \vec{F}, e \right\rangle} \text{ (Rs-Newlft)}$$

$$\dfrac{}{\left\langle H, R, \vec{F}, \texttt{endlft } \alpha; e \right\rangle \longrightarrow_D \left\langle H, R, \vec{F}, e \right\rangle} \text{ (Rs-Endlft)}$$

$$\dfrac{}{\left\langle H, R, \vec{F}, \texttt{fail} \right\rangle \longrightarrow_D \textbf{Fail}} \text{ (Rs-Fail)}$$

Fig. 5. Operational semantics of the source language.

3.1 Syntax of Types

The syntax of types is given as follows:

$$\tau \ (\text{types}) ::= \textbf{int} \mid \textbf{int ref}_B^{\alpha,r}$$
$$r \ (\text{ownership}) \in [0,1]$$
$$B \ (\text{borrowings}) ::= \emptyset \mid (\alpha, r)$$
$$\sigma \ (\text{function types}) ::= \forall \overrightarrow{\alpha} : \mathcal{L}. \ \langle \tau_1, \ldots, \tau_n \rangle \rightarrow \langle \tau_1', \ldots, \tau_n' \mid \tau \rangle$$
$$\mathcal{L} \ (\text{lifetime environment}) ::= \emptyset \mid \mathcal{L}, \alpha_1 \prec \alpha_2$$
$$\Gamma \ (\text{type environment}) ::= \emptyset \mid \Gamma, x : \tau$$

Our type system has the integer type **int** and the reference type of integer **int ref**$_B^{\alpha,r}$, where α, r indicates that the reference has the lifetime α and the fractional ownership r. The *borrowing* B specifies how much ownership is borrowed by variables during which lifetime. If $B = \emptyset$, no borrowing for other lifetime occurred. Otherwise, $B = (\beta, s)$, which indicates that the reference is lending ownership s to variables with lifetime β. For example, **int ref**$_{\beta,0.5}^{\alpha,0.5}$ is a reference with the lifetime α that has the ownership of 0.5 and lends the ownership of 0.5 to other variables during the lifetime β.

A *lifetime environment* $\mathcal{L}$ is a strict partial order on the set of valid lifetime variables, each element of which is denoted $\alpha \prec \beta$. We impose the constraint that when a reference type **int ref**$_{\beta,s}^{\alpha,r}$ exists, the lifetimes α and β are ordered $\beta \prec \alpha$ under the lifetime environment $\mathcal{L}$ at the point. A *type environment* Γ is a finite set of type bindings of the form $x : \tau$.

A *function type* σ takes the form $\forall \overrightarrow{\alpha} : \mathcal{L}. \ \langle \tau_1, \ldots, \tau_n \rangle \rightarrow \langle \tau_1', \ldots, \tau_n' \mid \tau \rangle$. This indicates that the i-th argument has the type τ_i before the function call and changes its type to τ_i' when the function returns. In addition, the function type has a direct return type τ. The function types are parameterized over lifetime variables $\overrightarrow{\alpha}$ with an ordering $\mathcal{L}$, and the lifetime variables appearing in τ, τ_i, τ_i' must be included in $\overrightarrow{\alpha}$.

Notations. Hereafter, we will identify the borrowing of $(\alpha,0)$ with $\emptyset$ and denote **int ref**$_\emptyset^{\alpha,r}$ by **int ref**$^{\alpha,r}$. Given a reference type $\tau = $ **int ref**$_B^{\alpha,r}$, we define $\textbf{Lft}(\tau) \overset{\text{def}}{:=} \alpha$ and $\textbf{Own}(\tau) \overset{\text{def}}{:=} r$; $\textbf{Lft}(\textbf{int})$ is undefined and we define $\textbf{Own}(\textbf{int}) \overset{\text{def}}{:=} 0$ for technical convenience. For $x : \tau \in \Gamma$, we write $\textbf{Lft}_\Gamma(x)$ and $\textbf{Own}_\Gamma(x)$ for $\textbf{Lft}(\tau)$ and $\textbf{Own}(\tau)$, respectively.

3.2 Typing Rules

The typing rules for expressions are defined in Figs. 6, 8 and 9. A type judgment for expressions is of the form $\Theta \mid \mathcal{L} \mid \Gamma \vdash e : \tau \rhd \mathcal{L}' \mid \Gamma'$ where Θ, called a *function type environment*, is a map from function names, ranging over by f, to function types. The judgment indicates that an expression e is well-typed with a type τ under the environments Θ, $\mathcal{L}$, and Γ, and further that the lifetime and type environments change to $\mathcal{L}'$ and Γ' after evaluating e.

$$\Theta \mid \mathcal{L} \mid \Gamma + \Gamma' + x{:}\tau \vdash x{:}\tau \triangleright \mathcal{L} \mid \Gamma' \quad \text{(T-VAR)} \qquad \Theta \mid \mathcal{L} \mid \Gamma \vdash \mathtt{fail}{:}\tau \triangleright \mathcal{L}' \mid \Gamma' \quad \text{(T-FAIL)}$$

$$\frac{\tau = \tau_x + \tau_y \qquad x \notin \mathrm{dom}(\Gamma') \qquad \Theta \mid \mathcal{L} \mid \Gamma, x : \tau_x, y : \tau_y \vdash e : \rho \triangleright \mathcal{L}' \mid \Gamma'}{\Theta \mid \mathcal{L} \mid \Gamma, y : \tau \vdash \mathtt{let}\ x = y\ \mathtt{in}\ e : \rho \triangleright \mathcal{L}' \mid \Gamma'} \quad \text{(T-LET)}$$

$$\frac{\Gamma \vdash o : \mathbf{int} \qquad \Theta \mid \mathcal{L} \mid \Gamma, x : \mathbf{int} \vdash e : \rho \triangleright \mathcal{L}' \mid \Gamma' \qquad x \notin \mathrm{dom}(\Gamma')}{\Theta \mid \mathcal{L} \mid \Gamma \vdash \mathtt{let}\ x = o\ \mathtt{in}\ e : \rho \triangleright \mathcal{L}' \mid \Gamma'} \quad \text{(T-ARITH)}$$

$$\frac{\Theta \mid \mathcal{L} \mid \Gamma, x : \mathbf{int} \vdash e_1 : \rho \triangleright \mathcal{L}' \mid \Gamma' \qquad \Theta \mid \mathcal{L} \mid \Gamma, x : \mathbf{int} \vdash e_2 : \rho \triangleright \mathcal{L}' \mid \Gamma'}{\Theta \mid \mathcal{L} \mid \Gamma, x : \mathbf{int} \vdash \mathtt{ifz}\ x\ \mathtt{then}\ e_1\ \mathtt{else}\ e_2 : \rho \triangleright \mathcal{L}' \mid \Gamma'} \quad \text{(T-IF)}$$

$$\frac{\begin{array}{c} \rho = [\vec{\beta}/\vec{\alpha}]\tau, \rho_i = [\vec{\beta}/\vec{\alpha}]\tau_i, \rho_i' = [\vec{\beta}/\vec{\alpha}]\tau_i' \qquad x \notin \mathrm{dom}(\Gamma') \\ \Theta(f) = \forall \vec{\alpha} : \mathcal{M}. \langle \tau_1, \ldots, \tau_n \rangle \to \langle \tau_1', \ldots, \tau_n' \mid \tau \rangle \qquad [\vec{\beta}/\vec{\alpha}]\mathcal{M} \subseteq \mathcal{L} \\ \Theta \mid \mathcal{L} \mid \Gamma, x : \rho, y_1 : \rho_1', \ldots, y_n : \rho_n' \vdash e : \xi \triangleright \mathcal{L}' \mid \Gamma' \end{array}}{\Theta \mid \mathcal{L} \mid \Gamma, y_1 : \rho_1, \ldots, y_n : \rho_n \vdash \mathtt{let}\ x = f\langle \vec{\beta} \rangle(y_1, \ldots, y_n)\ \mathtt{in}\ e : \xi \triangleright \mathcal{L}' \mid \Gamma'} \quad \text{(T-CALL)}$$

Fig. 6. Typing rules for standard expressions.

We tacitly assume that every type judgment is *well-formed*. A type judgment $\Theta \mid \mathcal{L} \mid \Gamma \vdash e : \tau \triangleright \mathcal{L}' \mid \Gamma'$ is well-formed if $\mathcal{L} \vdash_{\mathrm{WF}} \Gamma$, $\mathcal{L}' \vdash_{\mathrm{WF}} \Gamma'$, and $\mathcal{L} \vdash_{\mathrm{WF}} \tau$, which mean the type τ or every type used in Γ, Γ' should satisfy the order of the corresponding lifetime environment, $\mathcal{L}$ or $\mathcal{L}'$, and every reference type $\mathbf{int\,ref}^{\alpha,r}_{\beta,s}$ should satisfy $r + s \leq 1$. For example, if $\beta \prec \alpha \in \mathcal{L}$, then a type $\mathbf{int\,ref}^{\beta,r}_{\alpha,s}$ is invalid because the smaller lifetime β lends ownership to the larger lifetime α. (The necessity of the ordering is discussed in Remark 1).

We explain our type system using (T-LET) as a representative example. The let expression allows an *ownership transfer*. The expression $\mathtt{let}\ x = y\ \mathtt{in}\ e$ is well-typed only when the body e is typed under a type environment where x and y have types τ_x and τ_y that are obtained as split of τ, which is the type y originally had. The split expressed by the type addition $\tau = \tau_x + \tau_y$ (described below) intuitively means that some portion of y's ownership is passed to its new alias x. The condition $x \notin \mathrm{dom}(\Gamma')$ ensures that x does not escape its scope.

Type addition $\tau_1 + \tau_2$ is an essential feature in this type system to handle the ownership transfers. The type addition is defined by the rules in Fig. 7. The rules (A-SHARE) and (A-BORROW) express ownership distribution between two aliasing references but are conceptually different. *Sharing* distributes ownership between references of the same lifetime, whereas *borrowing* is done between references of different lifetimes. For example, $\mathbf{int\,ref}^{\alpha,1} = \mathbf{int\,ref}^{\alpha,0.5}_{\beta,0.5} + \mathbf{int\,ref}^{\beta,0.5}$ means that a reference with lifetime α is lending half of its ownership to a reference with lifetime β, where we have $\alpha \succ \beta$ due to the well-formedness condition.

$$\frac{}{\textbf{int} = \textbf{int} + \textbf{int}} \text{ (A-Int)} \qquad \frac{\tau_1 = \tau_2}{\tau_2 = \tau_1} \text{ (A-Ex)}$$

$$\frac{r_1 + r_2 \leqslant 1 \qquad s_1 + s_2 \leqslant 1}{\textbf{int ref}\,^{\alpha,r_1+r_2}_{\beta,s_1+s_2} = \textbf{int ref}\,^{\alpha,r_1}_{\beta,s_1} + \textbf{int ref}\,^{\alpha,r_2}_{\beta,s_2}} \text{ (A-Share)}$$

$$\frac{}{\textbf{int ref}\,^{\alpha,r+s}_{\varnothing} = \textbf{int ref}\,^{\alpha,r}_{\beta,s} + \textbf{int ref}\,^{\beta,s}_{\varnothing}} \text{ (A-Borrow)}$$

Fig. 7. Type addition rules.

$$\frac{\Theta \mid \mathcal{L} \mid \Gamma, x : \textbf{int ref}\,^{\alpha,1}_{\varnothing}, y : \textbf{int} \vdash e : \rho \rhd \mathcal{L}' \mid \Gamma' \qquad x \notin \text{dom}(\Gamma')}{\Theta \mid \mathcal{L} \mid \Gamma, y : \textbf{int} \vdash \texttt{let } x = \texttt{mkref } y \texttt{ in } e : \rho \rhd \mathcal{L}' \mid \Gamma'} \text{ (T-MkRef)}$$

$$\frac{\Theta \mid \mathcal{L} \mid \Gamma, x : \textbf{int}, y : \textbf{int ref}\,^{\alpha,r}_{B} \vdash e : \rho \rhd \mathcal{L}' \mid \Gamma'}{\Theta \mid \mathcal{L} \mid \Gamma, y : \textbf{int ref}\,^{\alpha,r}_{B} \vdash \texttt{let } x = \star y \texttt{ in } e : \rho \rhd \mathcal{L}' \mid \Gamma'} \text{ (T-Deref)}$$

$$\frac{\Theta \mid \mathcal{L} \mid \Gamma, y : \textbf{int} \vdash e : \rho \rhd \mathcal{L}' \mid \Gamma' \qquad \textbf{Own}_\Gamma(x) = 1}{\Theta \mid \mathcal{L} \mid \Gamma, y : \textbf{int} \vdash x := y; e : \rho \rhd \mathcal{L}' \mid \Gamma'} \text{ (T-Assign)}$$

Fig. 8. Typing rules for reference manipulations.

Type addition is extended to an operation on type environments, written $\Gamma + \Gamma'$, in a pointwise manner.[4]

Since we believe that most of the rules in Fig. 6 are self-explanatory, we only explain a few nontrivial points. In (T-Var), we allow some variables in the initial type environment or part of their ownership to be discarded. This allows us to meet the condition $x' \notin \text{dom}(\Gamma')$ that appears in rules such as (T-Let). The premise $\Gamma \vdash o : \textbf{int}$ in (T-Arith) expresses that each free variable x of an arithmetic expression o has type $\textbf{int}$ in Γ. The condition $[\vec{\beta}/\vec{\alpha}]\mathcal{M} \subseteq \mathcal{L}$ in (T-Call) means that the lifetime variables contained in the arguments' types must follow the function's constraint $\mathcal{M}$ under the lifetime environment $\mathcal{L}$. As already explained, τ_i and τ_i' of a function type represent how the type of the arguments change by calling the function. Types of y_i, therefore, need to match τ_i and τ_i' (up-to substitution) before and after the function call, respectively.

Figure 8 shows the typing rules for reference manipulations. A newly created reference has the full ownership as expressed by $x : \textbf{int ref}\,^{\alpha,1}_{\varnothing}$ in the premise of (T-MkRef). The rule (T-Assign) has the condition $\textbf{Own}(x) = 1$ to ensure that only a reference with full ownership is updated. Dereferencing can be done regardless of the ownership.[5]

The rule (T-Alias) allows us to transfer the ownership of references based on alias information. Rules (T-NewLft) and (T-EndLft) are rules for life-

---

[4] If a variable x only appears in one of the type environment, say $x \in \text{dom}(\Gamma)$ but $x \notin \text{dom}(\Gamma')$, $(\Gamma + \Gamma')(x)$ is given as $\Gamma(x)$.

[5] If we want to disallow dereferencing references of ownership 0 we may add a premise $r > 0$ to (T-Deref); it is a matter of preference.

$$\tau_x + \tau_y = \mathbf{int\,ref}_B^{\alpha,r} \qquad \mathbf{int\,ref}_B^{\alpha,r} = \rho_x + \rho_y$$

$$\frac{\Theta \mid \mathcal{L} \mid \Gamma, x : \rho_x, y : \rho_y \vdash e : \rho \rhd \mathcal{L}' \mid \Gamma'}{\Theta \mid \mathcal{L} \mid \Gamma, x : \tau_x, y : \tau_y \vdash \mathbf{alias}\,(x = y); e : \rho \rhd \mathcal{L}' \mid \Gamma'} \text{ (T-Alias)}$$

$$\frac{\Theta \mid \mathcal{M} \mid \Gamma \vdash e : \rho \rhd \mathcal{L}' \mid \Gamma' \qquad \mathcal{M} = \mathcal{L} \cup \{\alpha\} \qquad \alpha = \min \mathcal{M}}{\Theta \mid \mathcal{L} \mid \Gamma \vdash \mathbf{newlft}\,\alpha\,\mathbf{in}\,e : \rho \rhd \mathcal{L}' \mid \Gamma'} \text{ (T-NewLft)}$$

$$\frac{\Theta \mid \mathcal{L}{\uparrow}_\alpha \mid \Gamma{\uparrow}_\alpha \vdash e : \rho \rhd \mathcal{L}' \mid \Gamma' \qquad \alpha = \min \mathcal{L}}{\Theta \mid \mathcal{L} \mid \Gamma \vdash \mathbf{endlft}\,\alpha; e : \rho \rhd \mathcal{L}' \mid \Gamma'} \text{ (T-EndLft)}$$

Fig. 9. Typing rules for ghost instructions.

time introduction and termination, which do not exist in the type system of CONSORT. The expression $\mathbf{newlft}\,\alpha\,\mathbf{in}\,e$ is typed when e is typed under the lifetime environment $\mathcal{M}$ such that $\mathcal{M} = \mathcal{L} \cup \{\alpha\}$ and $\alpha = \min \mathcal{M}$, which mean that the newly introduced lifetime variable α becomes a minimal lifetime in typing of e. The operator ${\uparrow}_\alpha$ in (T-EndLft) removes all information concerning α from the environment and gives back the ownership that references with lifetime α have been borrowing. Concretely, $\Gamma{\uparrow}_\alpha$ is defined by

$$\mathbf{int}{\uparrow}_\alpha \overset{\text{def}}{:=} \mathbf{int} \qquad \mathbf{int\,ref}_B^{\beta,r}{\uparrow}_\alpha \overset{\text{def}}{:=} \begin{cases} \mathbf{int\,ref}_\emptyset^{\beta,r+s} & (\text{if } B = (\alpha, s)) \\ \mathbf{int\,ref}_B^{\beta,r} & (\text{otherwise}) \end{cases}$$

$$\emptyset{\uparrow}_\alpha \overset{\text{def}}{:=} \emptyset \qquad (\Gamma, x : \tau){\uparrow}_\alpha \overset{\text{def}}{:=} \begin{cases} \Gamma{\uparrow}_\alpha & (\text{if } \alpha = \mathbf{Lft}(\tau)) \\ \Gamma{\uparrow}_\alpha, x : (\tau{\uparrow}_\alpha) & (\text{if } \alpha \neq \mathbf{Lft}(\tau)) \end{cases}$$

whereas $\mathcal{L}{\uparrow}_\alpha$ is the subposet induced by $\mathcal{L} \setminus \{\alpha\}$. The premise $\alpha = \min \mathcal{L}$ says that we can only end a minimal lifetime. This ensures that only a lifetime with which references are not lending their ownership can be ended.

The typing rules for functions and programs are defined in Fig. 10.

Here, the rule (T-FunDef) stipulates the contract that a function cannot have free variables and should end all lifetimes introduced in the function body by requiring $\mathcal{L} \subseteq \vec{\alpha}$ and setting $\mathcal{L}$ to both initial and ending lifetime context. The rule (T-Prog) checks that all the function definitions and the main expression

$$\frac{\begin{array}{c} \Theta(f) = \forall \vec{\alpha} : \mathcal{L}. \langle \tau_1, \ldots, \tau_n \rangle \to \langle \tau_1', \ldots, \tau_n' \mid \tau \rangle \qquad \mathcal{L} \subseteq \vec{\alpha} \\ \Theta \mid \mathcal{L} \mid x_1 : \tau_1, \ldots, x_n : \tau_n \vdash e : \tau \rhd \mathcal{L} \mid x_1 : \tau_1', \ldots, x_n : \tau_n' \end{array}}{\Theta \vdash f \mapsto \langle \vec{\alpha} \rangle (x_1, \ldots, x_n) e} \text{ (T-FunDef)}$$

$$\frac{\begin{array}{c} \Theta \vdash d_1 \qquad \cdots \qquad \Theta \vdash d_n \\ \mathrm{dom}(\Theta) = \mathrm{dom}(\{d_1, \ldots, d_n\}) \qquad \Theta \mid \emptyset \mid \emptyset \vdash e : \tau \rhd \emptyset \mid \emptyset \end{array}}{\vdash \langle \{d_1, \ldots, d_n\}, e \rangle} \text{ (T-Prog)}$$

Fig. 10. Typing rules for functions and programs.

are well-typed. It also checks that the main expression does not have any free (lifetime) variables.

Figures 11 and 12 show the examples in Fig. 1 and Fig. 2 typed under our type system, respectively. The typings are mostly the same as those of CONSORT and RUSTHORN. The difference between Fig. 2 and Fig. 12 is that the reference x is not simply invalidated by the type system like in Rust but given a type of $x : \mathsf{int\,ref}\,_{\beta,1}^{\alpha,0}$. This demonstrates that our type system can express the lending references as a type, which is not possible in Rust. We will utilize this difference in a later example. Additionally, it is noteworthy that the typing of Fig. 11 is obtained just by giving the identical lifetime to both references. (Recall that CONSORT can be seen as a specific case of our type system).

```
let x = mkref 0 in
// x : int ref α,1
let y = x in
// x : int ref α,0,  y : int ref α,1
y := 1;
alias(x = y);
// x : int ref α,0.5,  y : int ref α,0.5
assert( *x + *y = 2 )
```

Fig. 11. Typed example of Fig. 1.

```
let x = mkref 0 in
// x : int ref α,1
let y = x in
// x : int ref α,0, y : int ref β,1
//     β,1
y := 1;
endlft β;
// x : int ref α,1
assert( *x = 1 )
```

Fig. 12. Typed example of Fig. 2.

Another example of typing is given in Fig. 13. This example is the program in Fig. 4 typed under our type system (extended with pair types). This program cannot be typed under CONSORT nor RUSTHORN (namely, Rust) because CONSORT cannot insert **alias** statements after the use of z since z is a dynamic alias of either x or y, and in Rust, we cannot touch x or y after the creation of p and q. Our type system can type this program by employing borrowing information and expressing partially lending references as a type of $\mathsf{int\,ref}\,_{\beta,0.5}^{\alpha,0.5}$.

3.3   Type Preservation

As usual, reduction preserves well-typedness, and, in our type system, well-typedness ensures the ownership invariant of fractional types:

Theorem 1. *Suppose that* $\vdash \langle D, e \rangle$ *and* $\langle \emptyset, \emptyset, \cdot, e \rangle \rightarrow_D \langle H, R, \overrightarrow{F}, e' \rangle$. *Then we have* $\Theta \mid \mathcal{L} \mid \Gamma \vdash e' : \tau \rhd \mathcal{L}' \mid \Gamma'$ *for some* $\Theta, \Gamma, \Gamma', \mathcal{L}, \mathcal{L}'$ *and* τ. *Moreover, for any* $a \in \mathrm{dom}(H)$, *we have* $\sum_{x \in \mathbf{Var}; R(x)=a} \mathbf{Own}_\Gamma(x) \le 1$.   □

The proof of this theorem is given in the full version [25]. In the theorem above, $\Theta, \mathcal{L}', \Gamma'$ and τ are fixed, whereas $\mathcal{L}$ and Γ differ at each step of the execution.[6]

---

[6] Strictly speaking, $\mathcal{L}', \Gamma'$ and τ also changes when a function is called.

```
// Θ(minmax) = ∀α, β : β ≺ α. ⟨int ref α,1, int ref α,1⟩
//            → ⟨int ref α,0.5 β,0.5, int ref α,0.5 β,0.5 | ⟨int ref β,0.5, int ref β,0.5⟩⟩
minmax(x, y) { if *x < *y then (x, y) else (y, x) }
// Θ(rand_choose) = ∀α, β : β ≺ α. ⟨int ref α,0.5 β,0.5, int ref α,0.5 β,0.5⟩
//                → ⟨int ref α,0, int ref α,0 | int ref α,0.5 β,0.5⟩
rand_choose(x, y) { if _ then x else y }
let x = mkref _ in          // x : int ref α,1
let y = mkref _ in          // y : int ref α,1
let (p, q) = minmax(x, y) in
// x : int ref α,0.5 β,0.5, y : int ref α,0.5 β,0.5, p : int ref β,0.5, q : int ref β,0.5
let z = rand_choose(x, y) in // x : int ref α,0, y : int ref α,0, z : int ref α,0.5 β,0.5
assert( *p <= *z && *z <= *q );
endlft β;                   // dispose p, q;  z : int ref α,1
z := 1
```

Fig. 13. Typed example of Fig. 4.

This is because, along the execution, we may introduce or terminate lifetime variables or change ownership of variables. What is important, however, is that Γ always gives an assignment of ownership so that the sum of the ownerships of references pointing to an address a does not exceed 1. This is a key property we exploit to show the soundness of the translation introduced in the next section.

Remark 1. We introduced the lifetime ordering because Theorem 1 does not hold if we have cyclic borrows. Figure 14 shows an example of such programs. □

```
let x = mkref 0 in // x : int ref α,1
let y = x in        // x : int ref α,0 β,1,  y : int ref β,1
let z = y in        // y : int ref β,0 α,1,  z : int ref α,1
endlft β            // x : int ref α,1,  z : int ref α,1
// ownership sum of x and z becomes 2
```

Fig. 14. An example where cyclic borrows occur.

4 Translation

This section introduces our translation from the source language into a standard functional language with non-deterministic assignments but without references. Our translation is mainly based on RUSTHORN [22]. We convert all references to a pair of current and prophecy values as described in Sect. 1 and exploit the prophecy technique when borrowings occur. Employing this translation, we can verify that a well-typed program in the source language will not fail by checking the safety of the translated program using existing verifiers that can efficiently verify functional programs such as WHY3 [10] or MOCHI [27].

We first define our target language, then formulate translation rules, and finally state its soundness.

4.1 Target Language

In this subsection, we introduce the target language of our translation.

Syntax. The syntax of the target language is as follows:

$$o \text{ (arithmetic terms)} ::= n \mid x \mid o_1 \, op \, o_2$$
$$t \text{ (terms)} ::= x \mid \text{let } x = o \text{ in } t \mid \text{let } x = y \text{ in } t \mid \text{let } x = f(y_1, \dots, y_n) \text{ in } t$$
$$\mid \text{let } x = \langle y, z \rangle \text{ in } t \mid \text{let } x = \text{fst } y \text{ in } t \mid \text{let } x = \text{snd } y \text{ in } t$$
$$\mid \text{let } x = \_ \text{ in } t \mid \text{assume} (x = y); t \mid \text{ifz } x \text{ then } t_1 \text{ else } t_2 \mid \text{fail}$$
$$d \text{ (function definitions)} ::= f \mapsto (x_1, \dots, x_n)t$$
$$P \text{ (programs)} ::= \langle \{d_1, \dots, d_n\}, t \rangle$$

We only describe the differences from the source language. First, this language uses a *non-deterministic value* $\_$, whose exact value is not determined until a corresponding **assume** instruction appears. For example, in Fig. 15, x and y can be any integer on their initialization. But after running the third and fourth lines, we have $x = 3$ and $y = 3$. Once the value of a non-deterministic value gets fixed, we can no longer change it. So if we add **assume** $(x = 5)$ at the end of this program, the program execution raises an error. Second, this language is free of references (or lifetimes). We no longer handle the heap memory in the operational semantics (described later), and thus, the verification problem for the target language is much more tractable than that for the source language. Third, the target language has pairs, created by $\langle x, y \rangle$ and decomposed by **fst** and **snd**. Pairs will be used to model references using prophecy variables.

Operational Semantics. We now introduce the operational semantics for our target language. The set of runtime values $\mathbf{Val}_{tgt}$ is defined recursively as follows:

$$\mathbf{Val}_{tgt} \ni v ::= n \mid \langle v_1, v_2 \rangle$$

```
let x = _ in  let y = _ in
assume(x = y);  assume(y = 3);
// 'assume(x = 5)' here causes an error
```

Fig. 15. Example demonstrating non-determinism.

A configuration (runtime state) of this language has the form $\langle \mathcal{D}, \mathcal{S}, \vec{\mathcal{F}}, t \rangle$, consisting of function definitions $\mathcal{D}$, a *set* of registers $\mathcal{S}$, a call stack $\vec{\mathcal{F}}$, and the currently reducing term t. The main difference to the source language is that a configuration has a set of registers $\mathcal{S}$ to handle the non-determinism. Each register $S \in \mathcal{S}$ is a partial mapping from **Var** to $\mathbf{Val}_{tgt}$. Important rules of the operational semantics are given by the rules in Fig. 16; the rules for other constructs are standard and are defined similarly to those for the source language. (See the full version [25] for the complete definition).

$$\frac{\mathcal{S}' = \{S\{x \mapsto S(y)\} \mid S \in \mathcal{S}\}}{\langle \mathcal{D}, \mathcal{S}, \vec{\mathcal{F}}, \textbf{let } x = y \textbf{ in } t \rangle \longrightarrow \langle \mathcal{D}, \mathcal{S}', \vec{\mathcal{F}}, t \rangle} \text{ (Rt-Let)}$$

$$\frac{\mathcal{S}' = \{S\{x \mapsto n\} \mid S \in \mathcal{S}, n \in \mathbb{Z}\}}{\langle \mathcal{D}, \mathcal{S}, \vec{\mathcal{F}}, \textbf{let } x = \_ \textbf{ in } t \rangle \longrightarrow \langle \mathcal{D}, \mathcal{S}', \vec{\mathcal{F}}, t \rangle} \text{ (Rt-LetNondet)}$$

$$\frac{\mathcal{S}' = \{S \in \mathcal{S} \mid S(x) = S(y)\}}{\langle \mathcal{D}, \mathcal{S}, \vec{\mathcal{F}}, \textbf{assume } (x = y); t \rangle \longrightarrow \langle \mathcal{D}, \mathcal{S}', \vec{\mathcal{F}}, t \rangle} \text{ (Rt-Assume)}$$

Fig. 16. Operational semantics of the target language (excerpt).

4.2   Translation Rules

Here, we define our type-directed translation. Figure 17 shows some notable translation rules; the translation is almost homomorphic for the other constructs (see the full version [25] for details). The translation relation is of the form $\mathcal{L} \mid \Gamma \vdash e{:}\tau \Rightarrow t$, meaning that e is translated to t under the type environment Γ. Note that if we ignore the "$\Rightarrow t$" part, the rules are essentially the same as the typing rules given in Sect. 3. We omitted the function type environment and the resulting environments "$\triangleright \mathcal{L}' \mid \Gamma'$" for simplicity. These environments are not needed to define the translation for the intraprocedual fragment of the source language. To ease the presentation, we focus on this fragment in the rest of Sect. 4.

As briefly explained, a reference x is translated into a pair $\langle x, x_o \rangle$, where the first and the second elements (are supposed to) represent the current value of x and the *future (or prophesized) value* of x, respectively. We treat the future value as if x has the value stored in x when x's lifetime ends. To help readers understand this idea, let us look at the rule (C-MkRef). A new reference created by **mkref** y is translated to a pair $\langle y, \_ \rangle$. The first element is y because the current value stored in the reference is the value y is bound to. We set a non-deterministic value for the second element because, at the time of creation, we do not know what the value stored in this reference will be when its lifetime

$$\frac{\mathcal{L} \mid \Gamma, x : \text{int ref}_\varnothing^{\alpha,1}, y : \text{int} \vdash e : \rho \Rightarrow t}{\mathcal{L} \mid \Gamma, y : \text{int} \vdash \text{let } x = \text{mkref } y \text{ in } e : \rho \Rightarrow \text{let } x = \langle y, \_\rangle \text{ in } t} \quad \text{(C-MkRef)}$$

$$\frac{\mathcal{L} \mid \Gamma, x : \text{int}, y : \text{int ref}_B^{\alpha,r} \vdash e : \rho \Rightarrow t \qquad r > 0}{\mathcal{L} \mid \Gamma, y : \text{int ref}_B^{\alpha,r} \vdash \text{let } x = \star y \text{ in } e : \rho \Rightarrow \text{let } x = \text{fst } y \text{ in } t} \quad \text{(C-Deref-Pos)}$$

$$\frac{\mathcal{L} \mid \Gamma, x : \text{int}, y : \text{int ref}_B^{\alpha,0} \vdash e : \rho \Rightarrow t}{\mathcal{L} \mid \Gamma, y : \text{int ref}_B^{\alpha,0} \vdash \text{let } x = \star y \text{ in } e : \rho \Rightarrow \text{let } x = \_ \text{ in } t} \quad \text{(C-Deref-Zero)}$$

$$\frac{\mathcal{L} \mid \Gamma \vdash e : \rho \Rightarrow t \qquad \text{Own}_\Gamma(x) = 1}{\mathcal{L} \mid \Gamma \vdash x := y; e : \rho \Rightarrow \text{let } x = \langle y, \text{snd } x\rangle \text{ in } t} \quad \text{(C-Assign)}$$

$$\frac{\begin{array}{c} \tau_x + \tau_y = \text{int ref}_B^{\alpha,r} \quad \text{int ref}_B^{\alpha,r} = \rho_x + \rho_y \\ \mathcal{L} \mid \Gamma, x : \rho_x, y : \rho_y \vdash e : \rho \Rightarrow t \quad t' = \mathfrak{T}_{\tau_x \to \rho_y}^{\tau_x \to \rho_x}(x,y,t) \end{array}}{\mathcal{L} \mid \Gamma, x : \tau_x, y : \tau_y \vdash \text{alias}(x = y); e : \rho \Rightarrow t'} \quad \text{(C-Alias)}$$

$$\frac{\mathcal{L}{\uparrow}_\alpha \mid \Gamma{\uparrow}_\alpha \vdash e : \rho \Rightarrow t \qquad \alpha = \min(\mathcal{L}) \qquad \Gamma \backslash \Gamma{\uparrow}_\alpha = \{x_1, \ldots, x_n\}}{\mathcal{L} \mid \Gamma \vdash \text{endlft } \alpha; e : \rho \Rightarrow \text{assume}_{1 \leqslant i \leqslant n}(\text{fst } x_i = \text{snd } x_i); t} \quad \text{(C-EndLft)}$$

Fig. 17. Translation rules (excerpt).

ends. In other words, we randomly *guess* the future value. We *check* if the guess was correct by confirming whether the first and second elements of the pair coincide when we drop this reference. These checks are conducted by inserting $\text{assume}(\text{fst } x_i = \text{snd } x_i)$ that is obtained by translating $\text{endlft } \alpha$ as expressed in the rule (C-EndLft). Here, x_i is a variable with the lifetime α.

Our translation obeys the following two principles:

- *Trustable XOR Zero (TXZ)*: If a reference x has non-zero ownership, then the first element of the translation of x indeed holds the current value.
- *Prophecy for Zero (PFZ)*: The translation of a reference with zero ownership holds a prophesized value of a borrowing reference.

TXZ resembles CONSORT's idea that only a type with non-zero ownership can have a non-trivial refinement predicate. The fact that we do not trust a reference with zero ownership is reflected in the rule (C-Deref-Zero), where dereferencing is translated into an assignment of a non-deterministic value. TXZ enables us to express an update of a reference with full ownership as an update of the first element of the pair as in (C-Assign). (Note that, by Theorem 1, aliases of a reference with ownership 1 must have ownership 0).

PFZ is the core of the RUSTHORN-style prophecy technique. A reference of ownership 0 can regain some ownership in two ways: by retrieving the lending ownership through endlft or by ownership transfer through alias annotations. PFZ handles the former case. (How the latter is handled will be explained below). Suppose a reference x is lending its ownership to y. PFZ says that $\text{fst } x = \text{snd } y$ in the translated program. When the lifetime of y ends, in the translated program, the current value of y must equal the prophesized value because of assume $(\mathbf{fst}\ y = \mathbf{snd}\ y)$. Hence, at this moment, we have $\mathbf{fst}\ x = \mathbf{snd}\ y = \mathbf{fst}\ y$, which means that x also holds the correct value. Therefore, TXZ is not violated when the ownership of x becomes non-zero.

We now explain the rule (C-ALIAS), which is the key rule to understand how our translation deals with ownership transfers. The rule (C-ALIAS) uses an auxiliary function $\mathfrak{T}^{\tau_x \to \rho_x}_{\tau_y \to \rho_y}(x, y, t)$ defined as follows:

$$
\mathfrak{T}^{\tau_x \to \rho_x}_{\tau_y \to \rho_y}(x, y, t) \overset{\text{def}}{:=}
\begin{cases}
t & (\mathbf{Own}(\tau_x) = \mathbf{Own}(\rho_x) = 0) \\
\mathbf{let}\ x = \langle \mathbf{fst}\ y, \mathbf{snd}\ x \rangle\ \mathbf{in}\ t & \left(\begin{matrix} \mathbf{Own}(\tau_y) \\ \mathbf{Own}(\rho_x), \mathbf{Own}(\rho_y) \end{matrix} > 0 \right) \\
\begin{matrix} \mathbf{let}\ y = \langle \mathbf{fst}\ x, \mathbf{snd}\ y \rangle\ \mathbf{in} \\ \mathbf{let}\ x = \langle \mathbf{snd}\ y, \mathbf{snd}\ x \rangle\ \mathbf{in} \\ t \end{matrix} & \left(\begin{matrix} \mathbf{Own}(\tau_x) > 0 \\ \text{and} \\ \mathbf{Own}(\rho_x) = 0 \end{matrix} \right) \\
\mathfrak{T}^{\tau_y \to \rho_y}_{\tau_x \to \rho_x}(y, x, t) & \text{(otherwise)}
\end{cases}
$$

The essential case is when $\mathbf{Own}(\tau_x) > 0$ and $\mathbf{Own}(\rho_x) = 0$, that is, when y takes all the ownership of x. In this case, x takes the future value of y to obey PFZ. At the same time, to ensure TXZ, y takes the old value of x so that y has the correct value. (Note that $\mathbf{Own}(\tau_y)$, the previous ownership of y, can be zero). Similarly, when $\mathbf{Own}(\tau_y), \mathbf{Own}(\rho_x), \mathbf{Own}(\rho_y) > 0$, the value y holds is passed to x to ensure TXZ ($\mathbf{Own}(\rho_y) > 0$ is necessary to avoid case overlapping). Figures 18 and 19 show the translated example of Fig. 11 and Fig. 12.

```
let x = (0, _) in
let y = (fst x, _) in
let x = (snd y, snd x) in
let y = (fst y + 1, snd y) in
let x = (fst y, snd x) in
assert(fst x + fst y = 2)
```

Fig. 18. Translated example of Fig. 11.

```
let x = (0, _) in
let y = (fst x, _) in
let x = (snd y, snd x) in
let y = (1, snd y) in
assume(fst y = snd y);
assert(fst x = 1)
```

Fig. 19. Translated example of Fig. 12.

Remark 2 (Difference from RUSTHORN*).* The main difference between our translation and RUSTHORN's method is that our method converts all references to pairs, whereas RUSTHORN only converts mutably borrowing references. This gap emerges because our type system enables an immutable reference to recover its ownership by alias; meanwhile, in Rust, once a mutable reference becomes immutable, it cannot be changed to mutable. Note that in Fig. 19, both references x and y become pairs unlike in RUSTHORN (Fig. 3).

4.3   Soundness

Our translation is sound, that is, if the translated program does not reach **Fail** neither does the original program.

As we are considering the intraprocedual fragment, we omit the stacks from configurations. For clarity, we write the reduction in the source language and the target language as $\langle H, R, e \rangle \xrightarrow{\text{src}} \langle H', R', e' \rangle$ and $\langle \mathcal{S}, t \rangle \xrightarrow{\text{tgt}} \langle \mathcal{S}', t' \rangle$, respectively.

Theorem 2 (Soundness of the translation). *Assume* $\emptyset \mid \emptyset \vdash e : \tau \Rightarrow t$ *and* $\langle \{\emptyset\}, t \rangle \xrightarrow{\text{tgt}}{\twoheadrightarrow}$ **Fail**, *then* $\langle \emptyset, \emptyset, e \rangle \xrightarrow{\text{src}}{\twoheadrightarrow}$ **Fail** *holds.*

We briefly explain the proof strategy below; the proof is given in the full version [25]. We define an (indexed) simulation $\overset{\text{sim}}{\rightsquigarrow}_{\mathcal{L},\Gamma}$ between the configurations of the two languages. The relation $\langle H, R, e \rangle \overset{\text{sim}}{\rightsquigarrow}_{\mathcal{L},\Gamma} \langle \mathcal{S}, t \rangle$ not only requires $\mathcal{L} \mid \Gamma \vdash e{:}\tau \Rightarrow t$ to hold, but also requires that TXZ and PFZ hold; without them the relation is too weak to be an invariant. Whenever $\langle H, R, e \rangle \xrightarrow{\text{src}} \langle H', R', e' \rangle$ we show that there exists $\langle \mathcal{S}', t' \rangle$ such that $\langle \mathcal{S}, t \rangle \xrightarrow{\text{tgt}}{\twoheadrightarrow} \langle \mathcal{S}', t' \rangle$ and $\langle H', R', e' \rangle \overset{\text{sim}}{\rightsquigarrow}_{\mathcal{L}',\Gamma'} \langle \mathcal{S}', t' \rangle$ *for some* $\mathcal{L}', \Gamma'$. Such environments can be chosen by an argument identical to the proof of type preservation (Theorem 1).

5   Preliminary Experiments

We conducted preliminary experiments to evaluate the effectiveness of our translation. In this experiment, we first typed the benchmark programs and translated them into OCaml programs by hand[7] (with slight optimizations). After that, we measured the time of the safety verification of the translated programs using MoCHi [27], a fully-automated verifier for OCaml programs. The experiments were conducted on a machine with AMD Ryzen 7 5700G 3.80 GHz CPU and 32 GB RAM. We used MoCHi of build a3c7bb9d as frontend solver, which used Z3 [23] (version 4.11.2) and HorSat2 [19] (version 0.95) as backend solvers.

We ran experiments on small but representative programs with mutable references and summerized the results in Table 1. Columns 'program' and 'safety' are the name of programs and whether they involve `fail` in assertions, respectively. When two programs have the same name but different safeties, they represent slightly modified variations of a single program with different safety. Columns 'ConSORT?' and 'RustHorn?' indicate that the program can be typed by Con-SORT or RustHorn, respectively. Columns B_o and B_t are the byte counts of the original and translated programs, respectively. The column 'time(sec)' shows how many seconds MoCHi takes to verify translated programs. These results show that our method works in practical time for representative programs, though it increases the program size by 2 to 4 by translation.

---

[7] This translation process can be automated if we require programmers to provide type annotations or if the type inference mentioned in Sect. 7 is worked out.

Below, we describe each benchmark briefly. 'consort-demo,' 'rusthorn-demo,' and 'minmax' are the sample programs in Figs. 1, 2 and 4. 'simple-loop' and 'shuffle-in-call', both included in CONSORT's benchmark, are the programs utilizing fractional ownership. 'just-rec' and 'inc-max' are from RUSTHORN, which exploits borrowable ownership. 'linger-dec' is the program with loop and dynamic allocating of references and adopted by both CONSORT and RUSTHORN benchmark. 'hhk2008' is a SEAHORN [13] test to check whether the verifier can find an invariant on the loop. See the full version [25] for benchmark programs.

Table 1. Experimental results.

program	safety	ConSORT?	RustHorn?	B_o	B_t	B_t/B_o	time(sec)
consort-demo	safe	✓	✗	76	285	3.75	< 0.1
rusthorn-demo	safe	✓	✓	68	276	4.06	0.8
borrow-merge	safe	✗	✗	216	756	3.5	1.0
simple-loop	safe	✓	✓	194	347	1.79	< 0.1
	unsafe			150	320	2.13	0.2
just-rec	safe	✓	✓	156	362	2.32	0.6
	unsafe			157	372	2.37	0.2
shuffle-in-call	safe	✓	✗	121	395	3.26	0.8
	unsafe			121	395	3.26	0.2
inc-max	safe	✗	✓	143	550	3.85	1.2
	unsafe			143	549	3.84	0.4
minmax	safe	✗	✗	251	920	3.67	0.2
	unsafe			251	920	3.67	0.6
linger-dec	safe	✓	✓	239	836	3.5	1.1
	unsafe			243	841	3.46	0.3
hhk2008	safe	✓	✓	321	630	1.96	2.7

6 Related Work

As introduced in Sect. 1, CONSORT [31] and RUSTHORN [22] are the direct ancestor of this work. We have combined the two approaches as borrowable fractional ownership types.

CREUSOT [9] and AENEAS [14] are verification toolchains for Rust programs based on a translation to functional programs. CREUSOT removes references and translates Rust programs to WHY3 [10] programs using RUSTHORN's prophecy technique. Our translation is closer to that of CREUSOT than that of RUSTHORN in that the target is a functional program rather than CHCs. But very unlike CREUSOT, our translation accommodates the CONSORT-style fractional ownership and alias annotations. AENEAS uses a different encoding to translate Rust programs to pure functional programs for interactive verification (in Coq, F*, etc.). However, they can't support advanced patterns like nested borrowing because their model of a mutable reference is not a first-class value, unlike RustHorn's prophecy-based approach.

There are other verification methods and tools that utilize some notion of ownership for imperative program verification, such as STEEL [11,30], VIPER [24], and REFINEDC [26]. Still, their approaches and design goals are quite different from ours in that they do *semi*-automated program verification in F* or low-level separation logic, requiring more user intervention, such as annotations of loop invariants and low-level proof hints about ownership.

SEAHORN [13] and JAYHORN [18] both introduce a fully automated verification framework for programs with mutable references. SEAHORN is for LLVM-based languages (C, C++, Rust, etc.) and JAYHORN is for Java. Both do not use ownership types but model the heap memory directly as a finite array. As a result, they are ineffective or imprecise for programs with dynamic memory allocation, as shown in the experiments of CONSORT and RUSTHORN.

The work by Suenaga and Kobayashi [29] and CYCLONE [15] used ownership types to ensure memory safety for a language with explicit memory deallocation. We expect that our borrowable fractional ownership types can also be used to improve their methods.

7   Conclusion and Future Work

We have presented a type system based on the new notion of borrowable fractional ownership, and a type-directed translation to reduce the verification of imperative programs to that of programs without mutable references. Our approach combines that of CONSORT [31] and RUSTHORN[22], enabling automated verification of a larger class of programs.

Future work includes automated type inference and an extension of the type system to allow nested references. Type inference would not be so difficult if we assume some lifetime annotations as in Rust. Concerning nested references, as a naive extension of reference types from $\mathbf{int\,ref}_B^{\alpha,r}$ to $\tau\,\mathbf{ref}_B^{\alpha,r}$ seems too restrictive, we plan to introduce a reference type of the form $\xi\backslash\tau/\rho\,\mathbf{ref}_{\beta,s}^{\alpha,r}$, where ξ is the borrowing type, τ is the current content type, and ρ is the lending type to others. The point is that, instead of just keeping the amount of ownership being borrowed, we should keep in what type a reference is being borrowed.

Acknowledgements. We would like to thank anonymous referees for their useful comments. This work was supported by JSPS KAKENHI Grant Number JP20H05703 and JP22KJ0561.

244   T. Nakayama et al.

References

1. Rust. https://www.rust-lang.org/. Accessed 20 Aug 2023
2. Abadi, M., Lamport, L.: The existence of refinement mappings. Theoret. Comput. Sci. **82**(2), 253–284 (1991). https://doi.org/10.1016/0304-3975(91)90224-P
3. Astrauskas, V., et al.: The Prusti project: formal verification for Rust. In: Deshmukh, J.V., Havelund, K., Perez, I. (eds.) NASA Formal Methods: 14th International Symposium, NFM 2022, Pasadena, CA, USA, 24–27 May 2022, Proceedings, pp. 88–108. Springer, Heidelberg (2022). https://doi.org/10.1007/978-3-031-06773-0_5
4. Bao, Y., Wei, G., Bračevac, O., Jiang, Y., He, Q., Rompf, T.: Reachability types: tracking aliasing and separation in higher-order functional programs. Proc. ACM Program. Lang. **5**(OOPSLA), 1–32 (2021). https://doi.org/10.1145/3485516
5. Boyapati, C., Lee, R., Rinard, M.: Ownership types for safe programming: preventing data races and deadlocks. In: Proceedings of the 17th ACM SIGPLAN Conference on Object-Oriented Programming, Systems, Languages, and Applications, OOPSLA 2002, pp. 211–230. Association for Computing Machinery, New York, NY, USA, November 2002. https://doi.org/10.1145/582419.582440
6. Boyland, J.: Checking interference with fractional permissions. In: Cousot, R. (ed.) SAS 2003. LNCS, vol. 2694, pp. 55–72. Springer, Heidelberg (2003). https://doi.org/10.1007/3-540-44898-5_4
7. Calcagno, C., Parkinson, M., Vafeiadis, V.: Modular safety checking for fine-grained concurrency. In: Nielson, H.R., Filé, G. (eds.) SAS 2007. LNCS, vol. 4634, pp. 233–248. Springer, Heidelberg (2007). https://doi.org/10.1007/978-3-540-74061-2_15
8. Crichton, W., Patrignani, M., Agrawala, M., Hanrahan, P.: Modular information flow through ownership. In: Proceedings of the 43rd ACM SIGPLAN International Conference on Programming Language Design and Implementation. ACM, New York, NY, USA, June 2022. https://doi.org/10.1145/3519939.3523445
9. Denis, X., Jourdan, J.H., Marché, C.: CREUSOT: a foundry for the deductive verification of Rust programs. In: Riesco, A., Zhang, M. (eds.) Formal Methods and Software Engineering: 23rd International Conference on Formal Engineering Methods, ICFEM 2022, Madrid, Spain, 24–27 October 2022, Proceedings, pp. 90–105. Springer, Heidelberg (2022). https://doi.org/10.1007/978-3-031-17244-1_6
10. Filliâtre, J.-C., Paskevich, A.: Why3—where programs meet provers. In: Felleisen, M., Gardner, P. (eds.) ESOP 2013. LNCS, vol. 7792, pp. 125–128. Springer, Heidelberg (2013). https://doi.org/10.1007/978-3-642-37036-6_8
11. Fromherz, A., et al.: Steel: proof-oriented programming in a dependently typed concurrent separation logic. Proc. ACM Program. Lang. **5**(ICFP), 1–30 (2021). https://doi.org/10.1145/3473590
12. Gäher, L., et al.: Simuliris: a separation logic framework for verifying concurrent program optimizations. Proc. ACM Program. Lang. **6**(POPL), 1–31 (2022). https://doi.org/10.1145/3498689
13. Gurfinkel, A., Kahsai, T., Komuravelli, A., Navas, J.A.: The SeaHorn verification framework. In: Kroening, D., Păsăreanu, C.S. (eds.) CAV 2015. LNCS, vol. 9206, pp. 343–361. Springer, Cham (2015). https://doi.org/10.1007/978-3-319-21690-4_20
14. Ho, S., Protzenko, J.: Aeneas: rust verification by functional translation. Proc. ACM Program. Lang. **6**(ICFP), 711–741 (2022). https://doi.org/10.1145/3547647
15. Jim, T., Morrisett, J.G., Grossman, D., Hicks, M.W., Cheney, J., Wang, Y.: Cyclone: a safe dialect of C. In: Proceedings of the General Track of the Annual

Conference on USENIX Annual Technical Conference, ATEC 2002, pp. 275–288. USENIX Association, USA (2002). http://dl.acm.org/citation.cfm?id=647057. 713871

16. Jung, R., Jourdan, J.H., Krebbers, R., Dreyer, D.: RustBelt: securing the foundations of the Rust programming language. Proc. ACM Program. Lang. **2**(POPL), 1–34 (2017). https://doi.org/10.1145/3158154

17. Jung, R., et al.: The future is ours: prophecy variables in separation logic. Proc. ACM Program. Lang. **4**(POPL), 1–32 (2019). https://doi.org/10.1145/3371113

18. Kahsai, T., Rümmer, P., Sanchez, H., Schäf, M.: JayHorn: a framework for verifying Java programs. In: Chaudhuri, S., Farzan, A. (eds.) CAV 2016. LNCS, vol. 9779, pp. 352–358. Springer, Cham (2016). https://doi.org/10.1007/978-3-319-41528-4_19

19. Kobayashi, N.: HorSat2: a saturation-based higher-order model checker (2015). http://www-kb.is.s.u-tokyo.ac.jp/~koba/horsat2. Accessed 6 Sept 2023

20. Lehmann, N., Geller, A.T., Vazou, N., Jhala, R.: Flux: liquid types for Rust. Proc. ACM Program. Lang. **7**(PLDI), 1533–1557 (2023). https://doi.org/10.1145/3591283

21. Matsakis, N.D., Klock, F.S.: The Rust language. In: Proceedings of the 2014 ACM SIGAda Annual Conference on High Integrity Language Technology, HILT 2014, pp. 103–104. Association for Computing Machinery, New York, NY, USA, October 2014. https://doi.org/10.1145/2663171.2663188

22. Matsushita, Y., Tsukada, T., Kobayashi, N.: RustHorn: CHC-based verification for Rust programs. ACM Trans. Program. Lang. Syst. **43**(4), 1–54 (2021). https://doi.org/10.1145/3462205

23. de Moura, L., Bjørner, N.: Z3: an efficient SMT solver. In: Ramakrishnan, C.R., Rehof, J. (eds.) TACAS 2008. LNCS, vol. 4963, pp. 337–340. Springer, Heidelberg (2008). https://doi.org/10.1007/978-3-540-78800-3_24

24. Müller, P., Schwerhoff, M., Summers, A.J.: Viper: a verification infrastructure for permission-based reasoning. In: Jobstmann, B., Leino, K.R.M. (eds.) VMCAI 2016. LNCS, vol. 9583, pp. 41–62. Springer, Heidelberg (2016). https://doi.org/10.1007/978-3-662-49122-5_2

25. Nakayama, T., Matsushita, Y., Sakayori, K., Sato, R., Kobayashi, N.: Borrowable fractional ownership types for verification (Extended Version) (2023). https://doi.org/10.48550/arXiv.2310.20430

26. Sammler, M., Lepigre, R., Krebbers, R., Memarian, K., Dreyer, D., Garg, D.: RefinedC: automating the foundational verification of C code with refined ownership types. In: Proceedings of the 42nd ACM SIGPLAN International Conference on Programming Language Design and Implementation. ACM, New York, NY, USA, June 2021. https://doi.org/10.1145/3453483.3454036

27. Sato, R., Unno, H., Kobayashi, N.: Towards a scalable software model checker for higher-order programs. In: Proceedings of the ACM SIGPLAN 2013 Workshop on Partial Evaluation and Program Manipulation, PEPM 2013, pp. 53–62. Association for Computing Machinery, New York, NY, USA (2013). https://doi.org/10.1145/2426890.2426900

28. Sergey, I., Clarke, D.: Gradual ownership types. In: Seidl, H. (ed.) ESOP 2012. LNCS, vol. 7211, pp. 579–599. Springer, Heidelberg (2012). https://doi.org/10.1007/978-3-642-28869-2_29

29. Suenaga, K., Kobayashi, N.: Fractional ownerships for safe memory deallocation. In: Hu, Z. (ed.) APLAS 2009. LNCS, vol. 5904, pp. 128–143. Springer, Heidelberg (2009). https://doi.org/10.1007/978-3-642-10672-9_11

30. Swamy, N., Rastogi, A., Fromherz, A., Merigoux, D., Ahman, D., Martínez, G.: SteelCore: an extensible concurrent separation logic for effectful dependently typed programs. Proc. ACM Program. Lang. **4**(ICFP), 1–30 (2020). https://doi.org/10.1145/3409003

31. Toman, J., Siqi, R., Suenaga, K., Igarashi, A., Kobayashi, N.: ConSORT: context- and flow-sensitive ownership refinement types for imperative programs. In: ESOP 2020. LNCS, vol. 12075, pp. 684–714. Springer, Cham (2020). https://doi.org/10.1007/978-3-030-44914-8_25

Runtime Verification

TP-DejaVu: Combining Operational and Declarative Runtime Verification

Klaus Havelund[1], Panagiotis Katsaros[2(✉)], Moran Omer[3], Doron Peled[3], and Anastasios Temperekidis[2]

[1] Jet Propulsion Laboratory, California Institute of Technology, Pasadena, USA
[2] Aristotle University of Thessaloniki, Thessaloniki, Greece
katsaros@csd.auth.gr
[3] Bar Ilan University, Ramat Gan, Israel

Abstract. Runtime verification (RV) facilitates monitoring the executions of a system, comparing them against a formal specification. A main challenge is to keep the incremental complexity of updating its internal structure, each time a new event is inspected, to a minimum. There is a tradeoff between achieving a low incremental complexity and the expressive power of the used specification formalism. We present an efficient RV tool that allows specifying properties of executions that include data, with the possibility to apply arithmetic operations and comparisons on the data values. In order to be able to apply efficient RV for specifications with these capabilities, we combine two RV methodologies: the first one is capable of performing arithmetic operations and comparisons based on the most recent events; the second is capable of handling many events with data and relating events that occur at arbitrary distance in the observed execution. This is done by two phase RV, where the first phase, which monitors the input events directly and is responsible to the arithmetic calculations and comparisons, feeds the second phase with modified events for further processing. This is implemented as a tool called TP-DEJAVU, which extends the DEJAVU tool.

1 Introduction

Runtime verification (RV) allows monitoring the execution of systems, checking them against a formal specification. RV is often restricted to checking one execution at a time and viewing a prefix of the execution at any given moment, providing a verdict, which is often *true* or *false*. With these limitations, RV is devoid of some of the complexity and computability restrictions of more comprehensive formal methods such as formal verification and model checking [9].

The research performed by the first author was carried out at Jet Propulsion Laboratory, California Institute of Technology, under a contract with the National Aeronautics and Space Administration. The research performed by the third and fourth authors was partially funded by Israeli Science Foundation grant 2454/23: "Validating and controlling software and hardware systems assisted by machine learning".

R. Dimitrova et al. (Eds.): VMCAI 2024, LNCS 14500, pp. 249–263, 2024.
https://doi.org/10.1007/978-3-031-50521-8_12

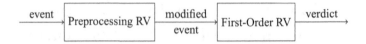

Fig. 1. Two phase monitoring.

Monitoring of executions with data against a first-order specification is challenging because of the need to keep and handle a large number of data values that appear in the observed prefix and the relationship between them. We are interested here in the capability of applying arithmetic operations and comparisons between data elements. In a general setting, a specification language should enable comparing different values that appear in distant events in the inspected execution. For example, checking that a new value that appears in the current event is exactly twice as big as some value observed in some previous event; this would require a comparison of the newly inspected value to the set of previously observed values. However, we have observed that in practice, in many cases the operations and comparisons required by the specification are limited, and can be performed using a fixed amount of memory. This includes in particular arithmetic operations that are based on past events within a limited distance, or on aggregate functions for calculating, e.g., sum, count, average, etc.

We introduce the TP-DEJAVU tool, for *Two Phase* DEJAVU. It combines two RV techniques: one that is capable of performing a rich collection of arithmetic, Boolean and string calculations, although restricted to using a fixed amount of memory elements, and based on the most recently seen or kept (through aggregation) values. The other one facilitates efficient processing of relational data but devoid of arithmetic operations and comparisons. This is done by applying a two phase RV processing (Fig. 1), where the first phase, which we also call the *preprocessing*, is responsible for the arithmetic manipulations, and the second part implements first-order based checks. The two parts interact through events sent by the first phase to the second one. These events can be modified from the original monitored events observed by the first phase.

RV algorithms typically maintain a *summary* that contains enough information for deciding on future verdicts without having to keep the full observed prefix. The implementation typically keeps a set of variables and updates them upon inspecting a new event. One can classify two main dichotomies for specifications for RV. The first one is *operational* specification. There, the specification describes the dynamics of changing the summary performed based on inspected events. This can be done by means of a set of assignments to summary variables, which are updated each time a new event is intercepted.

The second dichotomy is *declarative* specification, where *constraints* on the monitored sequence of events are given. An example of a declarative specification formalism is propositional temporal logic. It can express the relation between Boolean values that represent some inspected or measured properties (e.g. the light is on, a communication was successful) at different (not necessarily adjacent) time points during the monitored execution. First-order temporal logic extends this to also allow relating data values occurring in different events, e.g., requiring that a value is read from a file only if that file was opened and not closed yet. RV for declarative specification typically involves a translation into an internal operational form, involving updates to the summary performed when a new event occurs.

For past time *propositional* LTL, there is a direct and efficient translation from the specification [22], where the truth value of each subformula of the specification, based on the currently inspected prefix, is represented using a Boolean variable in the summary. RV algorithms may require the use of more general objects in the summary than Boolean variables. These variables can be numeric, e.g., integers or (fixed precision) reals. The *incremental complexity* of updating the summary is an important factor to consider for online RV algorithms, since these updates need to keep up with the speed of appearing events.

When moving from propositional to *first-order* temporal logic, instead of simple variables, one can use *relations* that represent subformulas [6,20]; for a given subformula η of the given specification, such a relation represents all the assignments to its free variables that satisfy η. The RV tool DEJAVU uses a BDD encoding of relations that represent assignments that satisfy the subformulas of the specification. To achieve high efficiency in time and space, exploiting BDD compactness [7], the individual objects of these relations are *bitstrings* that encode *enumerations* of the observed values rather than the values themselves. However, arithmetic operations and comparisons need to be performed based on the observed values themselves.

An arithmetic comparison between values, e.g., $<$, is implemented in DEJAVU by updating a relation $R_<$ that represents $<$ where $(v,w) \in R_<$ iff $v < w$. The relation $R_<$ is updated each time a new value v appears. This is done by comparing v against all the previously seen values. Abstractly, this can be thought of as adding to $R_<$ the tuples (v,w) when $v < w$ and (w,v) when $w < v$, for values w observed so far in the trace. Since making this update depends on the amount of values observed so far, this update can be time consuming, affecting the incremental complexity: it can grow unbounded with the length of the observed prefix. Furthermore, the need to perform arithmetic calculations and comparisons may tame down some representation optimization.

Our solution is to combine two approaches: the operational one to do some limited amount of arithmetic calculations and comparisons that are restricted to a finite amount of data, and the declarative one that is performed based on the QTL formalism of DEJAVU. We observed through the industrial case studies in the H2020 EU project FOCETA, from which we borrow some examples, that this is a powerful approach. On the other hand, we point out that some specifications (see Property 4 below) cannot be expressed using this combination.

The following examples show the complication involved in achieving a comparison between values within first-order temporal logic (the formal semantics of the logic is in Sect. 2.1), and gives a hint on where an alternative specification, based on combining operational and declarative components, may be useful.

1. $\forall x \, ((p(x) \wedge x > 7) \to \exists y \; \Diamond \; q(x, y))$. Here, the constraint $(x > 7)$ is imposed only on the *most recent* event. This can be checked by preprocessing the monitored event and checking the condition $x > 7$ as this event is intercepted.
2. $\forall x \, \forall y \, ((p(x) \wedge \ominus q(y) \wedge x < y) \to \Diamond \; r(x, y))$. The constraint $(x < y)$ is between the values observed within the current and the previous events. Remembering the value of y from the previous event, one can easily make this comparison. Using a finite amount of memory for past values will also work when $\ominus$ is replaced with some fixed number of the $\ominus$ operator.
3. $\forall x \, (p(x) \to (\forall y \, (\ominus \; \Diamond \; q(y) \to x > y) \wedge \exists z \, \ominus \; \Diamond \; q(z)))$. This property formalizes that when an event of the form $p(x)$ occurs, the value of x is bigger than any value y seen so far within an event of the form $q(y)$, and further, that at least one such a $q(z)$ event has occurred. A straightforward implementation, used in DEJAVU, builds a relation of tuples (x, y) where $x > y$, updating the relation as new values appear in events. A closer look reveals that we could have used a simpler way to monitor this property by keeping at each point the *maximal* value of y seen so far within $q(y)$ and comparing it with the new value x in a current event $p(x)$.
4. $\forall x \, (p(x) \to \exists y \exists z (\Diamond \; q(y) \wedge \; \Diamond \; q(z) \wedge y \neq z \wedge x = (y + z)/2))$. This case is difficult, since the property requires that a new p event has a value that is the average of two distinct values observed in previous q events. This requires remembering and comparing against all the previous values that appeared in previously observed q events. This case may necessitate comparison with an unboundedly growing number of past values. It cannot be handled by the particular two phase RV method suggested in this paper, which allows arithmetic computations limited to a finite amount of stored values.

As mentioned before, we split the specification into an operational part and a declarative part (cf. Fig. 1). The operational part, which specifies arithmetic operations and comparisons, is handled as a preprocessing stage. The declarative part is restricted not to include arithmetic components. The RV is then combined from the preprocessing of the intercepted events by the operational part, which generates modified (augmented) events for the declarative specification. The latter is handled by the old version of DEJAVU. Without the capability of arithmetic comparisons, DEJAVU is an extremely efficient tool for processing traces with data, see the experiments in [20]; using our two-phase approach, we enhance its capabilities to perform with similar efficiency, albeit the added expressiveness.

Related Work. Some early tools supported data comparison and computations as part of the logic [2,4]. The version of MONPOLY tool in [5] supports comparisons and aggregate operations such as sum and maximum/minimum within a first-order LTL formalism. It uses a database-oriented implementation. Other tools supporting automata based limited first-order capabilities include [10,26]. In [13], a framework that lifts the monitor synthesis for a propositional temporal logic to a temporal logic over a first-order theory, is described. A number of internal DSLs (libraries in a programming language) for RV have been developed [11,16,18], which offer the full power of the host programming language for writing monitors and therefore allow for arbitrary comparisons and computations on data to be performed. The concept of phasing monitoring such that one phase produces output to another phase has been explored in other frameworks, including early frameworks such as [24] with a fixed number of phases, but with propositional monitoring, and later frameworks with arbitrary user defined phases [4,17]. In particular, stream processing systems support this idea [12,23,25]. A more remotely related work on increasing the expressive power of temporal logic is the extension of DEJAVU with rules described in [21].

2   Combining Operational and Declarative RV

The structure of RV algorithms is typically simple and consists of capturing a monitored event and updating its internal memory, which we call a *summary*, as it summarises the needed information from the observed prefix of events. In case that a verdict is available, it is reported, sometimes also causing the verification procedure to terminate, if this kind of verdict is stable for the given specification (e.g., a negative verdict for a safety property).

2.1   Declarative Specification in Past Time First-Order Temporal Logic

The logic QTL, used by DEJAVU, and a core subset of the logic used by the MONPOLY tool [6], is a formalism that allows expressing properties of executions that include data. The restriction to past time allows interpreting the formulas on finite traces.

Syntax. The formulas of the QTL logic are defined using the following grammar, where p stands for a *predicate* symbol, a is a *constant* and x is a *variable*.

For simplicity of the presentation, we define here the QTL logic with unary predicates, but this is not due to a principal restriction, and in fact QTL supports predicates over multiple arguments, including zero arguments, corresponding to propositions. The DEJAVU system fully supports predicates over multiple arguments.

$$\varphi ::= true \mid p(a) \mid p(x) \mid (\varphi \wedge \varphi) \mid \neg\varphi \mid (\varphi \mathcal{S} \varphi) \mid \ominus \varphi \mid \exists x \, \varphi$$

A formula can be interpreted over multiple types (domains), e.g., natural numbers or strings. Accordingly, each variable, constant and parameter of a predicate is defined over a specific type. Type matching is enforced, e.g., for $p(a)$ ($p(x)$, respectively), the types of the parameter of p and of a (x, respectively) must be the same. We denote the type of a variable x by $type(x)$.

Propositional past time linear temporal logic is obtained by restricting the predicates to be parameterless, essentially Boolean propositions; then, no variables, constants and quantification are needed either.

Semantics. A QTL formula is interpreted over a *trace*, which is a finite sequence of *events*. Each event consists of a predicate symbol and parameters, e.g., $p(a)$,

$q(7)$. It is assumed that parameters belong to particular domains that are associated with (places in) the predicates. A more general semantics can allow each event to consist of a set of predicates with multiple parameters. However, this is *not* implemented in DEJAVU.

QTL subformulas have the following informal meaning: $p(a)$ is true if the last event in the trace is $p(a)$. The formula $p(x)$, for some variable x, holds if x is bound to a constant a such that $p(a)$ is the last event in the trace. The formula $(\varphi \, \mathcal{S} \, \psi)$, which reads as φ *since* ψ, means that ψ occurred in the past (including now) and since then (beyond that state) φ has been true. (The *since* operator is the past dual of the future time *until* modality in the commonly used future time temporal logic.) The property $\ominus \varphi$ means that φ is true in the trace that is obtained from the current one by omitting the last event. The formula $\exists x \, \varphi$ is true if there exists a value a such that φ is true with x bound to a. We can also define the following additional derived operators: $false = \neg true$, $(\varphi \vee \psi) = \neg(\neg\varphi \wedge \neg\psi)$, $(\varphi \rightarrow \psi) = (\neg\varphi \vee \psi)$, $\diamondsuit \varphi = (true \, \mathcal{S} \, \varphi)$ ("previously"), $\boxminus \varphi = \neg \diamondsuit \neg\varphi$ ("always in the past" or "historically"), and $\forall x \, \varphi = \neg\exists x \, \neg\varphi$.

Formally, let $free(\eta)$ be the set of free (i.e., unquantified) variables of $\eta \in sub(\varphi)$, i.e. η is a subformula of φ. Let γ be an assignment to the variables $free(\eta)$. We denote by $\gamma[v \mapsto a]$ the assignment that differs from γ only by associating the value a to v. We also write $[v \mapsto a]$ to denote the assignment that consists of a single variable v mapped to value a. Let σ be a trace of events of length $|\sigma|$ and i a natural number, where $i \leq |\sigma|$. Then $(\gamma, \sigma, i) \models \eta$ if η holds for the prefix of length i of σ with the assignment γ.

We denote by $\gamma|_{free(\varphi)}$ the restriction (projection) of an assignment γ to the free variables appearing in φ. Let ϵ be an empty assignment. In any of the following cases, $(\gamma, \sigma, i) \models \varphi$ is defined when γ is an assignment over $free(\varphi)$, and $i \geq 1$.

- $(\epsilon, \sigma, i) \models true$.
- $(\epsilon, \sigma, i) \models p(a)$ if $\sigma[i] = p(a)$.
- $([x \mapsto a], \sigma, i) \models p(x)$ if $\sigma[i] = p(a)$.
- $(\gamma, \sigma, i) \models (\varphi \wedge \psi)$ if $(\gamma|_{free(\varphi)}, \sigma, i) \models \varphi$ and $(\gamma|_{free(\psi)}, \sigma, i) \models \psi$.
- $(\gamma, \sigma, i) \models \neg\varphi$ if not $(\gamma, \sigma, i) \models \varphi$.
- $(\gamma, \sigma, i) \models (\varphi \, \mathcal{S} \, \psi)$ if for some $1 \leq j \leq i$, $(\gamma|_{free(\psi)}, \sigma, j) \models \psi$ and for all $j < k \leq i$, $(\gamma|_{free(\varphi)}, \sigma, k) \models \varphi$.
- $(\gamma, \sigma, i) \models \ominus\varphi$ if $i > 1$ and $(\gamma, \sigma, i - 1) \models \varphi$.
- $(\gamma, \sigma, i) \models \exists x \, \varphi$ if there exists $a \in type(x)$ such that $(\gamma[x \mapsto a], \sigma, i) \models \varphi$.

2.2   RV Monitoring First-Order Past LTL

We review the algorithm for monitoring first-order past LTL, implemented as part of DEJAVU [20]. Consider a classical algorithm for past time propositional LTL [22]. For propositional LTL, an event is a set of propositions. The summary consists of two vectors of bits. One vector, pre, keeps the Boolean (truth) value for each subformula, based on the trace observed so far *except* the last observed event. The other vector, now, keeps the Boolean value for each subformula based on that trace *including* the last event. When a new event e occurs, the vector now is copied to the vector pre; then a new version of the vector now is calculated based on the vector pre and the event e as follows:

- now($true$) = $True$
- now(p) = ($p \in e$)
- now(($\varphi \wedge \psi$)) = (now(φ) $\wedge$ now(ψ))
- now($\neg\varphi$) = $\neg$now(φ)
- now(($\varphi \; \mathcal{S} \; \psi$)) = (now($\psi$) $\vee$ (now(φ) $\wedge$ pre(($\varphi \; \mathcal{S} \; \psi$))))
- now($\ominus \varphi$) = pre(φ)

The *first-order* monitoring algorithm replaces the two vectors of bits by two vectors that represent *assignments*. The first vector, pre, contains, for each subformula η of the specification, a relation that represents the set of assignments to the free variables of η that satisfy η after the monitored trace seen so far *except its last event*. The second vector, now, contains the assignments that satisfy η given the complete monitored trace seen so far.

The updates in the first-order case replace negation with complementation (denoted using an overline), conjunction with intersection $\cap$ and disjunction with union $\cup$. The intuition behind the connection of the Boolean operators for the propositional logic and the set operators for QTL follows from redefining the semantics of QTL in terms of sets. Let $I[\varphi, \sigma, i]$ be the function that returns the *set* of assignments such that $(\gamma, \sigma, i) \models \varphi$ iff $\gamma|_{free(\varphi)} \in I[\varphi, \sigma, i]$. Then, e.g., $I[(\varphi \wedge \psi), \sigma, i] = I[\varphi, \sigma, i] \cap I[\psi, \sigma, i]$. For other cases and further details of set semantics, see [20].

The complementation, intersection and union are operators on relations. Note that the relations for different subformulas can be over different (but typically not disjoint) types of tuples. For example, one relation can be formed from a subformula with free variables x, y, z and the other one over y, z, w, say with x, w over strings and y, z over integers. Intersection actually operates as a database *join*, which matches tuples of its two arguments when they have the same values for their common variables. Union is actually a *co-join*. The symbol $\bot$ represents the relation with no tuple, and $\top$ represents the relation with all possible tuples over the given domains.

In order to form a finite representation of an unbounded domain, one can represent the values that were already observed in previous events, plus a special notation for all the values not yet seen, say "$\sharp$". This representation is updated each time a new value is observed in an input event. For example, the set $\{\sharp, 4, 9\}$ represents all the values that we have *not* seen so far in input events, and in addition, the values 4 and 9. This finite representation allows using full negation in monitored formulas.

The RV algorithm for a QTL formula φ is as follows:

1. Initially, for each $\eta \in sub(\varphi)$ of the specification φ, now(η) = $\bot$.
2. Observe a new event $p(a)$ as input;
3. Let pre:=now.
4. Make the following updates for the formulas $sub(\varphi)$, where
   if $\psi \in sub(\eta)$ then now(ψ) is updated before now(η).
   - now($true$) = $\top$

- $\text{now}(p(a)) = $ if current event is $p(a)$ then $\top$ else $\bot$
- $\text{now}(p(x)) = $ if current event is $p(a)$ then $\{a\}$ else $\bot$
- $\text{now}((\eta \wedge \psi)) = (\text{now}(\eta) \cap \text{now}(\psi))$
- $\text{now}(\neg \eta) = \overline{\text{now}(\eta)}$
- $\text{now}((\eta \mathcal{S} \psi)) = (\text{now}(\psi) \cup (\text{now}(\eta) \cap \text{pre}((\eta \mathcal{S} \psi))))$
- $\text{now}(\ominus \eta) = \text{pre}(\eta)$
- $\text{now}(\exists x\ \eta) = \text{now}(\eta)_x$

5. Goto step 2.

In DEJAVU, for each value a seen for the first time in an input trace one assigns an enumeration. Using hashing, subsequent occurrences of the value a obtain the same enumeration. This value is then converted to a *bitstring*. For a subformula η of the specification, $\text{now}(\eta)$ is a BDD representation of concatenations of such bitrstrings, corresponding to the tuples of values of the free variables of η that satisfy η. For further details on the DEJAVU implementation, see [20]. The DEJAVU tool uses keyword characters to express QTL formulas; it employs the following notation: forall and exists stand for $\forall$ and $\exists$, respectively, P, H, S and @ for $\diamondsuit$, $\boxminus$, $\mathcal{S}$ and $\ominus$, respectively, and |, & and ! for $\vee$, $\wedge$ and $\neg$, respectively.

2.3   The Operational RV

An operational specification can be expressed using updates to variables that are included in the summary. The next state, which is the updated summary, is obtained by performing the assignments based on the values of the previous summary and the parameters appearing within the new event. The input to the first phase of RV is, as in DEJAVU, a sequence of events $e_1.e_2.e_3 \ldots$, where each event consists of a predicate name with parameters, enclosed in parentheses and separated by commas, as in $q(a, 7)$. The syntax of the specification consists of two parts:

Initialization. This part starts with the keyword initiate. It provides values to the variables in the *initial* summary, i.e., the summary *before* any event occurs.

Update. Depending on the input predicate p, the variables of the summary are updated based on the values that are associated with the current event and the previous values of the variables in the previous summary, as described below. An updated event is then generated.

An update consists of a sequence of on clauses, starting with "on $p(x, y, \ldots)$)", where p is a predicate name and $x, y, \ldots$ are variables that are set to the positional arguments of p. Thus, for the event $p(3, \text{``}abc\text{''})$, $p(x, y)$ will cause x to be assigned to 3 and y to "abc". Such on *clause* is followed by a sequence of assignments to be executed when an event that matches the on clause occurs. The assignments can include Boolean operations, arithmetic operations (e.g., $*, /, +, -$), arithmetic comparisons ($<, \leq, >, \geq$) and string operations.

The construct *ite* stands for *if-then-else*. Its first parameter is a Boolean expression (obtained, e.g., from an arithmetic comparison). The second parameter corresponds to the value returned when the Boolean expression calculates to *true* (the *then* case). The third parameter is the value returned when the Boolean expression calculates to *false* (the *else* case). The second and third parameters must be of the same type. This can be also extended to string operators. The symbol @ prefixing a variable x means that we refer to the value of x in the *previous* summary. Thus, $x:=ite(v > @x, v, @x)$, where v is a value that appears in the current event, updates x to keep the maximal value of v that was observed. If the value of $@x$ is needed for the first event then x must be initialized. If a summary variable is not assigned to, in the on clause that matches the current event, then this variable retains its value from the previous summary. This helps to shorten the description of the property.

The operational phase can deal with different types; the types of variables are declared when they are assigned. We do not necessarily need to provide an on clause for each predicate that can appear in an event; if some predicate is not intercepted by an on clause in the operational phase, the event is forwarded unchanged to the declarative phase.

Example 1. In this example we have events that include two data items: a car vendor and speed. A *true* verdict is returned each time that a new vendor has broken the speed record *for the first time*. Note that the verdict for the inspected traces can change from *true* to *false* and back to *true* any number of times (the *false* verdict is *not* necessarily stable, as is the case for safety properties [1]).

```
initiate
      MaxSpeed: int := 0
on recorded(vendor: str, speed: int)
      NewRecord: bool := @MaxSpeed <speed
      MaxSpeed: int := ite(NewRecord == true, speed, @MaxSpeed)
      output fast(vendor, NewRecord)
exists x . (fast(x, "true") & ! @ P fast(x, "true"))
```

Example 2. The following example records from time to time the temperature in one of a collection of running cars. It calculates the temperature increase between the two successive startMeasure and endMeasure events. If the increase in temperature is more than 5 degrees (say Celsius), then it indicates an overheated car. The monitor reports if some car is overheated more than twice.

```
on startMeasure(Car: str, temp: float)
      output startMeasure(Car)
on endMeasure(Car: str, temp: float)
      Warming: bool := (temp − ⊖temp) > 5
      output warmAlert(Car, Warming)
exists x . (warmAlert(x, "true") & @
(startMeasure(x) & P (warmAlert(x, "true") &
@ P (warmAlert(x, "true")))))
```

Example 3. The next property deals with temperature setting commands sent to air conditioners (ac). In order to send a command, we must have turned the air conditioner on (and not off since). However, if the command is out of temperature bounds, specifically, below 17C or above 26C, then it is ignored as a faulty command, and there is no such requirement.

on set(ac: *str* , temp: *float*)
     WithinBound: *bool* := (temp >= 17 && temp <= 26)
     output set(ac, temp, WithinBound)
forall ac . ((**exists** temp . set(ac, temp, "true")) -> (! turn_off (ac) **S** turn_on(ac)))

2.4   Examples from Use Cases of Autonomous Systems

In this subsection, we detail specifications developed for learning-enabled autonomous systems (LEAS) as part of the H2020 European Union FOCETA project.

Prediction of Obstacle Behaviour in an Automated Valet Parking System. An automated valet parking (AVP) system is an L4 (level four) autonomous driving system. A user owning a vehicle equipped with the AVP functionality can stop the car in front of the parking lot entry area. Whenever the user triggers the AVP function, the vehicle will communicate with the infrastructure and park the car at designated regions. The system is expected to operate under mixed traffic, i.e., the parking lot will have other road users including walking pedestrians and other vehicles.

The AVP system implements object detection capabilities for sensing the surrounding objects and localisation functions for inferring the system's location on the map. It also features mission and path planning functions, as well as trajectory tracking and a prediction module that is based on the available information that predicts positions of traffic participants (i.e., obstacle list) in the future. The position is given in x, y coordinates. The prediction error is computed as the distance between an obstacle's actual position at cycle t and the predicted position for it at cycle $t-1$. The prediction error reported for an obstacle at any bounded within a certain value Epo. Also, a maximum accumulated error Emax exists for the system.

initiate
     LErr: *double* := 0
     SysErr: *double* := 0
     NewPred: *bool* := false
on mk_prediction(ru: *str* , px: *int* , py: *int*)
     LErr: *double* := 0
     NewPred: *bool* := true
     output predicted(ru)

```
on obstacle(ru: str, x: int, y: int)
    LErr: double :=
    ite (@NewPred == true, (((x−@px)^2)+((y−@py)^2)^0.5), 0)
    SysErr: double := @SysErr+LErr
    error : bool := (LErr > Epo) || (SysErr > Emax)
    NewPred: bool := false
    output valid(ru, error, L_err)
on exit(ru: str, p_err: double)
    SysErr: double := @SysErr − p_err
    output exit(ru)
```

forall ru . (exit (ru) | ((@ predicted (ru) ->
exists n . valid (ru, "false", n)) **S** entry(ru))

This specification states that a prediction can be made for the future position of any road user (ru) that has entered the parking lot area (event entry) and has not yet exited (event exit). Upon having a new prediction (NewPred), the prediction error (LErr) measured when ru's actual position is known (event obstacle) does not exceed Epo. Moreover, the accumulated error (SysErr) for all road users that are still moving in the parking lot area does not exceed Emax. When a ru leaves the parking space, the last reported prediction error (p_err) ceases to be considered in the accumulated error (SysErr).

Monitoring the Perception Function of an AVP System. The perception monitor takes as input from the perception subsystem the computed free space area (event new_fspace), obstacles (ru) and their localisation (event location). The dimensions of the free space are given as a rectangle defined by its diagonally opposite corners. The dimensions of localised rus are given through reporting their detected 2D bounding boxes. The perception monitor detects and alerts about inconsistencies in the perception process. Specifically, the output of free space detection and localisation of rus should never overlap, to ensure consistency of the overall perception function.

```
on new_fspace(fs_x1: int, fs_y1: int, fs_x2: int, fs_y2: int)
    output new_fspace(fs_x1, fs_y1, fs_x2, fs_y2)
on location(ru: str, bb_xa: int, bb_ya: int, bb_xb: int, bb_yb: int)
    overlap: bool := (bb_xb >= fs_x1) && (bb_xa <= fs_x2)
    && (bb_yb <= fs_y1) && (bb_ya >= fs_y2)
    output error(ru, overlap)
on exit_fspace (fs_x1: int, fs_y1: int, fs_x2: int, fs_y2: int)
    output exit_fspace (fs_x1, fs_y1, fs_x2, fs_y2)
```

exists x1 . **exists** y1 . **exists** x2 . **exists** y2 . (exit_fspace (x1,y1,x2,y2) |
(**forall** ru . ! error (ru, "true") & (! exit_fspace (x1,y1,x2,y2) **S** new_fspace(x1, y1, x2, y2)))

3   The TP-DejaVuTool and Experimental Results

TP-DejaVu is an extension of DejaVu [29], both written in Scala. It facilitates two phase RV processing. The first phase (the prepossessing) performs operational RV with syntax as described in Sect. 2.3. Upon each intercepted event, it performs summary update calculations and generates a modified event, which it passes to the declarative phase tool. The second phase is based on the DejaVu tool, and performs monitoring against a first-order specification. The source of the TP-DejaVu tool, with documentation and experiment files, including the examples described in this paper, appear in [28].

Next, we present experimental results on the relative efficiency of runtime monitoring Qtl specifications using DejaVu tool versus the efficiency of two phase RV using TP-DejaVu. In all cases, the two specifications output exactly the same results. Our benchmarks focused exclusively on the time and memory consumption during the evaluation phase, without the compilation time. The properties in our experiment were taken from Sect. 1, specifically properties 1–3 (property 4 cannot be translated into TP-DejaVu); their adaptations for TP-DejaVu are illustrated in Fig. 2.

Traces: For properties 1 and 2, the event sequences and their respective values were assigned at random. However, for property 3, we tried to construct traces such that DejaVu can process them within the given time limit of 1000 s. Out of every 100 events, one event is labelled as "p", while the rest are labelled "q". For our experiment, we utilized traces comprising 10K, 100K, 500K, 1M, and 5M events.

Results: Table 1 presents the results from our comparison experiments. These results highlight the advantages of a two phase RV over the singular phase launched by DejaVu when complex inequality operators are involved. Executions where the evaluation process exceeded 1000 s are marked with the symbol ∞. For property 1, both methods display a rise in execution times with increasing trace sizes. In this case, TP-DejaVu manages to be marginally faster but may need more memory. For Property 2 and 3, DejaVu was unable to evaluate any of the traces (except one), as evidenced by the consistently infinite execution times. On the other hand, TP-DejaVu managed to assess all traces success-

```
on p(x: int)
  in_bound: bool := x > 7
  output p(x, in_bound)

forall x . (p(x, "true") ->
    exists y q(x,y))
```

```
initiate
  prev_q: bool := false

on p(x: int)
  x_lt_y: bool := x < y
  prev_q: bool := false
  output p(x, x_lt_y)

on q(y: int)
  prev_q: bool := true
  output q(y)

forall x . forall y .
  ((p(x, "true") & @q(y)) ->
  P r(x, y))
```

```
initiate
    MaxY: int := −1

on p(x: int)
 xGTy: bool := x > MaxY
 output p(x, xGTy)

on q(y: int)
 NewMaxY: bool := @MaxY < y
 MaxY: int :=
 ite (NewMaxY, y, @MaxY)
 output q(y)

forall x . forall w .
  (p(x, w) -> (p(x, "true") &
  exists y . @P q(y)))
```

Fig. 2. TP-DejaVu properties corresponding to properties 1 (left), 2 (middle), and 3 (right).

fully. However, this came with a significant increase in memory consumption, especially evident with the larger trace files.

Table 1. TP-DEJAVU vs. DEJAVU.

Property	Method	Trace 10K	Trace 100K	Trace 500K	Trace 1M	Trace 5M
$P1$	DEJAVU	0.64 s 125.26 MB	1.31 s 335.52 MB	4.76 s 1.10 GB	8.85 s 1.88 GB	185.65 s 3.59 GB
	TP-DEJAVU	0.54 129.74 MB	0.96 s 311.24 MB	3.25 s 858.30 MB	5.44 s 1.21 GB	41.18 s 5.70 GB
$P2$	DEJAVU	∞ –	∞ –	∞ –	∞ –	∞ –
	TP-DEJAVU	0.56 s 135.18 MB	1.12 s 308.72 MB	4.12 s 1.17 GB	7.54 s 1.78 GB	56.78 s 3.55 GB
$P3$	DEJAVU	5.23 s 805.47 MB	∞ –	∞ –	∞ –	∞ –
	TP-DEJAVU	0.64 s 119.85 MB	0.90 s 326.38 MB	2.39 s 738.78 MB	4.08 s 1.11 GB	21.30 s 3.43 GB

Table 2. Examples and use cases of Two Phase RV.

Property	# Objects	Trace 10K	Trace 100K	Trace 500K	Trace 1M	Trace 5M
Example 1	10 cars	0.53 s 105.68 MB	0.92 s 157.34 MB	1.55 s 229.31 MB	2.35 s 292.72 MB	5.74 s 303.02 MB
Example 2	10 cars	0.67 s 106.88 MB	0.92 s 150.04 MB	1.52 s 263.37 MB	2.41 s 288.28 MB	11.12 s 291.95 MB
Example 3	10 ACs	0.58 s 108.30 MB	0.81 s 165.85 MB	1.68 s 240.72 MB	2.28 s 297.82 MB	7.22 s 300.19 MB
Use case 1	10 rus	0.72 s 114.17 MB	1.54 s 169.08 MB	1.95 s 292.89 MB	3.23 s 296.25 MB	8.81 s 315.48 MB
Use case 2	10 rus	0.64 s 134.00 MB	1.30 s 189.15 MB	3.37 s 340.44 MB	6.10 s 349.80 MB	22.72 s 1.11 GB

In Table 2, we provide experimental results highlighting the efficiency of the two-phase RV based on the examples detailed in Sects. 2.3 and 2.4. This includes Examples 1 to 3 and the two LEAS applications.

4   Conclusions and Further Work

We presented an RV approach based on two phases. The first, operational RV phase, takes care of algebraic (but also Boolean and string) calculations. The second, declarative RV phase, takes care of relational computation. The experimental results indicate that the two phase approach can gain considerable advantage in speed over incrementally encoding relations that represent comparisons between the observed data by the declarative RV, as is done in DEJAVU. It allows handling a rich number of algebraic operators and comparisons. On the other hand, it is limited to applying these operations to a fixed number of kept values, recently observed or aggregated from earlier events. An extension that allows storing unbounded collections of values can be considered. Currently, the operational phase of the TP-DEJAVU tool outputs a single event for every input event. An extension of the tool can relax this rule and can generate several or zero events, or even have several possibilities, depending on the data in the events.

Data-Availability Statement

The TP-DEJAVU tool is open source and publicly available at https://doi.org/10.5281/zenodo.8322559, as well as from the GitHub repository at https://github.com/moraneus/TP-DejaVu.

References

1. Alpern, B., Schneider, F.B.: Recognizing safety and liveness. Distrib. Comput. **2**(3), 117–126 (1987)
2. Barringer, H., Goldberg, A., Havelund, K., Sen, K.: Rule-based runtime verification. In: Steffen, B., Levi, G. (eds.) VMCAI 2004. LNCS, vol. 2937, pp. 44–57. Springer, Heidelberg (2004). https://doi.org/10.1007/978-3-540-24622-0_5
3. Barringer, H., Havelund, K.: TraceContract: a scala DSL for trace analysis. In: Butler, M., Schulte, W. (eds.) FM 2011. LNCS, vol. 6664, pp. 57–72. Springer, Heidelberg (2011). https://doi.org/10.1007/978-3-642-21437-0_7
4. Barringer, H., Rydeheard, D., Havelund, K.: Rule systems for run-time monitoring: from Eagle to RuleR. In: Sokolsky, O., Taşıran, S. (eds.) RV 2007. LNCS, vol. 4839, pp. 111–125. Springer, Heidelberg (2007). https://doi.org/10.1007/978-3-540-77395-5_10
5. Basin, D.A., Klaedtke, F., Marinovic, S., Zalinescu, E.: Monitoring of temporal first-order properties with aggregations. Formal Methods Syst. Des. **46**(3), 262–285 (2015)
6. Basin, D.A., Klaedtke, F., Müller, S., Zalinescu, E.: Monitoring metric first-order temporal properties. J. ACM **62**(2), 45 (2015)
7. Bryant, R.E.: Symbolic boolean manipulation with ordered binary-decision diagrams. ACM Comput. Surv. **24**(3), 293–318 (1992)
8. Burch, J.R., Clarke, E.M., McMillan, K.L., Dill, D.L., Hwang, L.J.: Symbolic model checking: 10^{20} states and beyond, LICS, pp. 428–439 (1990)
9. Clarke, E.M., Grumberg, O., Peled, D.A.: Model Checking, 1st edn, pp. 1–314, I–XIV. MIT Press (2001). ISBN 978-0-262-03270-4
10. Colombo, C., Gauci, A., Pace, G.J.: LarvaStat: monitoring of statistical properties (RV 2010), pp. 480–484 (2010)
11. Dams, D., Havelund, K., Kauffman, S.: A Python library for trace analysis. In: Dang, T., Stolz, V. (eds.) Proceedings of the Runtime Verification: 22nd International Conference, RV 2022, Tbilisi, 28–30 September 2022, pp. 264–273. Springer, Cham (2022). https://doi.org/10.1007/978-3-031-17196-3_15
12. D'Angelo, B., et al.: LOLA: runtime monitoring of synchronous systems. In: 12th International Symposium on Temporal Representation and Reasoning (TIME), pp. 166–174. IEEE (2005)
13. Decker, N., Leucker, M., Thoma, D.: Monitoring modulo theories. Softw. Tools Technol. Transf. **18**(2), 205–225 (2016)
14. Duckett, B., Havelund, K., Stewart, L.: Space telemetry analysis with PyContract. In: Haxthausen, A.E., Huang, W.-l., Roggenbach, M. (eds.) Applicable Formal Methods for Safe Industrial Products: Essays Dedicated to Jan Peleska on the Occasion of His 65th Birthday, pp. 272–288. Springer, Cham (2023). https://doi.org/10.1007/978-3-031-40132-9_17
15. Fowler, M., Parsons, R.: Domain-Specific Languages. Addison-Wesley (2010)

16. Gorostiaga, F., Sánchez, C.: HStriver: a very functional extensible tool for the runtime verification of real-time event streams. In: Huisman, M., Păsăreanu, C., Zhan, N. (eds.) FM 2021. LNCS, vol. 13047, pp. 563–580. Springer, Cham (2021). https://doi.org/10.1007/978-3-030-90870-6_30

17. Hallé, S., Villemaire, R.: Runtime enforcement of web service message contracts with data. IEEE Trans. Serv. Comput. 5(2), 192–206 (2012)

18. Havelund, K.: Data Automata in Scala, Theoretical Aspects of Software Engineering Conference (TASE), pp. 1–9. IEEE Computer Society (2014)

19. Havelund, K.: Rule-based runtime verification revisited. Int. J. Softw, Tools Technol. Transf. 17(2), 143–170 (2014). https://doi.org/10.1007/s10009-014-0309-2

20. Havelund, K., Peled, D., Ulus, D.: First-order temporal logic monitoring with BDDs. In: FMCAD 2017, pp. 116–123 (2017)

21. Havelund, K., Peled, D.: An extension of first-order LTL with rules with application to runtime verification. Int. J. Softw. Tools Technol. Transf. 23(4), 547–563 (2021)

22. Havelund, K., Roşu, G.: Synthesizing monitors for safety properties. In: Katoen, J.-P., Stevens, P. (eds.) TACAS 2002. LNCS, vol. 2280, pp. 342–356. Springer, Heidelberg (2002). https://doi.org/10.1007/3-540-46002-0_24

23. Kallwies, H., Leucker, M., Schmitz, M., Schulz, A., Thoma, D., Weiss, A.: TeSSLa – an ecosystem for runtime verification. In: Dang, T., Stolz, V. (eds.) RV 2022. LNCS 13498, pp. 314–324. Springer, Cham (2022). https://doi.org/10.1007/978-3-031-17196-3_20

24. Kim, M., Kannan, S., Lee, I., Sokolsky, O.: Java-MaC: a run-time assurance tool for Java. In: 1st International Workshop on Runtime Verification (RV), ENTCS, vol. 55, no. 2. Elsevier (2001)

25. Perez, I., Dedden, F., Goodloe, A.: Copilot 3. Technical report, NASA Langley Research Center (2020)

26. Reger, G, Cruz, H.C., Rydeheard, D.: MarQ: monitoring at runtime with QEA. In: Baier, C., Tinelli, C. (eds.) TACAS 2015. LNCS, vol. 9035, pp. 596–610. Springer, Heidelberg (2015). https://doi.org/10.1007/978-3-662-46681-0_55

27. Wolper, P.: Temporal logic can be more expressive. Inf. Control 56(1/2), 72–99 (1983)

28. TP-DejaVu Tool Source Code. https://doi.org/10.5281/zenodo.8322559

29. DejaVu Tool Source Code. https://github.com/havelund/dejavu

Synthesizing Efficiently Monitorable Formulas in Metric Temporal Logic

Ritam Raha[1,2] , Rajarshi Roy[3(✉)] , Nathanaël Fijalkow[2] ,
Daniel Neider[4,5] , and Guillermo A. Pérez[1]

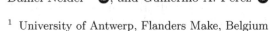

[1] University of Antwerp, Flanders Make, Belgium
`guillermoalberto.perez@uantwerpen.be`
[2] LaBRI, University of Bordeaux, Bordeaux, France
[3] Max Planck Institute for Software Systems, Kaiserslautern, Germany
`rajarshi@mpi-sws.org`
[4] TU Dortmund University, Dortmund, Germany
`daniel.neider@tu-dortmund.de`
[5] Center for Data Science and Security, University Alliance Ruhr,
Dortmund, Germany

Abstract. In runtime verification, manually formalizing a specification for monitoring system executions is a tedious and error-prone process. To address this issue, we consider the problem of automatically synthesizing formal specifications from system executions. To demonstrate our approach, we consider the popular specification language Metric Temporal Logic (MTL), which is particularly tailored towards specifying temporal properties for cyber-physical systems (CPS). Most of the classical approaches for synthesizing temporal logic formulas aim at minimizing the size of the formula. However, for efficiency in monitoring, along with the size, the amount of "lookahead" required for the specification becomes relevant, especially for safety-critical applications. We formalize this notion and devise a learning algorithm that synthesizes concise formulas having bounded lookahead. Our learning algorithm reduces the synthesis task to a series of satisfiability problems in Linear Real Arithmetic (LRA) and generates MTL formulas from their satisfying assignments. The reduction uses a novel encoding of a popular MTL monitoring procedure using LRA. Finally, we implement our algorithm in a tool called **TEAL** and demonstrate its ability to synthesize efficiently monitorable MTL formulas in CPS applications.

Keywords: Metric Temporal Logic · Runtime Verification · Specification Mining · Interpretability · Formal Methods

The first two authors contributed equally; the remaining authors are ordered alphabetically. This research was funded by the FWO G030020N project "SAILor", DFG grant no. 434592664 and the Research Center Trustworthy Data Science and Security (https://rc-trust.ai), one of the Research Alliance centers within the UA Ruhr (https://uaruhr.de).

R. Dimitrova et al. (Eds.): VMCAI 2024, LNCS 14500, pp. 264–288, 2024.
https://doi.org/10.1007/978-3-031-50521-8_13

1   Introduction

Runtime verification is a well-established method for ensuring the correctness of cyber-physical systems during runtime. Techniques in runtime verification are known to be more rigorous than conventional testing while not being as resource-intensive as exhaustive techniques of formal verification [20]. In the field of runtime verification, among other techniques, monitoring system executions against formal specifications during runtime is a widely used one. Over the years, numerous monitoring techniques have been proposed for a variety of specification languages [7,21,23,28].

In this work, we focus on Metric Temporal Logic (MTL) [37]—a specification language popularly employed for monitoring cyber-physical systems [29,43]. MTL is a real-time extension of Linear Temporal Logic (LTL) [51] augmented with timing constraints for temporal operators. MTL specifications are often easy to interpret due to their resemblance to natural language and, thus, also find applications in Artificial Intelligence [62]. While there are many possible semantics of MTL (e.g., discrete, dense-time pointwise, etc. [50]), we employ the dense-time continuous semantics as it is more natural and general than the counterparts [4,9]. We expand on MTL and other prerequisites in Sect. 2.

Virtually all verification techniques for MTL rely on the availability of a formal specification. However, manually writing specifications is a tedious and error-prone task [1]. Synthesizing functional, correct, and interpretable specifications that precisely express the design requirements has remained one of the major challenges in the adoption of formal techniques for verification [12,58].

To tackle the lack of formal specifications, there have been efforts to automatically synthesize specifications from system executions. Most of the existing works have targeted specification languages such as Linear Temporal Logic (LTL) [16,48,53] and Signal Temporal Logic (STL) [3,40,45,59], with few works for MTL [30,62]. Many of the works tend to synthesize specifications that are *concise* in size. Concise specifications are preferred over large ones because, based on the principle of Occam's razor, they are easier for humans to understand [57].

However, conciseness is not the only measure of interest for specifications, especially in the context of online monitoring. In online monitoring, specifically in *stream-based* runtime monitoring, a monitor reads an execution as a stream of data and verifies if a given specification is invariant (i.e., holds at all time points) in the execution. Many stream-based monitors [27,32,39] support MTL formulas. Typically, such monitors produce a stream of (Boolean) verdicts with some "latency", which depends on the lookahead of the formula. The lookahead required for an MTL formula is often formalized as its *future-reach* [29,31], which is the amount of time required to determine its satisfaction at any time point.

With the aim of reducing the latency for efficient online monitoring, we focus on automatically synthesizing MTL specifications based on two regularizers: size and future-reach. As input data, we rely on a sample S consisting of system executions that are observed for a finite duration. We consider the sample to be partitioned into a set P of positive (or desirable) executions and a set N of negative (or undesirable) executions.

We now formulate the central problem of synthesizing MTL formulas as follows: given a sample $S = (P, N)$ and a future-reach bound K, synthesize a minimal size MTL formula φ that (i) is *globally-separating* for S, in that, φ holds at all time points in the positive executions and does not hold at some time point in the negative executions, and (ii) the future-reach of φ is smaller than K. According to this problem, the prospective formula φ, being globally-separating, is invariant in the desirable executions and not in the undesirable executions. Such a property is typically preferred for specifications used in online monitoring [11].

Also, the future-reach bound forms a crucial part of this problem since, without a future-reach bound, the most concise prospective formula can have a large future-reach value (thus negatively affecting the latency required for online monitoring). To illustrate this, assume that we observe some simulations of an autonomous vehicle. During the simulations, we sample executions (shown below) of the vehicle every second for six seconds. We classify them as positive (denoted using u_i's) or negative (denoted using v_i's) based on whether the vehicle encountered a collision or not.

$$
\begin{array}{ccccccc}
 & 0 & 1 & 2 & 3 & 4 & 5 \\
u_1: & \{p, q\} & \{p\} & \{q\} & \{p, q\} & \{p\} & \{p\} \\
u_2: & \{q\} & \{\} & \{q\} & \{p\} & \{p\} & \{p, q\} \\
v_1: & \{p\} & \{q\} & \{\} & \{\} & \{\} & \{\} \\
v_2: & \{p\} & \{p, q\} & \{p\} & \{\} & \{p\} & \{\}
\end{array}
$$

In the executions, we use p to denote that there is no obstacle within a particular unsafe distance ahead of the vehicle and q to denote that the vehicle's brake is triggered. Our setting considers executions to be *continuous*. Thus, to ensure continuity of executions, in the above example, if p occurs at time point t, we interpret it as p holding during the entire interval $[t, t+1)$. We also assume that the executions last up to a final time point T, which is 6 for this example. Thus, for the execution u_1, p holds in the intervals $[0, 2)$ and $[3, 6)$.

In the sample, a minimal globally separating formula is $\varphi_1 = \mathbf{F}_{[0,3]}\, q$. The formula φ_1 being globally separating indicates that in all positive executions, the brake is triggered every three seconds (i.e., within the interval $[t, t + 3]$ for every time point t), irrespective of whether there is an obstacle within the unsafe distance. The formula φ_1 has size two and a future-reach of three seconds, meaning that any online monitor requires a three second lookahead window to check the satisfaction of φ_1. There is another formula $\varphi_2 = \neg p \rightarrow \mathbf{F}_{[0,1]}\, q$ that is globally separating for the sample. The formula φ_2 being globally-separating indicates that in all positive executions, for every time point t, if an obstacle is within the unsafe distance, then the brake is triggered within one second (i.e., within the interval $[t, t + 1]$). Although of size five, φ_2 has future-reach of one second and will be typically preferred over φ_1 for online monitoring in a safety-critical scenario. We present further details of the problem of synthesizing MTL formulas in Sect. 3.

For the problem of synthesizing MTL formulas, we first study whether a solution exists. It turns out that there exist samples S and future-reach bounds

K for which there might not exist any MTL formula that is both globally-separating for S and has future-reach within K. To check whether a prospective formula exists, we identify a simple characterization of S based on the future-reach K. Such a characterization results in an NP algorithm that can decide whether a prospective formula exists. Also, it provides an upper-bound, which is polynomial in the inputs S and K, on the size of the prospective formula if one exists. We mention the details of the existence check in Sect. 4.

To synthesize a concise prospective formula, we rely on a reduction to constraint satisfaction problems. In particular, we follow other works in formula synthesis [48,57] and devise an algorithm that encodes the synthesis problem in a series of satisfiability modulo theory (SMT) problems in Linear Real Arithmetic (LRA). To our knowledge, we design the first SMT-based algorithm that can synthesize MTL formulas of arbitrary syntactic structure.

Our SMT-based algorithm is highly modular and, thus, easily adaptable to other popular settings in formula synthesis. In particular, by modifying our algorithm, one can handle the typical *passive learning* problem [15,48] (i.e., the problem of synthesizing *separating* formulas without any future-reach bound), the *specification sketching* problem [38,42] (i.e., the problem of completing *partial* formulas), synthesizing formulas from noisy data [26,41] and so on.

Also, as a by-product of our SMT-based algorithm, we are able to analyze the complexity of deciding whether a prospective MTL formula of a *particular size* exists. While its exact complexity lower bounds are open, we show that this decision problem is in NP. The central SMT-based algorithm with all the theoretical results is in Sect. 5.

We also implement our algorithm using a popular SMT solver in a prototype named TEAL. We evaluate the ability of TEAL to synthesize MTL formulas typically employed for monitoring cyber-physical systems. We also empirically study the interplay between the size and future-reach of a formula. We present all the experimental results in Sect. 6.

Related Works. To our knowledge, there are only a limited number of works for synthesizing MTL formulas. One of them [62] infers MTL formulas as decision trees for representing task knowledge in Reinforcement Learning. Some other works [30,63] consider the parameter search problem for MTL where, given a parametric MTL formula (i.e., an MTL formula with missing temporal bounds), they infer the ranges of parameters where the formula holds/does not hold on a given system. Unlike our work, none of these works aims at synthesizing concise MTL specifications for monitoring tasks.

There are, nevertheless, numerous runtime monitoring procedures for MTL [4,10,17,22,29,33,39,60], clearly indicating the need for efficiently monitorable MTL specifications. Many of them also rely on the future-reach of a specification [10,29] or other similar measures (e.g., horizon [22], worst-case propagation delay [33], etc.) to quantify the efficiency of their monitoring procedure.

Interestingly, several works focus on synthesizing formulas in STL, an extension of MTL to reason about real-valued signals. Bartocci et al. [8] provide a comprehensive survey of the existing works on inferring STL. Many of them [3,34,35] solve the parameter search for STL, while others [13,14] learn decision trees over STL formulas, which typically do not result in concise formulas. There are few works [45,49] that do prioritize the conciseness of formulas during inference. These works cannot be directly applied to solve our problem for two main reasons. First, these works assume inputs to be *piecewise-affine continuous* signals. While the above assumption is natural for synthesizing STL formulas inference from real-valued signals, in our setting, we must rely on the assumption that our inputs are *piecewise-constant* signals, which is natural for Boolean-valued signals. Second, these works do not employ any measure, apart from conciseness, that directly influences the efficiency of runtime monitoring.

Finally, there are also several works on synthesizing concise formulas in Linear Temporal Logic (LTL) [16,48,53,56], Property Specification Language (PSL) [57], Computation Tree Logic [61], etc. These works are only designed to handle discrete executions (i.e., traces instead of signals) and, thus, cannot be applied directly to our setting.

2 Preliminaries

In this section, we introduce the basic notations used throughout the paper.

Signals and Prefixes. We represent continuous system executions as signals. A *signal* $x \colon \mathbb{R}_{\geq 0} \to 2^{\mathcal{P}}$ over a set of propositions $\mathcal{P}$ is an infinite time series that describes relevant system events over time. A prefix of a signal x restricted to domain $\mathbb{T} = [0, T), T \in \mathbb{R}_{\geq 0}$ is a function $x_{\mathbb{T}} \colon \mathbb{T} \to 2^{\mathcal{P}}$ where $x_{\mathbb{T}}(t) = x(t)$ for all $t \in \mathbb{T}$.

To synthesize MTL formulas, we rely on finite observations that are sequences of the form $\Omega = \langle (t_i, \delta_i) \rangle_{i \leq n}$, $n \in \mathbb{N}$ such that (i) $t_0 = 0$, (ii) $t_n < T$, and (iii) for all $i \leq n$, $\delta_i \subseteq \mathcal{P}$ is the set of propositions that hold at time point $t_i \in \mathbb{R}$. To construct well-defined signal prefixes, we approximate each observation Ω as a *piecewise-constant* signal prefix $x_{\mathbb{T}}^{\Omega}$ using interpolation as: (i) for all $i < n$, for all $t \in [t_i, t_{i+1})$, $x_{\mathbb{T}}(t) = \delta_i$; and (ii) for all $t \in [t_n, T)$, $x_{\mathbb{T}}(t) = \delta_n$. For brevity, we refer to signal prefixes simply as 'prefixes' when clear from the context.

Metric Temporal Logic. MTL is a logic formalism for specifying real-time properties of a system. We consider the following syntax of MTL:

$$\varphi := p \in \mathcal{P} \mid \neg p \mid \varphi_1 \wedge \varphi_2 \mid \varphi_1 \vee \varphi_2 \mid \varphi_1 \, \mathbf{U}_I \, \varphi_2 \mid \mathbf{F}_I \, \varphi \mid \mathbf{G}_I \, \varphi$$

where $p \in \mathcal{P}$ is a proposition, $\neg$ is the negation operator, $\wedge$ and $\vee$ are the conjunction and disjunction operators respectively, and $\mathbf{U}_I, \mathbf{F}_I$ and $\mathbf{G}_I$ are the timed-Until, timed-Finally and timed-Globally operators respectively. Here, I is a closed interval of non-negative real numbers of the form $[a, b]$ where $0 \leq a \leq b$[1].

---

[1] Since we infer MTL formulas with bounded lookahead, we restrict I to be bounded.

Note that the syntax is presented in *negation normal form*, meaning that the ¬ operator can only appear before a proposition.

Fig. 1. Syntax DAG of $(p \wedge \mathbf{G}_I q) \vee (\mathbf{F}_I p)$.

As a syntactic representation of an MTL formula, we rely on *syntax-DAGs*. A syntax-DAG is similar to the parse tree of a formula but with shared common subformulas. We define the size $|\varphi|$ of an MTL formula φ as the number of nodes in its syntax-DAG, e.g., the size of $(p \wedge \mathbf{G}_I q) \vee (\mathbf{F}_I p)$ is six as its syntax-DAG has six nodes, as shown in Fig. 1.

As mentioned already, we follow the continuous semantics of MTL. For the standard continuous semantics ($\models$) of MTL over infinite signals, we refer to the work of [50] and provide detailed descriptions in the extended version of the paper [54]. However, our setting demands a semantics of MTL over finite prefixes such that the synthesized formulas will be 'useful' while monitoring over infinite signals. Intuitively, we want an 'optimistic' semantics ($\models_f$) of an MTL formula φ over a prefix $\boldsymbol{x}_T$ such that $\boldsymbol{x}_T \models_f \varphi$ if there exists an infinite signal *extending* $\boldsymbol{x}_T$ that satisfies φ. In other words, $\boldsymbol{x}_T$ "carries no evidence against" the formula φ. Formally, we want the definition of $\models_f$ to satisfy the following lemma.

Lemma 1. *Given a prefix $\boldsymbol{x}_T$, let $ext(\boldsymbol{x}_T) = \{\boldsymbol{x} \mid \boldsymbol{x}_T \text{ is a prefix of } \boldsymbol{x}\}$ be the set of all infinite extensions of $\boldsymbol{x}_T$. Then given an MTL formula φ, $\boldsymbol{x}_T \models_f \varphi$ if there exists $\boldsymbol{x} \in ext(\boldsymbol{x}_T)$ such that $\boldsymbol{x} \models \varphi$.*

The proof of the above lemma proceeds via structural induction over φ, which we provide in the extended version of the paper [54].

For the above lemma to hold, we follow the idea of 'weak semantics' of MTL defined in [29][2] and interpret MTL over finite prefixes. Given a prefix $\boldsymbol{x}_T$, we inductively define when an MTL formula φ holds at time point $t \in \mathbb{T}$, i.e., $(\boldsymbol{x}_T, t) \models_f \varphi$, as follows:

$(\boldsymbol{x}_T, t) \models_f p \iff p \in \boldsymbol{x}_T(t);$

$(\boldsymbol{x}_T, t) \models_f \neg p \iff p \notin \boldsymbol{x}_T(t);$

$(\boldsymbol{x}_T, t) \models_f \varphi_1 \wedge \varphi_2 \iff (\boldsymbol{x}_T, t) \models_f \varphi_1 \text{ and } (\boldsymbol{x}_T, t) \models_f \varphi_2;$

$(\boldsymbol{x}_T, t) \models_f \varphi_1 \vee \varphi_2 \iff (\boldsymbol{x}_T, t) \models_f \varphi_1 \text{ or } (\boldsymbol{x}_T, t) \models_f \varphi_2;$

$(\boldsymbol{x}_T, t) \models_f \varphi_1 \mathbf{U}_{[a,b]} \varphi_2 \iff$

- $\exists t' \in [t+a, t+b] \cap \mathbb{T} \text{ s.t. } (\boldsymbol{x}_T, t') \models_f \varphi_2 \text{ and } \forall t'' \in [t, t'], (\boldsymbol{x}_T, t'') \models_f \varphi_1, \text{ or}$
- $T \leq t+b \text{ and } \forall t'' \in [t, T), (\boldsymbol{x}_T, t'') \models_f \varphi_1$

$(\boldsymbol{x}_T, t) \models_f \mathbf{F}_{[a,b]} \varphi \iff t+b \geq T \text{ or } \exists t' \in [t+a, t+b] \cap \mathbb{T} \text{ s.t. } (\boldsymbol{x}_T, t') \models_f \varphi;$

$(\boldsymbol{x}_T, t) \models_f \mathbf{G}_{[a,b]} \varphi \iff t+a \geq T \text{ or } \forall t' \in [t+a, t+b] \cap \mathbb{T}, (\boldsymbol{x}_T, t') \models_f \varphi$

We say that $\boldsymbol{x}_T$ satisfies φ if $(\boldsymbol{x}_T, 0) \models_f \varphi$. Also, for ensuring that our semantics complies with Lemma 1, we define $(\boldsymbol{x}_T, t) \models_f \varphi$ for all $t \geq T$ for any φ.

---

[2] Following Eisner et al. [24], Ho et al. [29] defined the weak semantics of MTL for the pointwise setting, which we adapt here for the continuous setting.

3   The Problem Formulation

Next, we formally introduce the various aspects of the central problem.

Sample. The input data consists of a set of labeled (piecewise-constant) prefixes. Formally, we rely on a sample $\mathcal{S} = (P, N)$ consisting of a set P of positive prefixes and a set N of negative prefixes such that $P \cap N = \emptyset$. We say an MTL formula φ is *globally-separating* (**G**-sep, for short) for $\mathcal{S}$ if it satisfies all the positive prefixes at each time point and does not satisfy negative prefixes at some time point[3]. Formally, given a sample $\mathcal{S}$, we define an MTL formula φ to be **G**-sep for $\mathcal{S}$ if (i) for all $\boldsymbol{x}_T \in P$ and for all $t \in [0, T)$, $(\boldsymbol{x}_T, t) \models_f \varphi$; and (ii) for all $\boldsymbol{y}_T \in N$, there exists $t \in [0, T)$ such that $(\boldsymbol{y}_T, t) \not\models_f \varphi$.

Future-Reach. To formalize the lookahead of an MTL formula φ, we rely on its future-reach $fr(\varphi)$ [29,31], which indicates how much of the future is required to determine the satisfaction of φ. It is defined inductively as follows:

$$fr(p) = fr(\neg p) = 0$$
$$fr(\varphi_1 \wedge \varphi_2) = fr(\varphi_1 \vee \varphi_2) = \max(fr(\varphi_1), fr(\varphi_2))$$
$$fr(\varphi_1 \, \mathbf{U}_{[a,b]} \, \varphi_2) = b + \max(fr(\varphi_1), fr(\varphi_2))$$
$$fr(\mathbf{F}_{[a,b]} \, \varphi) = fr(\mathbf{G}_{[a,b]} \, \varphi) = b + fr(\varphi)$$

To highlight that $fr(\varphi)$ quantifies the lookahead of φ, we observe the following lemma:

Lemma 2. *Let φ be an MTL formula such that $fr(\varphi) \leq K$ for some $K \in \mathbb{R}^{\geq 0}$. Also, let $\boldsymbol{x}$ and $\boldsymbol{y}$ be two signals such that $\boldsymbol{x}_{[0,K]} = \boldsymbol{y}_{[0,K]}$. Then, for all $T \in \mathbb{R}^{\geq 0}$, $\boldsymbol{x}_T \models_f \varphi$ if and only if $\boldsymbol{y}_T \models_f \varphi$.*

Intuitively, the above lemma states that a formula with future-reach $\leq K$ cannot distinguish between two signals that are identical up to time K. The lemma can be proved using structural induction over φ. For space constraints, we include the whole proof in the extended version of the paper [54].

The Problem. We now formally introduce the problem of synthesizing an MTL formula. In the problem, we ensure that the MTL formula is efficient for monitoring by allowing the system designer to specify a future-reach bound.

Problem 1. (SYNTL). Given a sample $\mathcal{S} = (P, N)$ and a future-reach bound K, find an MTL formula φ such that (i) φ is **G**-sep for $\mathcal{S}$; (ii) $fr(\varphi) \leq K$; (iii) for every MTL formula φ' such that φ' is **G**-sep for $\mathcal{S}$ and $fr(\varphi') \leq K$, $|\varphi| \leq |\varphi'|$.

Intuitively, the above optimization problem asks to synthesize a minimal size MTL formula that is **G**-sep for the input sample and has a future-reach within the input bound. Before we dive into the procedure for finding such an MTL formula, we first study if such an MTL formula even exists.

---

[3] Most stream-based monitors check if the specification holds at every time point [11].

4   Existence of a Solution

As alluded to in the introduction, for any given sample S and future-reach bound K, the existence of a prospective formula is not always guaranteed. For an illustration, consider the sample S with one positive prefix $\boldsymbol{x}_{\mathbb{T}} = \langle(0, \{q\}), (2, \{\})\rangle$ and one negative prefix $\boldsymbol{y}_{\mathbb{T}} = \langle(0, \{q\})\rangle$, and domain $\mathbb{T} = [0, 4)$. For this sample, there is no formula φ with $fr(\varphi) \leq 1$ that is **G**-sep. To see this, assume that a prospective formula φ exists. Since φ must be **G**-sep, $(\boldsymbol{x}_{\mathbb{T}}, 0) \models \varphi$. Next, observe that, for all time-points $t \in \mathbb{T}$, $\boldsymbol{y}_{\mathbb{T}}$ when restricted to time interval $[t, t+1] \cap \mathbb{T}$ appears identical to $\boldsymbol{x}_{\mathbb{T}}$ when restricted to time interval $[0, 1]$ since φ has future-reach is 1 (using Lemma 2). Thus, for all time-points $t \in \mathbb{T}$, $(\boldsymbol{y}_{\mathbb{T}}, t) \models \varphi$ violating that φ is **G**-sep.

What we show now is that one can check whether a prospective formula exists by relying on a simple characterization of the inputs S and K. Towards this, we introduce some terminology.

We define an *infix* of a prefix $\boldsymbol{x}_{\mathbb{T}}$ as the restriction of $\boldsymbol{x}_{\mathbb{T}}$ to a specific time interval. Specifically, given two time-points $t_1 \leq t_2 < T$ and a prefix $\boldsymbol{x}_{\mathbb{T}}$, infix $\boldsymbol{x}_{\mathbb{T}}^{[t_1, t_2]}$ is the function $\boldsymbol{x}_{\mathbb{T}}^{[t_1, t_2]} : [0, t_2 - t_1] \to 2^{\mathcal{P}}$ such that $\boldsymbol{x}_{\mathbb{T}}^{[t_1, t_2]}(t) = \boldsymbol{x}_{\mathbb{T}}(t + t_1)$ for all $t \in [0, t_2 - t_1]$.

Next, we define a characterization of a sample S based on the future-reach K, which we term as *K-infix-separable*. Intuitively, we say S to be *K-infix-separable* if there is a K-length infix $\boldsymbol{y}_{\mathbb{T}}^{[t_1, t_2]}$ for every negative prefix $\boldsymbol{y}_{\mathbb{T}}$ in S that is not an infix of any positive prefix in S. Formally, $S = (P, N)$ is *K-infix-separable* if for every negative prefix $\boldsymbol{y}_{\mathbb{T}} \in N$, there exists an infix $\boldsymbol{y}_{\mathbb{T}}^{[t_1, t_2]}$ with $t_2 - t_1 \leq K$ such that $\boldsymbol{y}_{\mathbb{T}}^{[t_1, t_2]} \neq \boldsymbol{x}_{\mathbb{T}}^{[t_1', t_2']}$ for any infix $\boldsymbol{x}_{\mathbb{T}}^{[t_1', t_2']}$ of any positive prefix $\boldsymbol{x}_{\mathbb{T}} \in P$.

We now state the result that enables checking the existence of a solution to Problem 1.

Lemma 3. *For a given sample S and future-reach bound K, there exists an MTL formula φ with $fr(\varphi) \leq K$ that is **G**-sep for S if and only if S is K-infix-separable.*

Proof. ($\Rightarrow$) For the forward direction, consider φ to be an MTL formula with $fr(\varphi) \leq K$ that is **G**-sep for S. Since φ is **G**-sep, for any arbitrary negative prefix, say $\bar{\boldsymbol{y}}_{\mathbb{T}}$, there must be a time-point, say $\bar{t} < T$, such that $(\bar{\boldsymbol{y}}_{\mathbb{T}}, \bar{t}) \not\models \varphi$. If $\bar{t} + K < T$, we show by contradiction that the infix $\bar{\boldsymbol{y}}_{\mathbb{T}}^{[\bar{t}, \bar{t}+K]}$ is not an infix in any positive prefix. In particular, if $\bar{\boldsymbol{y}}_{\mathbb{T}}^{[\bar{t}, \bar{t}+K]} = \boldsymbol{x}_{\mathbb{T}}^{[t, t+K]}$, then $(\boldsymbol{x}_{\mathbb{T}}, t) \not\models \varphi$ as φ cannot distinguish between signals that are identical up to time K (using Lemma 2). If $\bar{t} + K \geq T$, the semantics of MTL being weak, there is an $L < K$ with $\bar{t} + L < T$ such that for any $\boldsymbol{y} \in ext(\boldsymbol{y}_{\mathbb{T}}^{[0, \bar{t}+L]})$, $(\boldsymbol{y}, t) \not\models \varphi$ (using Lemma 1). Once again, we show by contradiction that the infix $\bar{\boldsymbol{y}}_{\mathbb{T}}^{[\bar{t}, \bar{t}+L]}$ is not an infix in any positive prefix. In particular, if $\bar{\boldsymbol{y}}_{\mathbb{T}}^{[\bar{t}, \bar{t}+L]} = \boldsymbol{x}_{\mathbb{T}}^{[t, t+L]}$, then for all $\boldsymbol{x} \in ext(\boldsymbol{x}_{\mathbb{T}}^{[0, t+L]})$ $(\boldsymbol{x}, t) \not\models \varphi$. Also, for any $\boldsymbol{x} \in ext(\boldsymbol{x}_{\mathbb{T}})$ $(\boldsymbol{x}, t) \not\models \varphi$, meaning $(\boldsymbol{x}_{\mathbb{T}}, t) \not\models \varphi$ (again, using Lemma 1).

($\Leftarrow$) For the other direction, consider S to be *K-infix-separable*. Using the definition of *K-infix-separable*, for any arbitrary negative prefix, say $\bar{\boldsymbol{y}}_{\mathbb{T}}$, we have an infix $\bar{\boldsymbol{y}}_{\mathbb{T}}^{[t_1, t_2]}$ with $t_2 - t_1 \leq K$ that is not an infix in any positive prefix. We construct a formula $\varphi_{\bar{\boldsymbol{y}}_{\mathbb{T}}}$ that explicitly specifies the propositions appearing in each interval of the infix $\bar{\boldsymbol{y}}_{\mathbb{T}}^{[t_1,t_2]}$ using $\mathbf{G}$ and $\wedge$ operators. Observe that $fr(\varphi_{\bar{\boldsymbol{y}}_{\mathbb{T}}}) \leq K$ since $t_2 - t_1 \leq K$ in $\bar{\boldsymbol{y}}_{\mathbb{T}}^{[t_1,t_2]}$. Now, the formula $\neg\varphi_{\boldsymbol{x}_{\mathbb{T}}}$ holds at all time-points in all positive prefixes, while it does not hold at time-point t_1 in $\bar{\boldsymbol{y}}_{\mathbb{T}}$. We finally construct the prospective formula as $\varphi = \bigwedge_{\boldsymbol{y}_{\mathbb{T}} \in N} \neg\varphi_{\boldsymbol{y}_{\mathbb{T}}}$ which is $\mathbf{G}$-sep for $\mathcal{S}$ and also, $fr(\varphi) \leq K$. $\qquad\square$

We now describe an NP algorithm to check whether a sample $\mathcal{S}$ is K-*infix-separable*. The crux of the algorithm is to guess, for each negative prefix $\boldsymbol{y}_{\mathbb{T}}$, an infix $\boldsymbol{y}_{\mathbb{T}}^{[t_1,t_2]}$ with $t_2 - t_1 \leq K$ and then check whether it is an infix of any positive prefix. The procedure of checking involves comparing $\boldsymbol{y}_{\mathbb{T}}^{[t_1,t_2]}$ against the possible infixes of the positive prefixes.

To describe the checking procedure in detail, let $\bar{\boldsymbol{y}}_{\mathbb{T}}^{[t_1,t_2]}$ be an infix of the negative prefix $\bar{\boldsymbol{y}}_{\mathbb{T}}$. Suppose we like to check whether $\bar{\boldsymbol{y}}_{\mathbb{T}}^{[t_1,t_2]}$ is an infix of the positive prefix $\bar{\boldsymbol{x}}_{\mathbb{T}}$. For this, we check $\bar{\boldsymbol{y}}_{\mathbb{T}}^{[t_1,t_2]} = \bar{\boldsymbol{x}}_{\mathbb{T}}^{[t,t+(t_2-t_1)]}$ with only those infixes where the observation time points of $\boldsymbol{x}_{\mathbb{T}}$ and $\boldsymbol{y}_{\mathbb{T}}$ coincide. Precisely, we check $\bar{\boldsymbol{y}}_{\mathbb{T}}^{[t_1,t_2]} = \bar{\boldsymbol{x}}_{\mathbb{T}}^{[t,t+(t_2-t_1)]}$ for all those infixes of $\bar{\boldsymbol{x}}_{\mathbb{T}}$ where $t'' - t = t' - t_1$, t'' and t' being the time points where $\boldsymbol{x}_{\mathbb{T}}$ and $\boldsymbol{y}_{\mathbb{T}}$ have been observed, respectively. This procedure is based on the fact that the changes in an infix occur only at the observation time points. Also, this procedure requires polynomial time in the number of observation time points of $\bar{\boldsymbol{x}}_{\mathbb{T}}$ and $\bar{\boldsymbol{y}}_{\mathbb{T}}$. We can now perform the procedure for each positive and negative prefix. Overall, we have the following result.

Lemma 4. *Given a sample $\mathcal{S}$ and future-reach bound K, checking whether $\mathcal{S}$ is K-infix-separable is in* NP.

5   An SMT-Based Algorithm

Our algorithm relies on an SMT-based approach inspired by the numerous constraint satisfaction-based approaches for synthesizing temporal logic formulas [2,15,48,57]. Roughly speaking, our algorithm constructs a series of formulas in Linear Real Arithmetic (LRA) and uses an optimized SMT solver to search for the desired solution. To expand on the specifics of our algorithm, we first familiarize the readers with LRA.

Linear Real Arithmetic (LRA). In LRA [6], given a set of real variables $\mathcal{Y}$, a *term* is defined recursively as either constant $c \in \mathbb{R}$, a real variable $y \in \mathcal{Y}$, a product $c \cdot y$ of a constant $c \in \mathbb{R}$ and a real variable $y \in \mathcal{Y}$, or a sum $t_1 + t_2$ of two terms t_1 and t_2. An *atomic formula* is of the form $t_1 \diamond t_2$ where $\diamond \in \{<, \leq, =, \geq, >\}$. An *LRA formula*, defined recursively, is either an atomic formula, the negation $\neg\Phi$ of an LRA formula Φ, or the disjunction $\Phi_1 \vee \Phi_2$ of two formulas Φ_1, Φ_2. We additionally include standard Boolean constants *true*, and *false* and Boolean operators $\wedge$, $\rightarrow$ and $\leftrightarrow$.

To assign meaning to an LRA formula, we rely on a so-called *interpretation* function $\iota\colon \mathcal{Y} \to \mathbb{R}$ that maps real variables to constants in $\mathbb{R}$. An interpretation

---

Algorithm 1. Overview of our algorithm

    Input: Sample $\mathcal{S}$, fr-bound K

1: **if** $\mathcal{S}$ is not K-*infix-separable* **then return** No prospective formula
2: $n \leftarrow 0$
3: **while** True **do**
4:      $n \leftarrow n + 1$
5:      Construct $\Phi^n_{\mathcal{S},K} := \Phi^{str}_{n,\mathcal{S},K} \wedge \Phi^{fr}_{n,\mathcal{S},K} \wedge \Phi^{sem}_{n,\mathcal{S},K}$
6:      **if** $\Phi^n_{\mathcal{S},K}$ is SAT **then**
7:          Construct φ^ι from a satisfying interpretation ι **return** φ^ι

ι can easily be lifted to a term t in the usual way and is denoted by $\iota(t)$. We now define when ι *satisfies* a formula φ, denoted by $\iota \models \varphi$, recursively as follows: $\iota \models t_1 \diamond t_2$ for $\diamond \in \{<, \leq, =, \geq, >\}$ if and only if $\iota(t_1) \diamond \iota(t_2)$ is *true*, $\iota \models \neg \Phi$ if $\iota \not\models \Phi$, and $\iota \models \Phi_1 \vee \Phi_2$ if and only if $\iota \models \Phi_1$ or $\iota \models \Phi_2$. We say that an LRA formula Φ is *satisfiable* if there exists an interpretation ι with $\iota \models \Phi$.

Despite being NP-complete, with the rise of the SAT/SMT revolution [44], checking the satisfiability of LRA formulas can be handled effectively by several highly-optimized SMT solvers [5,18,47].

Algorithm Overview. Our algorithm constructs a series of LRA formulas $\langle \Phi^n_{\mathcal{S},K} \rangle_{n=1,2,\ldots}$ to facilitate the search for a suitable MTL formula. The formula $\Phi^n_{\mathcal{S},K}$ has the following properties:

1. $\Phi^n_{\mathcal{S},K}$ is satisfiable if and only if there exists an MTL formula φ of size n such that φ is **G**-sep for $\mathcal{S}$ and $\mathit{fr}(\varphi) \leq K$.
2. from any satisfying interpretation ι of $\Phi^n_{\mathcal{S},K}$, one can construct an appropriate MTL formula φ^ι.

In our algorithm, sketched in Algorithm 1, we first check whether $\mathcal{S}$ is K-*infix-separable* (as described in Sect. 4) which informs us whether a prospective formula exists. We now check the satisfiability of $\Phi^n_{\mathcal{S},K}$ for increasing values of size n starting from 1. If $\Phi^n_{\mathcal{S},K}$ is satisfiable for some n, then our algorithm constructs a prospective MTL formula φ^ι from a satisfying interpretation ι returned by the SMT solver. This algorithm terminates because of checking whether a solution exists apriori and it returns a minimal formula because of the iterative search through MTL formulas of increasing sizes.

The crux of our algorithm lies in constructing the formula $\Phi^n_{\mathcal{S},K}$. Internally, $\Phi^n_{\mathcal{S},K} := \Phi^{str}_{n,\mathcal{S},K} \wedge \Phi^{fr}_{n,\mathcal{S},K} \wedge \Phi^{sem}_{n,\mathcal{S},K}$ is a conjunction of three subformulas, each with a distinct role. The subformula $\Phi^{str}_{n,\mathcal{S},K}$ encodes the structure of the prospective MTL formula. The subformula $\Phi^{fr}_{n,\mathcal{S},K}$ ensures that the future-reach of the prospective formula is less than or equal to K. Finally, the subformula $\Phi^{sem}_{n,\mathcal{S},K}$ ensures that the prospective formula is **G**-sep for $\mathcal{S}$. In what follows, we expand on the construction of each of the introduced subformulas. We drop the subscripts n, $\mathcal{S}$, and K from the subformulas when clear from the context.

Structural Constraints. Following Neider and Gavran [48], we symbolically encode the syntax-DAG of the prospective MTL formula using the formula Φ^{str}. For this, we first fix a naming convention for the nodes of the syntax-DAG of an MTL formula. For a formula of size n, we assign to each of its nodes an identifier from $\{1, \ldots, n\}$ such that the identifier of each node is larger than that of its children if it has any. Note that such a naming convention may not be unique. Based on these identifiers, we denote the subformula of φ rooted at Node i as $\varphi[i]$. In that case, $\varphi[n]$ is precisely the formula φ.

Next, to encode a syntax-DAG symbolically, we introduce the following variables[4]: (i) Boolean variables $x_{i,\lambda}$ for $i \in \{1, \ldots, n\}$ and $\lambda \in \mathcal{P} \cup \{\neg, \vee, \wedge, \mathbf{U}_I, \mathbf{F}_I, \mathbf{G}_I\}$; (ii) Boolean variables $l_{i,j}$ and $r_{i,j}$ for $i \in \{1, \ldots, n\}$ and $j \in \{1, \ldots, i\}$; (iii) real variables a_i and b_i for $i \in \{1, \ldots, n\}$. The variable $x_{i,\lambda}$ tracks the operator labeled in Node i, meaning, $x_{i,\lambda}$ is set to true if and only if Node i is labeled with λ. The variable $l_{i,j}$ (resp., $r_{i,j}$) tracks the left (resp., right) child of Node i, meaning, $l_{i,j}$ (resp., $r_{i,j}$) is set to true if and only if the left (resp., right) child of Node i is Node j. Finally, the variable a_i (resp., b_i) tracks the lower (resp., upper) bound of the interval I of a temporal operator (i.e., operators $\mathbf{U}_I$, $\mathbf{F}_I$ and $\mathbf{G}_I$), meaning that, if a_i (resp., b_i) is set to $a \in \mathbb{R}$ (resp., $b \in \mathbb{R}$), then the lower (resp., upper) bound of the interval of the operator in Node i is a (resp., b). While we introduce variables a_i and b_i for each node, they become relevant only for the nodes that are labeled with a temporal operator.

We now impose structural constraints on the introduced variables to ensure they encode valid MTL formulas. Exemplarily, we have the following constraint:

$$\left[\bigwedge_{1 \leq i \leq n} \bigvee_{\lambda \in \Lambda} x_{i,\lambda} \right] \wedge \left[\bigwedge_{1 \leq i \leq n} \bigwedge_{\lambda \neq \lambda' \in \Lambda} \neg x_{i,\lambda} \vee \neg x_{i,\lambda'} \right],$$

where $\Lambda = \mathcal{P} \cup \{\neg, \vee, \wedge, \mathbf{U}_I, \mathbf{F}_I, \mathbf{G}_I\}$. The above constraint ensures that each node is labeled by exactly one operator or one proposition. We also impose other structural constraints, such as each node can have at most two children, Node 1 must be a proposition, etc. Such structural constraints are similar to the ones proposed by Neider and Gavran [48]. We here additionally ensure that the $\neg$ operator appears only in front of propositions. Also, we ensure that the intervals of the temporal operators are proper using the constraint $\bigwedge_{1 \leq i \leq n} 0 \leq a_i \leq b_i < K$. We refer interested readers to the extended version of the paper [54] for all the constraints.

The subformula Φ^{str} is a conjunction of all the structural constraints we described. Using a satisfying interpretation ι of Φ^{str}, one can construct the syntax DAG of a unique MTL formula φ^ι.

Future-Reach Constraints. To symbolically compute the future-reach of the prospective formula φ, we encode the inductive definition of the future-reach, as described in Sect. 3 in an LRA formula. To this end, we introduce real variables

---

[4] We include Boolean variables in our LRA formulas since Boolean variables can always be simulated using real variables that are constrained to be either 0 or 1.

f_i for $i \in \{1, \ldots, n\}$ to encode the future-reach of the subformula $\varphi[i]$. Precisely, f_i is set to $f \in \mathbb{R}$ if and only if $fr(\varphi[i]) = f$.

To ensure the desired meaning of the f_i variables, we impose constraints like

$$\bigwedge_{1 \le i \le n, 1 \le j < i} [x_{i,\mathbf{F}_I} \wedge l_{i,j}] \to [f_i = f_j + b_i]. \tag{1}$$

This constraint expresses that if Node i contains the $\mathbf{F}_I$ operator where I is encoded using a_i and b_i, then the future-reach of $\varphi[i]$, i.e., $fr(\varphi[i]))$, must be the future-reach of $\varphi[j]$ plus b, i.e., $fr(\varphi[j]) + b$. For other operators, we impose similar constraints based on the definition of future-reach for that operator, described in Sect. 3. We refer the readers to the extended version of the paper [54] for the remaining future-reach constraints.

Finally, to enforce that the future-reach of the prospective MTL formula is within K, along with the constraints mentioned above, we have $f_n \le K$ in Φ^{fr}.

Semantic Constraints. To symbolically check whether the prospective formula is **G**-sep, we must encode the procedure of checking the satisfaction of an MTL formula into an LRA formula. To this end, we rely on the monitoring procedure devised by Maler and Nickovic [43] for efficiently checking when a signal satisfies an MTL formula. Since our setting is slightly different, we take a brief detour via the description of our adaptation of the monitoring algorithm.

Given an MTL formula φ and a signal prefix $\boldsymbol{x}_\mathrm{T}$, our monitoring algorithm computes the (lexicographically) ordered set $\mathcal{I}_\varphi(\boldsymbol{x}_\mathrm{T}) = \{I_1, \cdots, I_\eta\}$ of *maximal disjoint time intervals* $I_1, \cdots, I_\eta$ where φ holds on $\boldsymbol{x}_\mathrm{T}$. Mathematically speaking, the following property holds for the set $\mathcal{I}_\varphi(\boldsymbol{x}_\mathrm{T})$ we construct:

Lemma 5. *Given an MTL formula φ and a prefix $\boldsymbol{x}_\mathrm{T}$, for all $t \in \mathbb{T}$, $(\boldsymbol{x}_\mathrm{T}, t) \models_\mathrm{f} \varphi$ if and only if $t \in I$ for some $I \in \mathcal{I}_\varphi(\boldsymbol{x}_\mathrm{T})$.*

In our monitoring algorithm, we compute the set $\mathcal{I}_\varphi(\boldsymbol{x}_\mathrm{T})$ inductively on the structure of the formula φ. To describe the induction, we use the notation $\mathcal{I}_\varphi^\cup(\boldsymbol{x}_\mathrm{T}) = \bigcup_{I \in \mathcal{I}_\varphi(\boldsymbol{x}_\mathrm{T})} I$ to denote the union of the intervals in $\mathcal{I}_\varphi(\boldsymbol{x}_\mathrm{T})$. For the base case, we compute $\mathcal{I}_p(\boldsymbol{x}_\mathrm{T})$ for every $p \in \mathcal{P}$ by accumulating the time points $t \in [0, T)$ where $(\boldsymbol{x}_\mathrm{T}, t) \models_\mathrm{f} p$ into maximal disjoint time intervals. In the inductive step, we exploit the relations presented in Table 1 for the different MTL operators. In the table, we use the following notation: $[t_1, t_2) \ominus [a, b] = [t_1 - b, t_2 - a) \cap \mathbb{T}$ and $\mathcal{I}^c = \mathbb{T} - \mathcal{I}$. While the table presents the computation of $\mathcal{I}_\varphi^\cup(\boldsymbol{x}_\mathrm{T})$, we can obtain $\mathcal{I}_\varphi(\boldsymbol{x}_\mathrm{T})$ by simply partitioning $\mathcal{I}_\varphi^\cup(\boldsymbol{x}_\mathrm{T})$ into maximal disjoint intervals.

For an illustration of the relations present in Table 1, we consider the introductory example and compute $\mathcal{I}_{\varphi_2}(u_1)$ where $u_1 = \{p, q\}\{p\}\{q\}\{p, q\}$ is the first positive prefix, $\varphi_2 = p \vee \mathbf{F}_{[0,1]} q$, and $\mathbb{T} = [0, 6)$. First, we have $\mathcal{I}_p(u_1) = \{[0, 2), [3, 6)\}$ and $\mathcal{I}_q(u_1) = \{[0, 1), [2, 4)\}$. Now, we can compute $\mathcal{I}_{\mathbf{F}_{[0,1]} q}(u_1) = \{[0, 4), [5, 6)\}$ and then $\mathcal{I}_{p \vee \mathbf{F}_{[0,1]} q}(u_1) = \{[0, 6)\}$.

Table 1. The relations for inductive computation of $\mathcal{I}_\varphi^\cup(\boldsymbol{x}_\mathbb{T})$.

$$\mathcal{I}_{\neg p}^\cup(\boldsymbol{x}_\mathbb{T}) = \left(\mathcal{I}_p^\cup(\boldsymbol{x}_\mathbb{T})\right)^c$$

$$\mathcal{I}_{\varphi_1 \vee \varphi_2}^\cup(\boldsymbol{x}_\mathbb{T}) = \mathcal{I}_{\varphi_1}^\cup(\boldsymbol{x}_\mathbb{T}) \cup \mathcal{I}_{\varphi_2}^\cup(\boldsymbol{x}_\mathbb{T})$$

$$\mathcal{I}_{\varphi_1 \wedge \varphi_2}^\cup(\boldsymbol{x}_\mathbb{T}) = \mathcal{I}_{\varphi_1}^\cup(\boldsymbol{x}_\mathbb{T}) \cap \mathcal{I}_{\varphi_2}^\cup(\boldsymbol{x}_\mathbb{T})$$

$$\mathcal{I}_{\mathbf{F}_{[a,b]}\,\varphi}^\cup(\boldsymbol{x}_\mathbb{T}) = \left(\bigcup_{I \in \mathcal{I}_\varphi(\boldsymbol{x}_\mathbb{T})} I \ominus [a,b]\right) \cup [T-b, T)$$

$$\mathcal{I}_{\mathbf{G}_{[a,b]}\,\varphi}^\cup(\boldsymbol{x}_\mathbb{T}) = \left(\bigcup_{I \in (\mathcal{I}_\varphi(\boldsymbol{x}_\mathbb{T}))^c} I \ominus [a,b]\right)^c \cup [T-a, T)$$

$$\mathcal{I}_{\varphi\,\mathbf{U}_{[a,b]}\,\psi}^\cup(\boldsymbol{x}_\mathbb{T}) = \bigcup_{I_\varphi \in \mathcal{I}_\varphi(\boldsymbol{x}_\mathbb{T})} \bigcup_{I_\psi \in \mathcal{I}_\psi(\boldsymbol{x}_\mathbb{T})} \left(\left((I_\varphi \cap I_\psi) \ominus [a,b]\right) \cap I_\varphi\right) \cup I_\mathbb{T},$$

$$\text{where } I_\mathbb{T} = \begin{cases} [\max(T-b, t), T), & \text{if } \exists t \text{ s.t. } [t, T) \in \mathcal{I}_\varphi(\boldsymbol{x}_\mathbb{T}) \\ \emptyset, & \text{otherwise} \end{cases}$$

In the monitoring algorithm, the number of maximal intervals required in $\mathcal{I}_\varphi(\boldsymbol{x}_\mathbb{T})$ is upper-bounded by $\mathcal{M} = \mu|\varphi|$, where $\mu = \max(\{|\mathcal{I}_p(\boldsymbol{x}_\mathbb{T})| \mid p \in \mathcal{P}\})$, as also observed by Maler and Nickovic [43]. The computation of this bound can also be done inductively on the structure of φ.

Now, in the subformula $\varPhi^{sem}$, we symbolically encode the set $\mathcal{I}_\varphi(\boldsymbol{x}_\mathbb{T})$ of our prospective MTL formula φ. To this end, we introduce variables $t_{i,m,s}^l$ and $t_{i,m,s}^r$ where $i \in \{1, \ldots, n\}$, $m \in \{1, \ldots, \mathcal{M}\}$, and $s \in \{1, \ldots, |\mathcal{S}|\}$, s being an identifier for the sth prefix $\boldsymbol{x}_\mathbb{T}^s$ in $\mathcal{S}$. The variables $t_{i,m,s}^l$ and $t_{i,m,s}^r$ encode the mth interval of $\mathcal{I}_{\varphi[i]}(\boldsymbol{x}_\mathbb{T}^s)$ for the subformula $\varphi[i]$. In other words, $t_{i,m,s}^l = t_1$ and $t_{i,m,s}^r = t_2$ if and only if $[t_1, t_2)$ is the mth interval of $\mathcal{I}_{\varphi[i]}(\boldsymbol{x}_\mathbb{T}^s)$.

To ensure that the variables $t_{i,m,s}^l$ and $t_{i,m,s}^r$ have their desired meaning, we introduce constraints for each operator based on the relations defined in Table 1. We now present these constraints for the different MTL operators.

For the $\neg$ operator, we have the following constraints:

$$\bigwedge_{\substack{1 \le i \le n \\ 1 \le j < i}} x_{i,\neg} \wedge l_{i,j} \rightarrow \Big[\bigwedge_{1 \le s \le |\mathcal{S}|} comp_s(i,j)\Big],$$

where, for every $\boldsymbol{x}_\mathbb{T}^s$ in $\mathcal{S}$, $comp_s(i,j)$ encodes that $\mathcal{I}_{\varphi[i]}^\cup(\boldsymbol{x}_\mathbb{T}^s)$ is the complement of $\mathcal{I}_{\varphi[j]}^\cup(\boldsymbol{x}_\mathbb{T}^s)$. We construct $comp_s(i,j)$ as follows:

$$\texttt{ite}(t_{j,1,s}^l = 0, \tag{2}$$

$$\bigwedge_{1 \le m \le \mathcal{M}-1} t_{i,m,s}^l = t_{j,m,s}^r \wedge t_{i,m,s}^r = t_{j,m+1,s}^l, \tag{3}$$

$$t_{i,1,s}^l = 0 \wedge t_{i,1,s}^r = t_{j,1,s}^l \wedge$$

$$\bigwedge_{1 \le m \le \mathcal{M}-1} t_{i,m+1,s}^l = t_{j,m,s}^r \wedge t_{i,m+1,s}^r = t_{j,m+1,s}^l), \tag{4}$$

where $\texttt{ite}$ is a syntactic sugar for the "if-then-else" construct over LRA formulas, which is standard in many SMT solvers. Here, Condition 2 checks whether the left bound of the first interval of $\mathcal{I}_{\varphi[j]}(\boldsymbol{x}_{\mathbb{T}}^s)$, encoded by $t_{j,1,s}^l$, is 0. If that holds, as specified by Constraint 3, the left bound of the first interval of $\mathcal{I}_{\varphi[i]}(\boldsymbol{x}_{\mathbb{T}}^s)$, encoded by $t_{1,i,s}^l$, will be the right bound of the first interval of $\mathcal{I}_{\varphi[j]}(\boldsymbol{x}_{\mathbb{T}}^s)$, encoded $t_{1,j,s}^r$ and so on. If Condition 2 does not hold, as specified by Constraint 4, the left bound of the first interval of $\mathcal{I}_{\varphi[i]}(\boldsymbol{x}_{\mathbb{T}}^s)$ will start with 0, and so on.

As an example, for a prefix $\boldsymbol{x}_{\mathbb{T}}^s$ and $\mathbb{T} = [0,7)$, let $\mathcal{I}_{\varphi[j]}(\boldsymbol{x}_{\mathbb{T}}^s) = \{[0,4),[6,7)\}$. Then, Constraint 3 ensures that $\mathcal{I}_{\varphi[i]}(\boldsymbol{x}_{\mathbb{T}}^s) = \{[4,6)\}^5$. Conversely, if $\mathcal{I}_{\varphi[j]}(\boldsymbol{x}_{\mathbb{T}}^s) = \{[1,4),[6,7)\}$, then Constraints 4 ensures that $\mathcal{I}_{\varphi[i]}(\boldsymbol{x}_{\mathbb{T}}^s) = \{[0,1),[4,6)\}$.

For the $\vee$ operator, we have the following constraint:

$$\bigwedge_{\substack{1 \le i \le n \\ 1 \le j,j' < i}} x_{i,\vee} \wedge l_{i,j} \wedge r_{i,j'} \to \Big[\bigwedge_{1 \le s \le |\mathcal{S}|} union_s(i,j,j') \Big],$$

where, for every $\boldsymbol{x}_{\mathbb{T}}^s$ in $\mathcal{S}$, $union_s(i,j,j')$ encodes that $\mathcal{I}_{\varphi[i]}(\boldsymbol{x}_{\mathbb{T}}^s)$ consists of the maximal disjoint intervals obtained from the union of the intervals in $\mathcal{I}_{\varphi[j]}(\boldsymbol{x}_{\mathbb{T}}^s)$ and $\mathcal{I}_{\varphi[j']}(\boldsymbol{x}_{\mathbb{T}}^s)$. We construct $union_s(i,j,j')$ as follows:

$$\bigwedge_{\sigma \in [l,r]} \bigwedge_{1 \le m \le M} \left(\bigvee_{1 \le m' \le M} (t_{i,m,s}^\sigma = t_{j,m',s}^\sigma) \vee \bigvee_{1 \le m' \le M} (t_{i,m,s}^\sigma = t_{j',m',s}^\sigma) \right) \wedge \quad (5)$$

$$\bigwedge_{\sigma \in [l,r]} \bigwedge_{1 \le m \le M} \left(\bigvee_{1 \le m' \le M} (t_{i,m,s}^\sigma = t_{j,m',s}^\sigma) \iff \bigwedge_{1 \le m'' \le M} (t_{j,m',s}^\sigma \notin I_{j',m'',s}) \right) \wedge \quad (6)$$

$$\bigwedge_{\sigma \in [l,r]} \bigwedge_{1 \le m \le M} \left(\bigvee_{1 \le m' \le M} (t_{i,m,s}^\sigma = t_{j',m',s}^\sigma) \iff \bigwedge_{1 \le m'' \le M} (t_{j',m',s}^\sigma \notin I_{j,m'',s}) \right), \quad (7)$$

where $I_{k,m,s}$ denotes the interval encoded by bounds $t_{k,m,s}^l$ and $t_{k,m,s}^r$[6]. Here, Constraint 5 states that the left (resp., right) bound of each interval of $\mathcal{I}_{\varphi[i]}(\boldsymbol{x}_{\mathbb{T}}^s)$, encoded by $t_{i,m,s}^l$ (resp., $t_{i,m,s}^r$) corresponds to one of the left (resp., right) bounds of the intervals in $\mathcal{I}_{\varphi[j]}(\boldsymbol{x}_{\mathbb{T}}^s)$ or in $\mathcal{I}_{\varphi[j']}(\boldsymbol{x}_{\mathbb{T}}^s)$. Then, Constraint 6 states that for each interval I in $\mathcal{I}_{\varphi[j]}(\boldsymbol{x}_{\mathbb{T}}^s)$, the left (resp., right) bound of I should appear as the left (resp., right) bound of some interval in $\mathcal{I}_{\varphi[i]}(\boldsymbol{x}_{\mathbb{T}}^s)$ if and only if the left (resp., right) bound of I is not included in any of the intervals in $\mathcal{I}_{\varphi[j']}(\boldsymbol{x}_{\mathbb{T}}^s)$. Constraint 7 mimics the statement made by Constraint 6 but for the bounds of the intervals in $\mathcal{I}_{\varphi[j']}(\boldsymbol{x}_{\mathbb{T}}^s)$.

For an illustration, assume that $\mathcal{I}_{\varphi[j]}(\boldsymbol{x}_{\mathbb{T}}^s) = \{[1,4),[6,7)\}$ and $\mathcal{I}_{\varphi[j']}(\boldsymbol{x}_{\mathbb{T}}^s) = \{[3,5),[6,7)\}$ for a prefix $\boldsymbol{x}_{\mathbb{T}}^s$ and $T = 7$. Now, if $\varphi[i] = \varphi[j] \vee \varphi[j']$, then

---

[5] $|\mathcal{I}_{\varphi[i]}(\boldsymbol{x}_{\mathbb{T}}^s)|$ may differ for different $\varphi[i]$; we address this at the end of Sect. 5.

[6] In LRA, $t \notin [t_1,t_2)$ can be encoded as $t < t_1 \vee t \ge t_2$.

$\mathcal{I}_{\varphi[i]}(\boldsymbol{x}_{\mathbb{T}}^s) = \{[1,5), [6,7)\}$ based on the relation for ∨-operator in Table 1. Observe that all the bounds of the intervals in $\mathcal{I}_{\varphi[i]}(\boldsymbol{x}_{\mathbb{T}}^s)$, i.e., 1, 5, 6, and 7, are present as the bounds of the intervals in either $\mathcal{I}_{\varphi[j]}(\boldsymbol{x}_{\mathbb{T}}^s)$ or $\mathcal{I}_{\varphi[j']}(\boldsymbol{x}_{\mathbb{T}}^s)$. This fact is in accordance with Constraint 5. Also, the right bound of $[1,4)$ in $\mathcal{I}_{\varphi[j]}(\boldsymbol{x}_{\mathbb{T}}^s)$ does not appear as a bound of any intervals in $\mathcal{I}_{\varphi[i]}(\boldsymbol{x}_{\mathbb{T}}^s)$, as it is included in an interval in $\mathcal{I}_{\varphi[j']}(\boldsymbol{x}_{\mathbb{T}}^s)$, i.e., $4 \in [3,5)$. This is in accordance with Constraint 6.

Next, for the $\mathbf{F}_I$-operator where I is encoded using a_i and b_i, we have the following constraint:

$$\bigwedge_{\substack{1 \leq i \leq n \\ 1 \leq j < i}} x_{i,\mathbf{F}_I} \wedge l_{i,j} \rightarrow \Big[\bigwedge_{1 \leq s \leq |\mathcal{S}|} union_s'(i,k,k) \wedge \ominus_s^{[a_i,b_i]}(k,j) \Big].$$

based on the relation for the $\mathbf{F}_{[a,b]}$ operator in Table 1. We here rely on an intermediate set of intervals $\tilde{\mathcal{I}}_k$ encoded using some auxiliary variables $\tilde{t}_{k,m,s}^l$ and $\tilde{t}_{k,m,s}^r$ where $m \in \{1,\dots,\mathcal{M}\}$ and $s \in \{1,\dots,|\mathcal{S}|\}$. Also, we use the formula $\ominus_s^{[a_i,b_i]}(k,j)$ to encode that the intervals in $\tilde{\mathcal{I}}_k$ can be obtained by performing $I \ominus [a,b]$ to each interval I in $\mathcal{I}_{\varphi[j]}(\boldsymbol{x}_{\mathbb{T}}^s)$, where $a_i = a$ and $b_i = b$. Finally, the formula $union'(i,k,k)$ encodes that $\mathcal{I}_{\varphi[i]}(\boldsymbol{x}_{\mathbb{T}}^s)$ consists of the maximal disjoint intervals obtained from the union of the intervals in $\tilde{\mathcal{I}}_k$ and $\{[T-b,T)\}$.

The construction of $union'(i,k,k)$ is similar to that of $union(i,j,j')$ in that the constraints involved are similar to Constraints 5 to 7. For $\ominus_s^{[a_i,b_i]}(k,j)$, we have the following constraint:

$$\bigwedge_{1 \leq m \leq \mathcal{M}-1} \big[\tilde{t}_{k,m,s}^l = \max\{0, (t_{j,m,s}^l - b_i)\} \wedge \tilde{t}_{k,m,s}^r = \max\{0, (t_{j,m,s}^r - a_i)\} \big] \quad (8)$$

As an example, consider $\mathcal{I}_{\varphi[j]}(\boldsymbol{x}_{\mathbb{T}}^s) = \{[1,4), [6,7)\}$ for a prefix $\boldsymbol{x}_{\mathbb{T}}^s$ and $T = 7$. Now, if $\varphi[i] = \mathbf{F}_{[1,4]} \varphi[j]$, then first we have $\tilde{\mathcal{I}}_k = \{[0,3), [2,6)\}$ based on Constraint 8[7]. Next, we have $\mathcal{I}_{\varphi[i]}(\boldsymbol{x}_{\mathbb{T}}^s) = \{[0,7)\}$ which consists of the maximal disjoint intervals from $\tilde{\mathcal{I}}_k \cup \{[T-4,T)\} = \{[0,3), [2,6), [3,7)\}$ using $union'(i,k,k)$.

For the $\mathbf{U}_I$ operator, we have the following constraint:

$$\bigwedge_{\substack{1 \leq i \leq n \\ 1 \leq j,j' < i}} x_{i,\mathbf{U}_I} \wedge l_{i,j} \wedge r_{i,j'} \rightarrow \Big[\bigwedge_{1 \leq s \leq |\mathcal{S}|} int_s(k_1,j,j') \wedge \ominus_s^{[a_i,b_i]}(k_2,k_1) \quad (9)$$

$$\wedge cond\text{-}int_s(k_3,k_2,j) \wedge union_s'(i,k_3,k_3) \big] \quad (10)$$

Here, we introduce three intermediate set of intervals $\tilde{\mathcal{I}}_{k_1}$, $\tilde{\mathcal{I}}_{k_2}$ and $\tilde{\mathcal{I}}_{k_3}$ encoded using auxiliary variables $\tilde{t}_{k_i,m,s}^l$ and $\tilde{t}_{k_i,m,s}^r$ where $i \in \{1,2,3\}$, $m \in \{1,\dots,\mathcal{M}\}$ and $s \in \{1,\dots,|\mathcal{S}|\}$. Similar to the constraints for the ∨ operator, we denote an interval in $\tilde{\mathcal{I}}_{k_i}$ as $I_{k_i,m,s}$ where $I_{k_i,m,s} = [\tilde{t}_{k_i,m,s}^l, \tilde{t}_{k_i,m,s}^r)$.

---

[7] While the intervals in $\tilde{\mathcal{I}}_k$ may not be disjoint, $union'(i,k,k)$ ensures that $\mathcal{I}_{\varphi[i]}(\boldsymbol{x}_{\mathbb{T}}^s)$ consists of only maximal disjoint intervals.

In the above constraint, $int_s(k_1, j, j')$ encodes that $\tilde{\mathcal{I}}_{k_1}$ consists of the maximal disjoint intervals obtained from the intersection of the intervals in $\mathcal{I}_{\varphi[j]}(\boldsymbol{x}_\mathbb{T}^s)$ and $\mathcal{I}_{\varphi[j']}(\boldsymbol{x}_\mathbb{T}^s)$. One can construct $int_s(k_1, j, j')$ using formulas $union_s(k_1, j, j')$ and $comp_s(k_1, j, j')$ based on De Morgan's law, $A \cap B = (A^c \cup B^c)^c$. Next, $\ominus_s^{[a_i, b_i]}(k_2, k_1)$ encodes that the intervals in $\tilde{\mathcal{I}}(k_2)$ are obtained by performing $I \ominus [a, b]$, where $a_i = a$ and $b_i = b$, to each interval I in $\tilde{\mathcal{I}}_{k_1}$ using Constraint 8. The next formula $cond-int_s(k_3, k_2, j)$ denotes the conditional intersection of intervals in $\tilde{\mathcal{I}}(k_2)$ and $\mathcal{I}_{\varphi[j]}(\boldsymbol{x}_\mathbb{T}^s)$. Precisely, it encodes that the mth interval in $\tilde{\mathcal{I}}(k_3)$, i.e., $I_{k_3, m, s}$, is obtained by performing the intersection of the mth interval in $\tilde{\mathcal{I}}(k_2)$, i.e., $I_{k_2, m, s}$, and the m'^{th} interval in $\mathcal{I}_{\varphi[j]}(\boldsymbol{x}_\mathbb{T}^s)$, i.e., $I_{j, m', s}$ if $I_{k_1, m, s}$ (note $I_{k_2, m, s} = I_{k_1, m, s} \ominus [a, b]$, by construction) is a subset of $I_{j, m', s}$. This can be achieved by encoding $cond-int_s(k_3, k_2, j)$ as the following constraint:

$$\bigwedge_{1 \leq m \leq \mathcal{M}} \bigwedge_{1 \leq m' \leq \mathcal{M}} (I_{k_1, m, s} \subseteq I_{j, m', s}) \rightarrow I_{k_3, m, s} = I_{k_2, m, s} \cap I_{j, m', s}$$

Both the subset check and the intersection of intervals allow simple encodings in LRA. Finally, the formula $union_s(i, k_3, k_3)$ encodes that $\mathcal{I}_{\varphi[i]}(\boldsymbol{x}_\mathbb{T}^s)$ consists of the maximal disjoint intervals obtained from the union of the intervals in $\tilde{\mathcal{I}}_{k_3}$.

For an illustration, assume that $\mathcal{I}_{\varphi[j]}(\boldsymbol{x}_\mathbb{T}^s) = \{[1, 3), [5, 8)\}$ and $\mathcal{I}_{\varphi[j']}(\boldsymbol{x}_\mathbb{T}^s) = \{[4, 6), [7, 9)\}$ for a prefix $\boldsymbol{x}_\mathbb{T}^s$ and $T = 9$. Now, let $\varphi[i] = \varphi[j] \mathbf{U}_{[0,3]} \varphi[j']$. Then, according to the computation in Table 1, $\mathcal{I}_{\varphi[i]}(\boldsymbol{x}_\mathbb{T}^s) = \{[5, 8)\}$.

Following Constraint 9, first we have $\tilde{\mathcal{I}}_{k_1} = \{[5, 6), [7, 8)\}$, obtained by the intersection of intervals in $\mathcal{I}_{\varphi[j]}(\boldsymbol{x}_\mathbb{T}^s)$ and $\mathcal{I}_{\varphi[j']}(\boldsymbol{x}_\mathbb{T}^s)$. Then, we have $\tilde{\mathcal{I}}_{k_2} = \{[2, 6), [4, 8)\}$, obtained by applying Minkowski minus with $a = 0$ and $b = 3$ on the intervals in $\tilde{\mathcal{I}}_{k_1}$. Next, we have $\tilde{\mathcal{I}}_{k_3} = \{[5, 6), [5, 8)\}$, obtained by the conditional intersection of the intervals in $\tilde{\mathcal{I}}_{k_2}$ and $\mathcal{I}_{\varphi[j]}(\boldsymbol{x}_\mathbb{T}^s)$. Notice that both the intervals in $\tilde{\mathcal{I}}_{k_1}$ are subsets of the interval $[5, 8)$ in $\mathcal{I}_{\varphi[j]}(\boldsymbol{x}_\mathbb{T}^s)$ and, thus, we intersect the intervals in $\tilde{\mathcal{I}}_{k_2}$ with $[5, 8)$. Finally, we have $\mathcal{I}_{\varphi[i]}(\boldsymbol{x}_\mathbb{T}^s) = \{[5, 8)\}$, obtained by the union of intervals of $\tilde{\mathcal{I}}_{k_3}$. The overall computation complies with the actual semantics of the $\mathbf{U}_I$ operator.

For Constraint 9, we briefly discuss an intricacy that results from the tricky computation of the intervals for $\mathbf{U}_I$, as also identified by [43] in Fig. 3(a). In particular, if one considered the (standard) intersection (i.e., $int_s(k_3, k_2, j)$) instead of the conditional intersection (i.e., $cond-int_s(k_3, k_2, j)$) in Constraint 9, it lead to a wrong result. In the above example, one can see that the (standard) intersection would have resulted in the incorrect set $\mathcal{I}_{[}(\boldsymbol{x}_\mathbb{T}^s)\varphi[i]] = \{[2, 3), [5, 8)\}$.

For the $\mathbf{G}_I$ and the $\wedge$ operators, we similarly encode the relations described in Table 1 by reusing the formulas $comp_s(i, j)$, $union_s(i, j, j')$, and $\ominus_s^{[a_i, b_i]}(k, j)$. We present the exact constraints in the extended version of the paper [54]. We now assert the correctness of the formulas encoding the set operations as follows:

Lemma 6. *The formulas $comp_s(i, j)$, $union_s(i, j, j')$, $\ominus_s^{[a_i, b_i]}(k, j)$, $int_s(i, j, j')$ and $cond-int_s(i, j, j')$ correctly encode the complement, union, $\ominus$, intersection and conditional intersection operations on sets of intervals, resp.*

The proof of the lemma is presented in the extended version of the paper [54].

It is worth noting that although the number of intervals in $\mathcal{I}_{\varphi[i]}(x_{\mathbb{T}}^s)$ for each subformula $\varphi[i]$ is bounded by $\mathcal{M}$, it may not contain the same number of intervals. For instance, $\mathcal{I}_p(x_{\mathbb{T}}^s) = \{[0,1), [6,7)\}$ has two intervals, while, assuming $T = 7$, $\mathcal{I}_{\neg p}(x_{\mathbb{T}}^s) = \{[1,6)\}$ has only one interval.

To circumvent this, we introduce some variables $num_{i,s}$ for $i \in \{1, \ldots, n\}$ and $s \in \{1, \ldots, |\mathcal{S}|\}$ to track of the number of intervals in $\mathcal{I}_{\varphi[i]}(x_{\mathbb{T}}^s)$ for each subformula $\varphi[i]$ for each prefix $x_{\mathbb{T}}^s$. We now impose $\bigwedge_{1 \leq i \leq n, 1 \leq m \leq \mathcal{M}} [m > num_{i,s}] \rightarrow [t_{i,m,s}^l = T \wedge t_{i,m,s}^r = T]$. This ensures that all the unused variables $t_{i,m,s}^\sigma$ for each Node i and prefix $x_{\mathbb{T}}^s$ in $\mathcal{S}$ are all set to T. We also use the $num_{i,s}$ variables in the constraints for easier computation of $\mathcal{I}_{\varphi[i]}(x_{\mathbb{T}}^s)$ for each operator. We include this in our implementation but omit it here for ease of presentation.

Finally, to ensure that the prospective formula φ is **G**-sep for $\mathcal{S}$, we add:

$$\bigwedge_{x_{\mathbb{T}}^s \in P} [(t_{n,1,s}^l = 0) \wedge (t_{n,1,s}^r = T)] \wedge \bigwedge_{x_{\mathbb{T}}^s \in N} [(t_{n,1,s}^l \neq 0) \vee (t_{n,1,s}^r \neq T)].$$

This constraint says that $\mathcal{I}_{\varphi[n]}(x_{\mathbb{T}}^s) = \{[0,T)\}$ for all the positive prefixes $x_{\mathbb{T}}^s$, while $\mathcal{I}_{\varphi[n]}(x_{\mathbb{T}}^s) \neq \{[0,T)\}$ for any negative prefixes $x_{\mathbb{T}}^s$.

We now state the correctness result for the overall algorithm, Algorithm 1, as follows:

Theorem 1 (Correctness). *Given a sample $\mathcal{S}$ and a future-reach bound K, Algorithm 1 terminates and outputs a minimal MTL formula φ such that φ is globally separating for $\mathcal{S}$ and $fr(\varphi) \leq K$, if such a formula exists.*

Proof. The termination of Algorithm 1 is guaranteed by the decision procedure of checking whether $\mathcal{S}$ is K-*infix-separable* (Sect. 4). The minimality of the synthesized formula is due to the iterative search of formulas of increasing size and the correct encoding of $\Phi_{\mathcal{S},K}^n$. The correctness of $\Phi_{\mathcal{S},K}^n$ follows from the correctness of computation of the sets $\mathcal{I}_\varphi(x_{\mathbb{T}})$ using Lemma 5 and the correctness of the encoding of set operations described in Lemma 6.   □

Our algorithm solves the optimization problem SYNTL by constructing LRA formulas. To analyze the computational hardness of SYNTL, we consider its corresponding decision problem SYNTL$_d$: given a sample $\mathcal{S}$, a future-reach bound K and size bound B (in unary), does there exist an MTL formula φ such that φ is **G**-sep for $\mathcal{S}$, $fr(\varphi) \leq K$, and $|\varphi| \leq B$. Following our algorithm, we can encode the SYNTL$_d$ problem in an LRA formula $\Phi = \bigvee_{n \leq B} \Phi_{\mathcal{S},K}^n$, where $\Phi_{\mathcal{S},K}^n$ is as described in Algorithm 1. One can check that the size of Φ is $\mathcal{O}(|\mathcal{S}||K|B^3\mathcal{M}^3)$. Since LRA satisfiability is NP-complete [19], we directly get the following:

Theorem 2. SYNTL$_d$ *is in* NP.

Remark 1. While the exact complexity lower bound for SYNTL$_d$ is unknown, we conjecture that SYNTL$_d$ is NP-hard. Our hypothesis stems from the fact that similar decision problems are already NP-hard for simple fragments of LTL [25]. However, these hardness results do not directly extend to SYNTL$_d$ since LTL and MTL are different in terms of expressivity and succinctness.

6   Experiments

In this section, we answer the following research questions to assess the performance of our algorithm for synthesizing MTL formulas.

RQ1: Can our algorithm synthesize concise formulas with small future-reach?
RQ2: How does lowering the future-reach bound affect the size of the formulas?
RQ3: How does our algorithm scale for different sample sizes?

To answer the research questions above, we have implemented a prototype TEAL[8] [55] of our algorithm in Python 3 using Z3 [46] as the SMT solver. To our knowledge, TEAL is the only tool for synthesizing minimal MTL formulas for runtime monitoring (see related works). In TEAL, we implement a heuristic on top of Algorithm 1. We set the maximum number of intervals $\mathcal{M}$ in sets $\mathcal{I}_{\varphi[i]}(\boldsymbol{x}_{\mathbb{T}})$ to be $\mu+2$ where $\mu = \max(\{|\mathcal{I}_p(\boldsymbol{x}_{\mathbb{T}})| \mid p \in \mathcal{P}\})$. This heuristic improves the runtime of TEAL significantly since most **G**-sep formulas φ never require the worst-case upper bound[9] of $\mathcal{M} = \mu|\varphi|$. To ensure that TEAL returns a correct MTL formula with this heuristic, we implement a verifier based on the inductive computation of $\mathcal{I}_{\varphi}(\boldsymbol{x}_{\mathbb{T}})$ from Table 1. In our experiments, the verifier ensured that all of the synthesized MTL formulas were correct. One can fine-tune the heuristic based on the sample and the expected MTL formulas.

As typically done in the literature of synthesizing formulas [2,48,53], we evaluate TEAL on benchmarks generated synthetically from MTL formulas. To obtain useful MTL formulas, we identify a number of MTL patterns, listed in Table 2, commonly used for monitoring cyber-physical systems. For instance, the time-sensitive requirement of an electronically controlled steering (ECS) system "operational checks like RAM verification must be done every 20 secs" can be monitored globally using the *bounded recurrence* formula $\mathbf{F}_{[0,20]}$ operational_check [36]; the requirement of an autonomous vehicle (from the introductory example) "brake should be triggered until within 2 secs the vehicle has no obstacle in an unsafe distance" can be monitored globally using the *bounded until* formula brake $\mathbf{U}_{[0,2]}$ no_obstacle.

In our experiments, we construct MTL formulas from the patterns in Table 2 by replacing time interval $[t_1, t_2]$ with different values. Now, to generate a sample

Table 2. Typical MTL patterns used for monitoring cyber-physical systems.

Bounded Recurrence:	$Globally(\mathbf{F}_{[t_1,t_2]}\, p)$
Bounded Response:	$Globally(p \rightarrow \mathbf{F}_{[t_1,t_2]}\, q)$
Bounded Invariance:	$Globally(p \rightarrow \mathbf{G}_{[t_1,t_2]}\, q)$
Bounded Until:	$Globally(p\, \mathbf{U}_{[t_1,t_2]}\, q)$

---

[8] TEAL is available at https://github.com/ritamraha/Teal.
[9] The operators $\mathbf{F}_I$, $\mathbf{G}_I$, $\wedge$, and $\neg$ increase the number of required intervals by at most one. Only the $\vee$ operator can double it in the worst-case.

from an MTL formula φ, we generated a set of random prefixes and then classified them into positive or negative depending on whether φ holds at all time-points of the prefix or not. We conducted all the experiments on a single core of an AMD EPYC 7702 64-core CPU (at 2GHz) using up to 10GB of RAM. The timeout was set to be 5400 secs for all the experiments.

To address **RQ1**, we ran TEAL on a benchmark suite generated from nine MTL formulas obtained from the three MTL patterns in Table 2 by replacing t_1 with 0 and t_2 with 1,2, and 3. The suite consists of 36 samples for each pattern (12 samples for each formula), with the number of prefixes ranging from 10 to 40 and the length of prefixes (i.e., the number of sampled time points) ranging from 4 to 6. For each sample $\mathcal{S}$, we set the future-reach bound K to be $fr(\varphi)$, where φ is the formula from which $\mathcal{S}$ was generated.

Table 3. Summary of the synthesized formulas.

Formula pattern	Successful runs		Timed out	Avg Size	Avg Time (in sec)
	Matched	Not Matched			
Bounded Recurrence	36	0	0	2	17.5
Bounded Response	25	5	6	3.7	1860.3
Bounded Invariance	15	7	14	3.6	1397.2
Bounded Until	32	4	0	2.9	362.4

We depict the summary of the results for RQ1 in Table 3. For each run, we noted the formula synthesized, its size, and the total time taken. We also noted whether the synthesized formula matched the pattern of the original formula using which the sample was generated. We observed that the synthesized formulas matched the pattern of the original formula in 87.1% of the cases in which TEAL did not time out. This shows that the randomly generated samples captured the behavior of the original formula rather well, enabling a fair evaluation of TEAL.

Further, we observed that the size of the synthesized formula is always less than or equal to that of the original formula. This demonstrates that TEAL always finds a concise formula for a given future-reach bound, answering RQ1 in positive.

To address **RQ2**, we investigate how the size of the synthesized formula changed over varying future-reach bounds. For this, we ran TEAL on the same benchmark suite from RQ1 but, this time, by varying the future-reach bound K from 1 to 4. We investigate the average size of the minimal formula we get over the generated 144 samples for each future-reach bound.

We observed that for future-reach bounds K of 1, 2, 3, and 4, the average size of the synthesized minimal formulas were 3.904, 3.734, 3.370, and 3.361, respectively. Thus, with an increase in K, the average size of the minimal formula decreased. This is because an increase in K allows a bigger search space of formulas. However, the decrease in size with the increase in future-reach bound

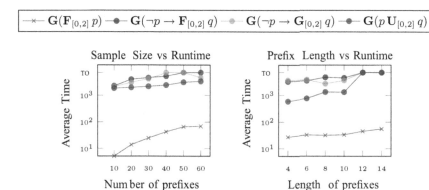

Fig. 2. Runtime change with respect to the number of prefixes and prefix lengths.

is not significant. This asserts the need for a future-reach bound for synthesizing formulas and confirms the efficacy of our algorithm.

To address **RQ3**, we ran TEAL on a benchmark suite generated from MTL formulas which originate from the MTL patterns in Table 2, setting $t_1 = 0$ and $t_1 = 2$. The suite consists of 36 samples for each formula, with the number of prefixes varying from 10 to 60 and the length of prefixes varying from 4 to 14. We set the future-reach bound K to be two.

Figure 2 illustrates the runtime variation of TEAL in two cases: increasing the number of prefixes fixing the length of them and increasing the length of prefixes fixing the number of them. We observe that to synthesize a larger formula, the time required grows significantly. This trend can be noticed in both cases.

7   Discussion and Conclusion

We have presented a novel SMT-based algorithm for automatically synthesizing MTL specifications from finite system executions. To be useful for efficient monitoring, we ensure that the synthesized formulas are both concise and have low future-reach. We have shown that our algorithm can synthesize concise formulas from benchmarks generated from commonly used MTL patterns.

Our algorithm is tailored to synthesize MTL formulas, which are particularly useful for monitoring. However, with minor modifications to our SMT encoding, we can adapt our algorithm to several different settings in formula synthesis. Moreover, we believe, we can improve the scalability of our algorithm by omitting the U_I-operator, since it is typically believed to be less interpretable.

From a practical point of view, an interesting future direction will be to lift our techniques to automatically synthesize STL formulas for verification. A naive adaptation our algorithm to work for STL has been explained in [52]. However, for scaling this technique, one must employ clever heuristics and optimizations.

Data Availability Statement. The artifact for the paper is publicly available at Zenodo [55]. The artifact contains the tool TEAL, along with the benchmarks used for the experiments in the paper. The artifact also contains all the necessary instructions to use TEAL and to reproduce the experiments used in the paper.

References

1. Ammons, G., Bodík, R., Larus, J.R.: Mining specifications. In: Launchbury, J., Mitchell, J.C. (eds.) Conference Record of POPL 2002: The 29th SIGPLAN-SIGACT Symposium on Principles of Programming Languages, Portland, OR, USA, 16–18 January 2002, pp. 4–16. ACM (2002). https://doi.org/10.1145/503272.503275
2. Arif, M.F., Larraz, D., Echeverria, M., Reynolds, A., Chowdhury, O., Tinelli, C.: SYSLITE: syntax-guided synthesis of PLTL formulas from finite traces. In: FMCAD, pp. 93–103. IEEE (2020)
3. Asarin, E., Donzé, A., Maler, O., Nickovic, D.: Parametric identification of temporal properties. In: Khurshid, S., Sen, K. (eds.) RV 2011. LNCS, vol. 7186, pp. 147–160. Springer, Heidelberg (2012). https://doi.org/10.1007/978-3-642-29860-8_12
4. Baldor, K., Niu, J.: Monitoring dense-time, continuous-semantics, metric temporal logic. In: Qadeer, S., Tasiran, S. (eds.) RV 2012. LNCS, vol. 7687, pp. 245–259. Springer, Heidelberg (2013). https://doi.org/10.1007/978-3-642-35632-2_24
5. Barbosa, H., et al.: cvc5: a versatile and industrial-strength SMT solver. In: TACAS 2022. LNCS, vol. 13243, pp. 415–442. Springer, Cham (2022). https://doi.org/10.1007/978-3-030-99524-9_24
6. Barrett, C.W., Sebastiani, R., Seshia, S.A., Tinelli, C.: Satisfiability modulo theories. In: Handbook of Satisfiability. Frontiers in Artificial Intelligence and Applications, vol. 336, pp. 1267–1329. IOS Press (2021)
7. Bartocci, E., Deshmukh, J., Donzé, A., Fainekos, G., Maler, O., Ničković, D., Sankaranarayanan, S.: Specification-based monitoring of cyber-physical systems: a survey on theory, tools and applications. In: Bartocci, E., Falcone, Y. (eds.) Lectures on Runtime Verification. LNCS, vol. 10457, pp. 135–175. Springer, Cham (2018). https://doi.org/10.1007/978-3-319-75632-5_5
8. Bartocci, E., Mateis, C., Nesterini, E., Nickovic, D.: Survey on mining signal temporal logic specifications. Inform. Comput. **289**, 104957 (2022). https://doi.org/10.1016/j.ic.2022.104957. https://www.sciencedirect.com/science/article/pii/S0890540122001122
9. Basin, D., Klaedtke, F., Zălinescu, E.: Algorithms for monitoring real-time properties. In: Khurshid, S., Sen, K. (eds.) RV 2011. LNCS, vol. 7186, pp. 260–275. Springer, Heidelberg (2012). https://doi.org/10.1007/978-3-642-29860-8_20
10. Basin, D., Krstić, S., Traytel, D.: Almost event-rate independent monitoring of metric dynamic logic. In: Lahiri, S., Reger, G. (eds.) RV 2017. LNCS, vol. 10548, pp. 85–102. Springer, Cham (2017). https://doi.org/10.1007/978-3-319-67531-2_6
11. Basin, D.A., Krstic, S., Traytel, D.: AERIAL: almost event-rate independent algorithms for monitoring metric regular properties. In: RV-CuBES. Kalpa Publications in Computing, vol. 3, pp. 29–36. EasyChair (2017)
12. Bjørner, D., Havelund, K.: 40 years of formal methods. In: Jones, C., Pihlajasaari, P., Sun, J. (eds.) FM 2014. LNCS, vol. 8442, pp. 42–61. Springer, Cham (2014). https://doi.org/10.1007/978-3-319-06410-9_4
13. Bombara, G., Belta, C.: Offline and online learning of signal temporal logic formulae using decision trees. ACM Trans. Cyber-Phys. Syst. **5**(3) (2021). https://doi.org/10.1145/3433994

14. Bombara, G., Vasile, C.I., Penedo, F., Yasuoka, H., Belta, C.: A decision tree approach to data classification using signal temporal logic. In: Proceedings of the 19th International Conference on Hybrid Systems: Computation and Control, HSCC 2016, pp. 1–10. Association for Computing Machinery, New York (2016). https://doi.org/10.1145/2883817.2883843
15. Camacho, A., McIlraith, S.A.: Learning interpretable models expressed in linear temporal logic. In: ICAPS, pp. 621–630. AAAI Press (2019)
16. Camacho, A., McIlraith, S.A.: Learning interpretable models expressed in linear temporal logic. Proceedings of the International Conference on Automated Planning and Scheduling, vol. 29(1), 621–630 (May 2021). https://doi.org/10.1609/icaps.v29i1.3529,https://ojs.aaai.org/index.php/ICAPS/article/view/3529
17. Chattopadhyay, A., Mamouras, K.: A verified online monitor for metric temporal logic with quantitative semantics. In: Deshmukh, J., Ničković, D. (eds.) RV 2020. LNCS, vol. 12399, pp. 383–403. Springer, Cham (2020). https://doi.org/10.1007/978-3-030-60508-7_21
18. Cimatti, A., Griggio, A., Schaafsma, B.J., Sebastiani, R.: The MathSAT5 SMT solver. In: Piterman, N., Smolka, S.A. (eds.) TACAS 2013. LNCS, vol. 7795, pp. 93–107. Springer, Heidelberg (2013). https://doi.org/10.1007/978-3-642-36742-7_7
19. Barrett, C., Tinelli, C.: Satisfiability modulo theories. In: Handbook of Model Checking, pp. 305–343. Springer, Cham (2018). https://doi.org/10.1007/978-3-319-10575-8_11
20. Dang, T., Stolz, V. (eds.): Runtime Verification - 22nd International Conference, RV 2022, Tbilisi, Georgia, September 28–30, 2022, Proceedings. LNCS, vol. 13498. Springer (2022)
21. Deshmukh, J.V., Donzé, A., Ghosh, S., Jin, X., Juniwal, G., Seshia, S.A.: Robust online monitoring of signal temporal logic. In: RV. LNCS, vol. 9333, pp. 55–70. Springer (2015)
22. Dokhanchi, A., Hoxha, B., Fainekos, G.: On-Line monitoring for temporal logic robustness. In: Bonakdarpour, B., Smolka, S.A. (eds.) RV 2014. LNCS, vol. 8734, pp. 231–246. Springer, Cham (2014). https://doi.org/10.1007/978-3-319-11164-3_19
23. Donzé, A., Ferrère, T., Maler, O.: Efficient robust monitoring for STL. In: Sharygina, N., Veith, H. (eds.) CAV 2013. LNCS, vol. 8044, pp. 264–279. Springer, Heidelberg (2013). https://doi.org/10.1007/978-3-642-39799-8_19
24. Eisner, C., Fisman, D., Havlicek, J., Lustig, Y., McIsaac, A., Van Campenhout, D.: Reasoning with temporal logic on truncated paths. In: Hunt, W.A., Somenzi, F. (eds.) CAV 2003. LNCS, vol. 2725, pp. 27–39. Springer, Heidelberg (2003). https://doi.org/10.1007/978-3-540-45069-6_3
25. Fijalkow, N., Lagarde, G.: The complexity of learning linear temporal formulas from examples. In: ICGI. Proceedings of Machine Learning Research, vol. 153, pp. 237–250. PMLR (2021)
26. Gaglione, J.-R., Neider, D., Roy, R., Topcu, U., Xu, Z.: Learning linear temporal properties from noisy data: A MaxSAT-based approach. In: Hou, Z., Ganesh, V. (eds.) ATVA 2021. LNCS, vol. 12971, pp. 74–90. Springer, Cham (2021). https://doi.org/10.1007/978-3-030-88885-5_6
27. Gorostiaga, F., Sánchez, C.: HLola: a very functional tool for extensible stream runtime verification. In: TACAS 2021. LNCS, vol. 12652, pp. 349–356. Springer, Cham (2021). https://doi.org/10.1007/978-3-030-72013-1_18
28. Havelund, K., Peled, D.: Runtime verification: from propositional to first-order temporal logic. In: Colombo, C., Leucker, M. (eds.) RV 2018. LNCS, vol. 11237, pp. 90–112. Springer, Cham (2018). https://doi.org/10.1007/978-3-030-03769-7_7

29. Ho, H.-M., Ouaknine, J., Worrell, J.: Online monitoring of metric temporal logic. In: Bonakdarpour, B., Smolka, S.A. (eds.) RV 2014. LNCS, vol. 8734, pp. 178–192. Springer, Cham (2014). https://doi.org/10.1007/978-3-319-11164-3_15

30. Hoxha, B., Dokhanchi, A., Fainekos, G.: Mining parametric temporal logic properties in model-based design for cyber-physical systems. Int. J. Softw. Tools Technol. Transf. **20**(1), 79–93 (2018)

31. Hunter, P., Ouaknine, J., Worrell, J.: Expressive completeness for metric temporal logic. In: 2013 28th Annual ACM/IEEE Symposium on Logic in Computer Science, pp. 349–357 (2013). https://doi.org/10.1109/LICS.2013.1

32. Kane, A., Chowdhury, O., Datta, A., Koopman, P.: A case study on runtime monitoring of an autonomous research vehicle (ARV) system. In: Bartocci, E., Majumdar, R. (eds.) RV 2015. LNCS, vol. 9333, pp. 102–117. Springer, Cham (2015). https://doi.org/10.1007/978-3-319-23820-3_7

33. Kempa, B., Zhang, P., Jones, P.H., Zambreno, J., Rozier, K.Y.: Embedding online runtime verification for fault disambiguation on Robonaut2. In: Bertrand, N., Jansen, N. (eds.) FORMATS 2020. LNCS, vol. 12288, pp. 196–214. Springer, Cham (2020). https://doi.org/10.1007/978-3-030-57628-8_12

34. Kong, Z., Jones, A., Belta, C.: Temporal logics for learning and detection of anomalous behavior. IEEE Trans. Autom. Control **62**(3), 1210–1222 (2017). https://doi.org/10.1109/TAC.2016.2585083

35. Kong, Z., Jones, A., Medina Ayala, A., Aydin Gol, E., Belta, C.: Temporal logic inference for classification and prediction from data. In: Proceedings of the 17th International Conference on Hybrid Systems: Computation and Control, HSCC 2014. pp. 273–282. Association for Computing Machinery, New York (2014). https://doi.org/10.1145/2562059.2562146

36. Konrad, S., Cheng, B.H.C.: Real-time specification patterns. In: Roman, G., Griswold, W.G., Nuseibeh, B. (eds.) 27th International Conference on Software Engineering (ICSE 2005), 15–21 May 2005, St. Louis, Missouri, USA, pp. 372–381. ACM (2005). https://doi.org/10.1145/1062455.1062526

37. Koymans, R.: Specifying real-time properties with metric temporal logic. Real Time Syst. **2**(4), 255–299 (1990)

38. Lemieux, C., Park, D., Beschastnikh, I.: General LTL specification mining (T). In: ASE, pp. 81–92. IEEE Computer Society (2015)

39. Lima, L., Herasimau, A., Raszyk, M., Traytel, D., Yuan, S.: Explainable online monitoring of metric temporal logic. In: TACAS (2). LNCS, vol. 13994, pp. 473–491. Springer (2023). https://doi.org/10.1007/978-3-031-30820-8_28

40. Linard, A., Tumova, J.: Active learning of signal temporal logic specifications. In: 2020 IEEE 16th International Conference on Automation Science and Engineering (CASE), pp. 779–785 (2020). https://doi.org/10.1109/CASE48305.2020.9216778

41. Luo, W., Liang, P., Du, J., Wan, H., Peng, B., Zhang, D.: Bridging ltlf inference to GNN inference for learning ltlf formulae. In: AAAI, pp. 9849–9857. AAAI Press (2022)

42. Lutz, S., Neider, D., Roy, R.: Specification sketching for linear temporal logic. In: ATVA. LNCS, vol. 14216, pp. 26–48. Springer (2023). https://doi.org/10.1007/978-3-031-45332-8_2

43. Maler, O., Nickovic, D.: Monitoring temporal properties of continuous signals. In: Lakhnech, Y., Yovine, S. (eds.) FORMATS/FTRTFT -2004. LNCS, vol. 3253, pp. 152–166. Springer, Heidelberg (2004). https://doi.org/10.1007/978-3-540-30206-3_12

44. Meel, K.S., Strichman, O. (eds.): 25th International Conference on Theory and Applications of Satisfiability Testing, SAT 2022, 2–5 August 2022, Haifa, Israel, LIPIcs, vol. 236. Schloss Dagstuhl - Leibniz-Zentrum für Informatik (2022)
45. Mohammadinejad, S., Deshmukh, J.V., Puranic, A.G., Vazquez-Chanlatte, M., Donzé, A.: Interpretable classification of time-series data using efficient enumerative techniques. In: HSCC 2020: 23rd ACM International Conference on Hybrid Systems: Computation and Control, Sydney, New South Wales, Australia, 21–24 April 2020, pp. 9:1–9:10. ACM (2020). DOI: https://doi.org/10.1145/3365365.3382218
46. de Moura, L., Bjørner, N.: Z3: an efficient SMT solver. In: Ramakrishnan, C.R., Rehof, J. (eds.) TACAS 2008. LNCS, vol. 4963, pp. 337–340. Springer, Heidelberg (2008). https://doi.org/10.1007/978-3-540-78800-3_24
47. de Moura, L.M., Bjørner, N.S.: Satisfiability modulo theories: introduction and applications. Commun. ACM **54**(9), 69–77 (2011)
48. Neider, D., Gavran, I.: Learning linear temporal properties. In: Bjørner, N.S., Gurfinkel, A. (eds.) 2018 Formal Methods in Computer Aided Design, FMCAD 2018, Austin, TX, USA, 30 October - 2 November 2018. pp. 1–10. IEEE (2018). https://doi.org/10.23919/FMCAD.2018.8603016
49. Nenzi, L., Silvetti, S., Bartocci, E., Bortolussi, L.: A robust genetic algorithm for learning temporal specifications from data. In: McIver, A., Horvath, A. (eds.) QEST 2018. LNCS, vol. 11024, pp. 323–338. Springer, Cham (2018). https://doi.org/10.1007/978-3-319-99154-2_20
50. Ouaknine, J., Worrell, J.: Some recent results in metric temporal logic. In: Cassez, F., Jard, C. (eds.) FORMATS 2008. LNCS, vol. 5215, pp. 1–13. Springer, Heidelberg (2008). https://doi.org/10.1007/978-3-540-85778-5_1
51. Pnueli, A.: The temporal logic of programs. In: FOCS, pp. 46–57. IEEE Computer Society (1977)
52. Raha, R.: Learning and verifying temporal specifications for cyber-physical systems. Ph.D. thesis, University of Antwerp, Belgium (2023). https://hdl.handle.net/10067/1986580151162165141
53. Raha, R., Roy, R., Fijalkow, N., Neider, D.: Scalable anytime algorithms for learning fragments of linear temporal logic. In: TACAS 2022. LNCS, vol. 13243, pp. 263–280. Springer, Cham (2022). https://doi.org/10.1007/978-3-030-99524-9_14
54. Raha, R., Roy, R., Fijalkow, N., Neider, D., Perez, G.A.: Synthesizing efficiently monitorable formulas in metric temporal logic (2023). https://arxiv.org/abs/2310.17410
55. Raha, R., Roy, R., Fijalkow, N., Neider, D., Perez, G.A.: TEAL: Synthesizing Efficiently Monitorable Formulas in Metric Temporal Logic (Oct 2023). https://doi.org/10.5281/zenodo.10046302
56. Riener, H.: Exact synthesis of LTL properties from traces. In: FDL, pp. 1–6. IEEE (2019)
57. Roy, R., Fisman, D., Neider, D.: Learning interpretable models in the property specification language. In: IJCAI, pp. 2213–2219. https://www.ijcai.org/Proceedings/2020/306 (2020)
58. Rozier, K.Y.: Specification: the biggest bottleneck in formal methods and autonomy. In: Blazy, S., Chechik, M. (eds.) VSTTE 2016. LNCS, vol. 9971, pp. 8–26. Springer, Cham (2016). https://doi.org/10.1007/978-3-319-48869-1_2
59. Silvetti, S., Nenzi, L., Bortolussi, L., Bartocci, E.: A robust genetic algorithm for learning temporal specifications from data. CoRR. arXiv: 1711.06202 (2017)

60. Thati, P., Rosu, G.: Monitoring algorithms for metric temporal logic specifications. In: Havelund, K., Rosu, G. (eds.) Proceedings of the Fourth Workshop on Runtime Verification, RV@ETAPS 2004, Barcelona, Spain, 3 April 2004. vol. 113, pp. 145–162. Elsevier (2004). https://doi.org/10.1016/j.entcs.2004.01.029

61. Wasylkowski, A., Zeller, A.: Mining temporal specifications from object usage. Autom. Softw. Eng. **18**(3–4), 263–292 (2011)

62. Xu, Z., Topcu, U.: Transfer of temporal logic formulas in reinforcement learning. In: Kraus, S. (ed.) Proceedings of the Twenty-Eighth International Joint Conference on Artificial Intelligence, IJCAI 2019, Macao, China, 10–16 August 2019, pp. 4010–4018. (2019). https://doi.org/10.24963/ijcai.2019/557, https://www.ijcai.org

63. Yang, H., Hoxha, B., Fainekos, G.: Querying parametric temporal logic properties on embedded systems. In: Nielsen, B., Weise, C. (eds.) ICTSS 2012. LNCS, vol. 7641, pp. 136–151. Springer, Heidelberg (2012). https://doi.org/10.1007/978-3-642-34691-0_11

Security and Privacy

Automatic and Incremental Repair for Speculative Information Leaks

Joachim Bard[1]([✉]), Swen Jacobs[1], and Yakir Vizel[2]

[1] CISPA Helmholtz Center for Information Security, Saarbrücken, Germany
{joachim.bard,jacobs}@cispa.de
[2] Technion, Haifa, Israel
yvizel@cs.technion.ac.il

Abstract. We present CURESPEC, the first model-checking based framework for *automatic repair* of programs with respect to information leaks in the presence of side-channels and speculative execution. CURESPEC is based on formal models of attacker capabilities, including observable side channels, inspired by the SPECTRE-PHT attacks. For a given attacker model, CURESPEC is able to either prove that the program is secure, or *detect* potential side-channel vulnerabilities and *automatically insert mitigations* such that the resulting code is provably secure. Moreover, CURESPEC can provide a *certificate* for the security of the program that can be independently checked. We have implemented CURESPEC in the SeaHorn framework and show that it can effectively repair security-critical code, for example the AES encryption from the OpenSSL library.

1 Introduction

Speculative execution is an indispensable performance optimization of modern processors: by predicting how branching (and other) conditions will evaluate and speculatively continuing computation, it can avoid pipeline stalls when data from other computations is still missing. When this data arrives, in case of a correct guess the results of the computations can be committed. Otherwise they have to be discarded, and the correct results computed. However, even when the results are not committed to registers that are available at the software level, speculative computations may leave traces in the microarchitecture that can leak through *side channels*, as demonstrated by the family of SPECTRE attacks [5,10,24]. For example, the cache is usually not cleaned up after a misspeculation, enabling *timing attacks* that can discover data used during speculation.

Since the discovery of SPECTRE, several countermeasures have been developed [22–24,35]. As neither speculation nor side channels can be removed from current hardware without sacrificing significant amounts of computing power, the problem is usually dealt with at the software level. Mitigations for SPECTRE

Joachim Bard carried out this work as PhD candidate at Saarland University, Germany.

R. Dimitrova et al. (Eds.): VMCAI 2024, LNCS 14500, pp. 291–313, 2024.
https://doi.org/10.1007/978-3-031-50521-8_14

292   J. Bard et al.

usually prevent information leaks during speculative execution by prohibiting "problematic" instructions from being executed speculatively.

Most of the existing mitigations prevent some SPECTRE attacks, but are known to be incomplete [10], i.e., the modified code may remain vulnerable, in circumstances that may or may not be known. In addition, there have been approaches that use formal methods to obtain code that is *guaranteed* to be resilient against clearly defined types of SPECTRE attacks, and formal notions of speculative non-interference [19] and speculative constant-time security [11] have been proposed that can give guarantees against side-channel attacks under speculation. However, these methods all have certain shortcomings: either they require manual modification of the code if potential leaks are found [11,13,19], they do not precisely state the security guarantees of automatically hardened code [34], or they are based on a security type system, which are known to be rather difficult to extend to different assumptions (e.g., attacker models) or guarantees [32].

Motivating Example. Consider the three programs in Fig. 1. The original program P is shown in (a). It accesses a public array a at position i after checking that the access to a is in bounds. In speculative execution, this bounds-check can be ignored, which enables to read arbitrary program memory and store it (albeit temporarily) into k. The information leak appears when k is used in another memory access to public array b, making k observable to an attacker through a cache-based timing attack.

Our repair approach is based on a transformation of P with the following goals: (i) capture computations that are executed speculatively; (ii) identify possible information leaks under speculation; and (iii) enable the prevention of speculation using fences[1].

```
1 if (i < size_a) {
2   k = a[i] * 512;
3   tmp = b[k];
4 }
```

```
1 bool spec = false;
2 if (*) {
3   spec = spec | !(i < size_a);
4   k = a[i] * 512;
5   assert(!spec);
6   tmp = b[k];
7 }
```

```
1  bool spec = false;
2  bool fence2 = true;
3  bool fence3 = false;
4  if (*) {
5    spec = spec | !(i < size_a);
6    assume(!(fence2 && spec));
7    k = a[i] * 512;
8    assume(!(fence3 && spec));
9    assert(!spec);
10   tmp = b[k];
11 }
```

(a) Original program

(b) Speculative execution semantics

(c) Speculative execution and fence semantics

Fig. 1. Example program and its version with speculative and fence semantics.

Figure 1(b) presents a modification P_s of P, demonstrating the first point. We assume that speculative executions can start at conditional statements, i.e.,

---

[1] Other methods, e.g. *speculative load hardening* (SLH, as appears in https://tinyurl.com/3nybax4u), can also be used as a mitigation in our repair algorithm.

the processor may ignore the condition and take the wrong branch. Therefore, we replace the branching condition by a non-deterministic operator $*$ that can return either true or false (Line 2). Moreover, an auxiliary variable spec is added in order to identify whether an execution of P_s corresponds to a speculative or a non-speculative execution of P. Namely, spec = true at some point in an execution of P_s iff the corresponding execution of P is possible *only* under speculation. In Fig. 1(b), spec is assigned true in Line 3 if the negation of the branching condition holds.

To detect information leaks under speculative execution, we assume that there is a set VInst of memory accesses that should not be performed under speculation, and for such memory accesses we add an assertion spec = false. Assuming that the nested array read in Line 3 of P (Fig. 1(a)) is in VInst, the transformed program P_s (Fig. 1(b)) contains such an assertion at Line 5.

To enable prevention of speculation, in Fig. 1(c) we add auxiliary variables fence$i$ for every line i in P with an instruction, and initialize them with truth values that determine whether speculation should be stopped before reaching line i of P. We model the fact that fences stop speculation by adding assumptions that at line i of P we cannot have fence$i$ = true and spec = true simultaneously (Line 6 and Line 8 of Fig. 1(c)). In this example, speculation can only start in Line 1 of P and fence2 = true stops speculation before Line 2 of P, implying that the vulnerable instruction in Line 3 of P is not reachable under speculation.

Note that in this example we assume that fence variables have fixed truth values, reducing the problem to a safety verification problem. In our repair algorithm, we start with a program where all variables fence$i$ are initialized to false, and allow the algorithm to manipulate initial values of the fence$i$ in order to find a version of the program that is secure against a given type of SPECTRE attacks (determined by our choice of vulnerable memory accesses VInst). Upon termination, our algorithm returns a version of P that has been made secure by adding fence instructions.

To formalize the ideas presented on this example, in Sect. 2 we will introduce a formal model for the standard semantics of a program P, then introduce a semantics that includes speculative executions in Sect. 3, and finally present our automatic repair approach in Sect. 4, which we evaluate in Sect. 5.

Our Contribution. In this paper, we present CURESPEC, the first model checking based framework for *automatic repair* of programs with respect to information leaks that are due to speculative execution. Applying CURESPEC to a given program results in a program with a *certified* security guarantee. In addition to the program, CURESPEC takes as input a set of instructions that may leak secret information to the attacker if executed under speculation, which we consider our *threat model*. This makes our technique directly applicable to the SPECTRE-PHT attacks[2], and easily extensible to any attacks for which a threat model in this form can be computed. For a given threat model, CURESPEC is able to either

---

[2] "PHT" refers to the pattern history table, which is the means of speculation that is abused by the attacker in this case. It is responsible for speculation on conditional branches. Our approach supports all four sub-types of SPECTRE-PHT.

prove that the program is secure, or *detect* potential side-channel vulnerabilities. In case vulnerabilities are detected, CURESPEC *automatically inserts mitigations* that remove these vulnerabilities, and proves that the modified code is *secure*. Since CURESPEC is based on model checking, it provides a *certificate* for the security guarantee in the form of an inductive invariant.

CURESPEC is a framework with two main parts: (i) a reduction of the problem of finding information leaks that are due to speculation to a safety verification problem; and (ii) a model checking based algorithm for detection and repair of possible vulnerabilities. For (i), we build on previous results that introduced formal modeling of speculative execution semantics [7], and extend this formal model to enable not only the detection of possible leaks, but also their automatic repair. For (ii), we extend the well-known IC3/PDR approach [9,21,25] to a repair algorithm for our problem. When the underlying model checking algorithm discovers a possible leak, CURESPEC modifies the program to eliminate this vulnerability. Then it resumes the verification process, and eliminates further vulnerabilities until the program is secure. An important feature of our technique is that the modified code is not verified from scratch, but the model checking algorithm maintains its state and re-uses the information obtained thus far. Finally, when CURESPEC proves safety of the (possibly repaired) program, the underlying PDR-algorithm produces an inductive invariant, which is a formal certificate of the desired security property.

We implemented CURESPEC in SEAHORN [20], a verification framework for C programs, and evaluated it on the "standard" test cases for SPECTRE-PHT vulnerabilities, as well as different parts of OpenSSL and HACL*, demonstrating its practicality.

To summarize, in this paper we provide the first method that *combines* the following aspects:

1. formal verification of programs with respect to information leaks under speculation, parameterized by a threat model;
2. automatic repair by insertion of mitigations that stop speculative execution;
3. formal guarantees for the repaired code in form of inductive invariants; and
4. an implementation that scales to practical encryption algorithms.

2 Preliminaries

2.1 Model Checking Programs

We consider first-order logic modulo a theory T and denote it by $FOL(T)$. We adopt the standard notation and terminology, where $FOL(T)$ is defined over a signature Σ of constant, predicate and function symbols, some of which may be interpreted by T. In this paper T is the theory of Linear Integer Arithmetic and Arrays (LIA). We use true and false to denote the constant truth values.

Transition Systems. We define transition systems as formal models of programs. Let X be a set of variables, used to represent program variables. A *state* is a valuation of X. For a state σ and $a \in X$, we denote by $\sigma(a)$ the value of a in σ. We write $\theta(X)$ to represent a formula over X in $FOL(\mathcal{T})$. $\theta(X)$ is called a *state formula* and represents a set of states.

A *transition system* is a tuple $M = \langle X, Init(X), Tr(X, X') \rangle$ where $Init(X)$ and $Tr(X, X')$ are quantifier-free formulas in $FOL(\mathcal{T})$. $Init$ represents the initial states of the system and Tr represents the (total) transition relation. We write $Tr(X, X')$ to denote that Tr is defined over variables $X \cup X'$, where X is used to represent the pre-state of a transition, and $X' = \{a' \mid a \in X\}$ is used to represent the post-state. A *path* in a transition system is a sequence of states $\pi := \sigma_0, \sigma_1, \ldots$, such that for $i \geq 1$: $(\sigma_{i-1}, \sigma'_i) \models Tr$. We also consider the case where a path is a finite sequence of states such that every two subsequent states have a transition. We use $\pi[i]$ to refer to the i-th state of π, namely σ_i. We use $\pi^{[0..n]}$ to refer to the prefix $\sigma_0, \sigma_1, \ldots, \sigma_n$, and π^n to the suffix $\sigma_n, \ldots$ of π. A path (or a prefix of a path) is called an *execution* of M when $\sigma_0 \models Init$. Given two paths $\pi_1 = \sigma_0, \sigma_1 \ldots$ and $\pi_2 = \sigma'_0, \sigma'_1 \ldots$, then $\pi = \pi_1^{[0..n]}\pi_2$ is a path if $(\sigma_n, \sigma'_0) \models Tr$.

A formula $\varphi(X, X')$ such that for every valuation I of X there is exactly one valuation I' of X' such that $(I, I') \models \varphi(X, X')$ is called a *state update function*. For $Y \subseteq X$, we denote by $\mathsf{id}(Y, Y')$ the state update function $\bigwedge_{a \in Y} a' = a$. While $\mathsf{id}(Y, Y')$ is a formula over $Y \cup Y'$, for readability we use $\mathsf{id}(Y)$.

Safety Verification. A *safety problem* is a tuple $\langle M, Bad(X) \rangle$, where $M = \langle X, Init, Tr \rangle$ is a transition system and Bad is a quantifier-free formula in $FOL(\mathcal{T})$ representing a set of bad states. A safety problem has a *counterexample of length* n if there exists an execution $\pi := \sigma_0, \ldots, \sigma_n$ with $\sigma_n \models Bad$. The safety problem is *SAFE* if it has no counterexample, of any length. It is *UNSAFE* otherwise.

A *safe inductive invariant* is a formula $Inv(X)$ such that (i) $Init(X) \rightarrow Inv(X)$, (ii) $Inv(X) \wedge Tr(X, X') \rightarrow Inv(X')$, and (iii) $Inv(X) \rightarrow \neg Bad(X)$. If such a safe inductive invariant exists, then the safety problem is SAFE.

In this work we use SPACER [25] as a solver for a given safety problem. SPACER is based on the Property Directed Reachability (PDR) algorithm [9,21]. Algorithm 1 presents SPACER as a set of rules, following the presentation style of [21]. We only give a brief overview of PDR and SPACER and highlight the details needed later in the paper for CURESPEC. Given a safety problem, SPACER tries to construct an inductive invariant, or find a counterexample. In order to construct an inductive invariant, SPACER maintains a sequence of formulas $F_0, F_1, \ldots, F_N$, with the following properties: (i) $F_0 \rightarrow Init$; (ii) $\forall 0 \leq j < N \cdot F_j \rightarrow F_{j+1}$; (iii) $\forall 0 \leq j < N \cdot F_j(X) \wedge Tr(X, X') \rightarrow F_{j+1}(X')$; and (iv) $\forall 0 \leq j < N \cdot F_j \rightarrow \neg Bad$.

F_j is an over-approximation of the states reachable in j steps or less. Additionally, SPACER maintains a set REACH of states that are known to be reachable. REACH is an under-approximation of the reachable states.

PDR performs a backward traversal of the states space. The traversal is performed starting from states that violate Bad and constructing a suffix of a counterexample backwards, trying to either show that a state that can reach Bad is

Input: A safety problem $\langle X, Init(X), Tr(X, X'), Bad(X) \rangle$.
Assumptions: $Init$, Tr and Bad are quantifier free.
Data: A queue $\mathcal{Q}$ of potential counterexamples, where $c \in \mathcal{Q}$ is a pair $\langle m, j \rangle$, m
   is a cube over state variables, $j \in \mathbb{N}$. A level N. A sequence $F_0, F_1, \ldots$.
   An invariant F_∞. A set of reachable states REACH.
Output: $(SAFE, F_\infty)$, where F_∞ is a safe inductive invariant,
or $(UNSAFE, \pi)$, where π is a counterexample
Initially: $\mathcal{Q} = \emptyset$, $N = 0$, $F_0 = Init$, $\forall j \geq 1 \cdot F_j = true$, $F_\infty = true$.
Require: $Init \rightarrow \neg Bad$
repeat
> **Safe** If $F_\infty \rightarrow \neg Bad$ **return** $(SAFE, F_\infty)$.
> **Cex** If $\langle m, j \rangle \in \mathcal{Q}$ and $m \cap (\text{REACH}) \neq \emptyset$, then there is a path π'
>   from m to $Bad(X)$ consisting of states in $\mathcal{Q}$ and a finite execution π^*
>   from $Init(X)$ to m, respectively. **return** $(UNSAFE, \pi^*\pi')$.
> **Unfold** If $F_N \rightarrow \neg Bad$, then set $N \leftarrow N + 1$.
> **Candidate** If for some m, $m \rightarrow F_N \wedge Bad$, then add $\langle m, N \rangle$ to $\mathcal{Q}$.
> **Predecessor** If $\langle m, j+1 \rangle \in \mathcal{Q}$ and there are m_0 and m_1 s.t.
> $m_1 \rightarrow m$, $m_0 \wedge m_1'$ is satisfiable, and $m_0 \wedge m_1' \rightarrow F_j \wedge Tr \wedge m'$, then
>   add $\langle m_0, j \rangle$ to $\mathcal{Q}$.
> **NewLemma** For $0 \leq j < N$: given $\langle m, j+1 \rangle \in \mathcal{Q}$ and a clause φ s.t. $\varphi \rightarrow \neg m$,
>   if $(\text{REACH}) \rightarrow \varphi$, and $\varphi \wedge F_j \wedge Tr \rightarrow \varphi'$, then add φ to F_k, for $k \leq j+1$.
> **ReQueue** If $\langle m, j \rangle \in \mathcal{Q}$, and $F_{j-1} \wedge Tr \wedge m'$ is unsatisfiable, then
>   add $\langle m, j+1 \rangle$ to $\mathcal{Q}$.
> **Push** For $1 \leq j$ and a clause $(\varphi \vee \psi) \in F_j \setminus F_{j+1}$,
>   if $(\text{REACH}) \rightarrow \varphi$ and $\varphi \wedge F_j \wedge Tr \rightarrow \varphi'$, then
>   add φ to F_k, for each $k \leq j+1$.
> **MaxIndSubset** If there is $j > N$ s.t. $F_{j+1} \subseteq F_j$, then
> $F_\infty \leftarrow F_i$, and $\forall k \geq j \cdot F_j \leftarrow F_\infty$.
> **Successor** If $\langle m, j+1 \rangle \in \mathcal{Q}$ and exist m_0, m_1 s.t.
> $m_0 \wedge m_1'$ are satisfiable and $m_0 \wedge m_1' \rightarrow (\text{REACH}) \wedge Tr \wedge m'$, then
>   add m_1 to REACH.
> **ResetQ** $\mathcal{Q} \leftarrow \emptyset$.
> **ResetReach** REACH $\leftarrow Init$.

until ∞;

Algorithm 1: The rules of the SPACER procedure.

reachable (**Candidate**, **Predecessor** and **Cex** rules) or prove that such states
are unreachable (**NewLemma** and **Push** rules). During this process the trace
of over-approximations $F_0, F_1, \ldots, F_N$ is constructed and reachable states are
discovered. We later show how these are used in CURESPEC. For more details
about PDR and SPACER the interested reader is referred to [9,21,25].

2.2   Standard Program Semantics

We assume a program P is represented in a low-level language (e.g. LLVM
bit-code) with standard semantics, and includes standard low-level instructions
such as unary and binary operations, conditional and unconditional branches,

load and store for accessing memory. In addition, it includes the instructions $\mathsf{assume}(b)$ and $\mathsf{assert}(b)$ used for safety verification.

Let $i \in \mathbb{N}$ be a line in the program P to be encoded. For simplicity we refer to the instruction at line i as i, and write $i \in P$. We assume there is a special program variable $\mathsf{pc} \in X$, called the *program counter*, defined over the domain $\mathbb{N} \cup \{\bot\}$. Let $i \in P$ be an instruction to be encoded. If i is a conditional branch instruction, it is encoded by a *conditional state update function* of the form

$$\tau_i(X, X'):=\mathsf{pc} = i \rightarrow (cond_i(X) \ ? \ \mathsf{pc}' = \vartheta_i(X) : \mathsf{pc}' = \varepsilon_i(X)) \wedge \mathsf{id}(X \setminus \{\mathsf{pc}\}),$$

where $cond_i(X)$ is the condition represented by a state formula, and $\vartheta_i(X)$ and $\varepsilon_i(X)$ are expressions that each evaluate to a value in $\mathbb{N}$, representing the first instruction of one of the two conditional branches. Accordingly, pc is updated to $\vartheta_i(X)$ when the condition holds, and otherwise it is updated to $\varepsilon_i(X)$.

All other instructions are encoded by an *unconditional state update function* of the form $\tau_i(X, X'):=\mathsf{pc} = i \rightarrow \varphi_i(X, X')$, where $\varphi_i(X, X')$ is a state update function[3]. Instructions are either conditional or unconditional depending on their corresponding state update function. We denote by $C \subseteq P$ the set of conditional instructions, i.e., if $i \in C$, then τ_i is a conditional state update function.

The semantics for the verification instruction $\mathsf{assume}(cond_i(X))$ is captured by a state update function, which is encoded by

$$\tau_i(X, X'):=\mathsf{pc} = i \rightarrow ((cond_i(X) \ ? \ \mathsf{pc}' = \vartheta_i(X) \ : \ \mathsf{pc}' = \mathsf{pc}) \wedge \mathsf{id}(X \setminus \{\mathsf{pc}\})).$$

This encoding requires that at line i the condition $cond_i(X)$ holds. If it does not hold, then the transition relation is stuck in an infinite loop and the program does not progress. Similarly, $\mathsf{assert}(cond(X))$ is captured by a state update function, which is encoded by

$$\tau_i(X, X'):=\mathsf{pc} = i \rightarrow ((cond_i(X) \ ? \ \mathsf{pc}' = \vartheta_i(X) \ : \ \mathsf{pc}' = \bot) \wedge \mathsf{id}(X \setminus \{\mathsf{pc}\}))$$

For **assert**, if the condition $cond_i(X)$ holds, the program continues. Otherwise, pc is set to $\bot$. This special case allows us to create a safety verification problem by defining the bad states to be those where $\mathsf{pc} = \bot$. To ensure the resulting transition relation is total, we add a state update function that makes sure that if a state where $\mathsf{pc} = \bot$ is ever reached, this state is stuttering:

$$\tau_\bot(X, X'):=pc = \bot \rightarrow \mathsf{id}(X)$$

To conclude, given a program P, we obtain a symbolic representation of the transition relation by conjoining the formulas for all lines of the program including $\tau_\bot$, i.e., $Tr(X, X'):=(\bigwedge_i \tau_i(X, X')) \wedge \tau_\bot(X, X')$. We call $M = \langle X, Init, Tr \rangle$ the *standard transition system* of P, and the corresponding safety problem is then defined by M and a set of bad states, given as $Bad(X):=\mathsf{pc} = \bot$.

---

[3] Note that pc is updated also by φ_i. We therefore assume that unconditional state update functions accompany every instruction that is not a conditional branch.

Remark 1. Let P be a program, $\langle M, Bad \rangle$ the corresponding safety problem, and $\pi = \sigma_0, \sigma_1, \ldots, \sigma_n$ an execution of M. Then, for every $1 \leq j \leq n$ such that $(\sigma_{j-1}, \sigma'_j) \models Tr$ there exists $i \in P$ such that $(\sigma_{j-1}, \sigma'_j) \models \tau_i$. We denote this as $i \in \pi$. Moreover, if $\sigma_n \models Bad$ then there exists an assertion violation in P.

3   Modeling Speculative Execution Semantics

When analyzing the *functionality* of a program, speculative execution can be ignored since the results of a computation that is based on *misspeculation* do not alter the program's state: if the condition of a branch turns out to be wrong after speculative execution, any of its results (visible in the microarchitecture, but not on the program level) are discarded and computation backtracks to the correct branch. However, the data used in such a computation may still leak through side channels. Therefore, when analyzing *information leaks through side channels*, the formal model must include speculative execution semantics and take into account possible observations based on misspeculation.

In this section, we first discuss the threat models we consider (Sect. 3.1), and introduce a formal speculative execution semantics (Sect. 3.2). Then, we formally define our notion of security and show how it can be reduced to a safety problem (Sect. 3.3).

3.1   Threat Models for Speculative Information Leaks

Verifying secure information flow deals with proving that confidential data does not leak through public outputs during the execution of a system [17]. More generally, we want to show that confidential data is not *observable* by an attacker.

A well-known class of attacks that can cause information leaks are *timing attacks*. These attacks use observations about the run-time of a system in order to infer secret data. More precisely, in order for a program to be secure against timing attacks, any two executions of a system that agree on their public inputs and outputs should be indistinguishable w.r.t. some measure of time (e.g. time to execute a program, latency in memory access, etc.).

Most variants of the SPECTRE attack fall within this class of timing attacks. In particular, they are based on observations that are due to code being executed speculatively. It is important to note that speculative execution does not change the architectural state of the CPU, but may have side-effects that are observable by an attacker (e.g. modifications to the cache).

Since the focus of this paper is timing attacks that can incur due to SPECTRE-PHT, we make the standard assumption that the attacker can distinguish executions if they differ in the values of pc, i.e., their control-flow, or in the execution of instructions with observable side-effects under speculation (e.g. memory accesses that modify the cache [24], or vector instructions that introduce a delay under certain conditions [30]). In the following we assume that these *vulnerable instructions*[4] are given in a set VInst $\subseteq P$. Note that compared to the work in [7], this means that our threat model is parameterized in the set VInst and strictly generalizes the fixed threat model used there.

3.2 Formal Model of Speculative Execution Semantics

Our goal is to check for information leaks *under speculative execution*. To this end, let us formalize the speculative semantics of the program P.

Let $M = \langle X, Init, Tr \rangle$ be the standard transition system (Sect. 2.1) of P. To model fences, let $\mathcal{F}:=\{\text{fence}_i \mid i \in P\}$ be a set of Boolean auxiliary variables, and let $B \subseteq P$ be the set of fenced instructions, i.e., with $\text{fence}_i = \text{true}$. We want to define a transition system that includes speculative execution semantics as $\hat{M}:=\langle \hat{X}, \hat{Init}, \hat{Tr} \rangle$ where $\hat{X}:=X \cup \{\text{spec}\} \cup \mathcal{F}$, where spec is of sort $\mathbb{N}_0$.

The initial states of $\hat{M}$ are defined as

$$\hat{Init}(\hat{X}):=Init \wedge (\text{spec} = 0) \wedge \bigwedge_{i \in B} \text{fence}_i \wedge \bigwedge_{i \in P \setminus B} \neg\text{fence}_i$$

where spec is initialized to 0, and auxiliary variables in $\mathcal{F}$ are initialized to true if the corresponding instruction is fenced, and otherwise to false.

To define $\hat{Tr}$, recall that speculation starts if the wrong branch is taken for some $i \in C$. At the first such position, spec becomes positive and remains positive for the rest of the execution. In order to formally model this behavior we modify the state update functions in the following manner.

Conditional Instructions. The state update function for each conditional instruction $i \in C$ (as it appears in Sect. 2.2) is defined as follows:

$$\hat{\tau}_i(\hat{X}, \hat{X}'):=\text{pc} = i \rightarrow \Big(\text{fence}_i \wedge \text{spec} > 0 \text{ ? } id(\hat{X}) : id(\hat{X} \setminus \{\text{spec}, \text{pc}\}) \wedge$$
$$\Big[((\text{spec}' = ((\neg cond(X) \vee \text{spec} > 0) \text{ ? } \text{spec} + 1 : 0)) \wedge \text{pc}' = \vartheta_i(X)) \vee$$
$$((\text{spec}' = ((cond(X) \vee \text{spec} > 0) \text{ ? } \text{spec} + 1 : 0)) \wedge \text{pc}' = \varepsilon_i(X))\Big]\Big)$$

Note that $\hat{\tau}_i$ is stuck in an infinite loop in case $\text{spec} > 0$ and fence_i is set to true. Otherwise, if the respective branch condition does not hold, the value of spec has to become positive. If spec is already positive, it remains positive and is incremented. Overall, spec can only be positive in a given execution iff at least one branch condition of the execution is not met.

Unconditional Instructions. For unconditional instructions, we must take into account the new auxiliary variables, as well as the fence assumptions that

---

[4] In practice, identifying these instructions depends on our assumptions on the attacker capabilities in the given setting, and may involve static analysis techniques such as taint tracking. For our experimental evaluation in Sect. 5, we consider a SPECTRE model with a powerful attacker that can observe (in addition to the control-flow) any memory access, i.e., every instruction that contains an array access $a[i]$ is in VInst.

prevent speculation. Given an unconditional instruction $\tau_i(X, X'):=pc = i \rightarrow \varphi_i(X, X')$ we define $\hat{\tau}_i$ in the following manner:

$$\hat{\tau}_i(\hat{X}, \hat{X}'):=pc = i \rightarrow \big(\text{fence}_i \wedge \text{spec} > 0 \ ? \ \text{id}(\hat{X}) :$$
$$\varphi_i(X, X') \wedge \text{id}(\mathcal{F}) \wedge \text{spec}' = (\text{spec} > 0 \ ? \ \text{spec} + 1 : 0)\big)$$

Again, $\hat{\tau}_i$ enters an infinite loop in case $\text{spec} > 0$ and fence_i is set to true.

Speculation Bound. In order to allow a realistic modeling of speculative execution, we consider a model that only allows a bounded speculation window. This modeling comes from the fact that, in any given microarchitecture, the Reorder Buffer (ROB) used to allow out-of-order execution is limited in the number of instructions it can occupy.

Therefore, we assume that for a given micro-architecture there exists a parameter $\Bbbk$ that is a bound on speculative executions. In order to enforce this bound, we constrain the transition relation $\hat{Tr}$ such that if spec ever reaches $\Bbbk$, $\hat{M}$ is stuck in an infinite loop (this can be viewed as a global assumption).

This results in the following two cases for the transition relation:

$$\hat{Tr}_{<\Bbbk}(\hat{X}, \hat{X}'):=(\text{spec} < \Bbbk) \wedge \bigwedge_{i \in P} \hat{\tau}_i(\hat{X}, \hat{X}') \wedge \tau_\perp(\hat{X}, \hat{X}')$$

$$\hat{Tr}_{\geq\Bbbk}(\hat{X}, \hat{X}'):=(\text{spec} >= \Bbbk) \wedge \text{id}(\hat{X})$$

Let $\hat{Tr}(\hat{X}, \hat{X}'):=\hat{Tr}_{<\Bbbk}(\hat{X}, \hat{X}') \vee \hat{Tr}_{\geq\Bbbk}(\hat{X}, \hat{X}')$. Then the *speculative transition system* of P is the system $\hat{M} = \langle \hat{X}, \hat{Init}, \hat{Tr} \rangle$.

3.3   Reducing Speculative Non-interference to a Safety Property

Now, we can define our notion of information-flow security formally. Let $o : \text{VInst} \rightarrow 2^X$ be a mapping that assigns to every instruction in VInst the set of variables for which the attacker learns their values if the instruction is executed. Then, the *observation* in a state σ with $\sigma(\text{pc}) = i$ is defined as $obs(\sigma) = \{i, o(i)\}$ if $i \in \text{VInst}$, and $obs(\sigma) = \{i\}$ otherwise. Then, $obs(\pi)$ for an execution $\pi = \sigma_0, \sigma_1, \dots$ of M is the sequence of observations $obs(\sigma_0), obs(\sigma_1), \dots$.

Denote by $\pi(\sigma)$ the unique execution of M that starts in σ. Note that for $\hat{M}$, executions starting in a given $\hat{\sigma}$ are not unique. Let $\hat{\Pi}(\hat{\sigma})$ be the set of executions of $\hat{M}$ that start in $\hat{\sigma}$, and denote by $\text{pc}(\hat{\pi})$ for $\hat{\pi} = \hat{\sigma}_0, \hat{\sigma}_1, \dots$ the sequence $i_0, i_1, \dots$ such that $\hat{\sigma}_j(\text{pc}) = i_j$ for all j. Observe that for two executions $\hat{\pi}_1$ and $\hat{\pi}_2$ of $\hat{M}$, $\text{pc}(\hat{\pi}_1) = \text{pc}(\hat{\pi}_2)$ if and only if they agree on the starting point of speculation and on the choice of conditional branches during speculation. For a state $\hat{\sigma}$ of $\hat{M}$, let $\hat{\sigma}|_X$ be the projection of $\hat{\sigma}$ on X, which can also be considered as a state of M.

Definition 1. Let M be the standard transition system of a program P, let $H \subseteq X$ be a set of high-security variables and $L:=X \setminus H$ a set of low-security variables with $\text{spec} \in L$ and $\mathcal{F} \subseteq L$. Let $\hat{M}$ be the speculative transition system of P.

We say that P satisfies *Speculative Non-Interference (SNI)* if for any two initial states $\hat{\sigma}_1, \hat{\sigma}_2$ with $\hat{\sigma}_1(x) = \hat{\sigma}_2(x)$ for all $x \in L$ and $obs(\pi(\hat{\sigma}_1|_X)) = obs(\pi(\hat{\sigma}_2|_X))$, we have $obs(\hat{\pi}_1) = obs(\hat{\pi}_2)$ for any $\hat{\pi}_1 \in \hat{\Pi}(\hat{\sigma}_1), \hat{\pi}_2 \in \hat{\Pi}(\hat{\sigma}_2)$ with $pc(\hat{\pi}_1) = pc(\hat{\pi}_2)$.

Note that this definition of SNI is a slight generalization of the original definition [19]. Note further that SNI can only be violated if there exists a pair of initial states that agree on the observations in their non-speculative executions, but do not agree on the observations on their speculative executions, for some fixed choice of when and how to speculate. Since we assume that additional side-channel observations of the attacker are only possible when a vulnerable instruction in VInst is executed, SNI can only be violated if instructions in VInst are *reachable under speculation*, which is a standard safety property.

This safety property can be modeled by adding assertion instructions to $\hat{M}$. In particular, we add an assertion instruction $\mathbf{assert}(\mathsf{spec} == 0)$ before every instruction $i \in \texttt{VInst}$, reflected by defining

$$\hat{\tau}_{a_i}(\hat{X}, \hat{X}') := \mathsf{pc} = a_i \rightarrow ((\mathsf{spec} = 0) \ ? \ \mathsf{pc}' = i \ : \ \mathsf{pc}' = \bot) \wedge \mathsf{id}(\hat{X} \setminus \{\mathsf{pc}\})$$

and by modifying all expressions $\vartheta_j(X)$ and $\varepsilon_j(X)$ in $\hat{Tr}_{<k}(\hat{X}, \hat{X}')$ such that whenever they would evaluate to some $i \in \texttt{VInst}$, they now evaluate to a_i, and similarly for state update functions $\varphi_j(X, X')$. I.e., an instruction $j \in P$ that precedes an instruction $i \in \texttt{VInst}$ now needs to precede the corresponding assertion a_i.

Let $\overline{Tr}_{<k}(\hat{X}, \hat{X}')$ be the result of applying these changes to $\hat{Tr}_{<k}(\hat{X}, \hat{X}')$. Then we consider the system with the transition relation defined as

$$\overline{Tr}(\hat{X}, \hat{X}') := \left(\overline{Tr}_{<k}(\hat{X}, \hat{X}') \wedge \bigwedge_{i \in \texttt{VInst}} \hat{\tau}_{a_i}(\hat{X}, \hat{X}') \right) \vee \hat{Tr}_{\geq k}(\hat{X}, \hat{X}').$$

Now, we can encode SNI of P as the safety problem $\langle \overline{M}, Bad(\hat{X}) \rangle$, where $\overline{M} = \langle \hat{X}, Init, \overline{Tr} \rangle$ and $Bad(\hat{X}) := \mathsf{pc} = \bot$ (i.e., the definition of bad states is the same as in Sect. 2.2). We call $\overline{M}$ the *speculative transition system with assertions* of P.

Properties of the Speculative Execution Semantics. We give some useful properties of the above semantics. Given an execution $\pi = \sigma_0, \sigma_1, \dots$ of a transition system, let π^{na} be the subsequence of π that contains exactly those σ_j where $\sigma_j(\mathsf{pc})$ does not refer to an **assert**-statement. We say that a transition system $M_1 = \langle X_1, Init_1, Tr_1 \rangle$ *is simulated by* a transition system $M_2 = \langle X_2, Init_2, Tr_2 \rangle$, denoted $M_1 \leq_{\mathsf{sim}} M_2$, if for every execution π_1 of M_1 there exists an execution π_2 of M_2 such that $\pi_2^{\mathsf{na}}|_{X_1} = \pi_1^{\mathsf{na}}$.

Lemma 1. *Let P be a program, $M = \langle X, Init, Tr \rangle$ its standard transition system, and $\overline{M}$ its speculative transition system with assertions. Then $M \leq_{\mathsf{sim}} \overline{M}$.*

Proof. See proof of Lemma 5. □

Lemma 2. *Let $\pi = \sigma_0, \sigma_1, \ldots, \sigma_k$ be an execution of $\overline{M}$ such that $\sigma_k \models Bad$. Then the following conditions hold: (i) $\sigma_0 \models$ spec $= 0$ (ii) $\sigma_k \models$ spec > 0 (iii) There exists a unique $0 \leq j < k$ such that $\sigma_j \models$ spec $= 0$ and $\sigma_{j+1} \models$ spec > 0.*

Proof. From $\sigma_0 \models \hat{Init}$ follows (i). Since $\sigma_k \models Bad$ we have $\sigma_k(\text{pc}) = \bot$. In order for pc to get $\bot$, there needs to be a point where spec $\neq 0$ holds (see $\hat{\tau}_{a_i}$). Since spec can never decrease starting from 0 in σ_0, spec > 0 at that point and (iii) follows. (ii) holds because $\tau_\bot$ never changes spec afterwards.   □

Lemma 3. *Let P be a program, $M = \langle X, Init, Tr \rangle$ its standard transition system, and $VInst \subseteq P$. Let $\overline{M} = \langle \hat{X}, \hat{Init}, \overline{Tr} \rangle$ be the speculative transition system with assertions of P. If $\overline{M}$ is SAFE wrt. Bad, then P satisfies SNI.*

Proof. We have already argued that violations of speculative non-interference are only possible if instructions in VInst are executed under speculation. Since this is exactly the definition of *Bad*, the statement follows immediately.   □

The careful reader will have observed that safety of $\overline{M}$ is a sufficient, but not a necessary condition for P satisfying SNI. I.e., our check over-approximates actual information leaks, and P being UNSAFE does not necessarily mean that SNI is violated. Note that this is a very common design choice in security verification that is e.g. also used in approaches based on taint tracking.

Moreover, a refinement of our approach can achieve full precision in the following way: when a counterexample π is found in $\overline{M}$, actual violation of SNI can be encoded into an SMT query φ_π such that SNI is violated iff φ_π is satisfiable (cp. [19]). If φ_π is unsatisfiable, verification has to continue, searching for other potential counterexamples (very much like in the repair method introduced later in Sect. 4). For simplicity, in the following we do not consider this refined approach.

4   Automatic Repair Under Speculative Execution

In this section we describe CURESPEC, an automatic, model checking based repair algorithm, that fixes SPECTRE-PHT related information leaks under speculative execution. CURESPEC receives a program P and a set of vulnerable instructions VInst. If CURESPEC detects that an instruction $i \in$ VInst is executed under speculation, it repairs P by analyzing the leaking execution and adding a fence instruction that disables speculation in program locations that enabled the leak. This process is iterative and continues until CURESPEC proves that P is secure.

Given a program P, its speculative transition system $\hat{M}$, and the safety problem $\langle \overline{M}, Bad \rangle$ (as defined in Sect. 3, w.r.t. a set of vulnerable instructions VInst). Recall that Algorithm 1 can determine if $\overline{M}$ is SAFE or UNSAFE wrt. *Bad*. When Algorithm 1 returns SAFE, then $\overline{M}$ satisfies speculative non-interference with respect to VInst. Otherwise, a counterexample describing a speculative execution that may leak information is returned and Algorithm 1 terminates.

CURESPEC builds upon Algorithm 1 and extends it such that if $\overline{M}$ is UNSAFE, instead of terminating, repair is applied. The repair process is iterative: When a leak is detected, it is analyzed, a fence is added to mitigate the leak, and verification is reapplied on the repaired program. However, CURESPEC is *incremental* in the sense that it maintains the state of Algorithm 1 as much as possible and reuses it when verification re-executes. To this end, CURESPEC includes all rules of Algorithm 1 excluding the **Cex** rule, and including two additional rules as described in Algorithm 2: **SpecLeak** and **AddFence**.

SpecLeak. This rule replaces the rule **Cex** from Algorithm 1 and prevents the algorithm from terminating when a counterexample is found. Instead, the leaking execution is stored in the variable π.

AddFence. This rule is responsible for the repair. The trace π is analyzed, and based on the misspeculation that leads to an information leak, a fence that makes π unfeasible is added (by letting $\text{fence}_k = \text{true}$ in $\hat{Init}$ for some k). Note that the trace of F_i and the invariant F_∞ are unchanged which ensures incrementality of the overall algorithm.

The **Safe** rule is amended to additionally return the list of added fences.

Input: A safety problem $\langle \overline{M}, Bad(\hat{X}) \rangle$ with $\overline{M} = \langle \hat{X}, \hat{Init}, \overline{Tr} \rangle$.
Assumptions: $\hat{Init}$, $\overline{Tr}$ and Bad are quantifier free.
Data: A queue $\mathcal{Q}$ of potential counterexamples, where $c \in \mathcal{Q}$ is a pair $\langle m, j \rangle$, m is a cube over state variables, $j \in \mathbb{N}$. A level N. A trace $F_0, \ldots, F_N$. An invariant F_∞. A set of reachable states REACH.
Output: A list of added fences $\mathcal{L}$ and a safe inductive invariant F_∞
Initially: $\mathcal{Q} = \emptyset$, $N = 0$, $F_0 = \hat{Init}$, $\forall j \geq 1 \cdot F_j = true$, $F_\infty = true$, $\pi = \langle \rangle$, $\mathcal{L} = \emptyset$.
Require: $\hat{Init} \rightarrow \neg Bad$
repeat

> **SpecLeak** If $\langle m, j \rangle \in \mathcal{Q}$ and $m \cap \text{REACH} \neq \emptyset$, then there is a path π' of $\overline{M}$ from m to Bad and a finite execution π^* of $\overline{M}$ ending in m, respectively. Set $\pi \leftarrow \pi^* \pi'$.
> **AddFence** If $\pi \neq \langle \rangle$, choose $j \leq k \leq N$ where $\sigma_k \models \text{spec} > 0$. Modify $\hat{Init}$ s. t. $\hat{Init} \models \text{fence}_k$. $\mathcal{L} \leftarrow \mathcal{L} \cup \{\text{fence}_k\}$. $\mathcal{Q} \leftarrow \emptyset$, REACH $\leftarrow \hat{Init}$, $\pi \leftarrow \langle \rangle$.

until ∞;

Algorithm 2: Repair algorithm CURESPEC

4.1 Analyzing the Leaking Execution π

When Algorithm 2 detects a potential leak in $\overline{M}$, **SpecLeak** stores this execution in π. In **AddFence**, Lemma 2 is used to identify where speculation starts in π. Based on that, we define the following:

Definition 2. Let $\pi = \sigma_0, \ldots, \sigma_N, \ldots$ be an execution of $\overline{M}$ such that $\sigma_N \models$ *Bad*, and $0 < k \leq N$ such that $\sigma_{k-1} \models$ spec $= 0$ and $\sigma_k \models$ spec > 0. Then $\pi^{[0..k-1]} = \sigma_0, \ldots, \sigma_{k-1}$ is called the *non-speculating prefix* of π, and $\pi^k = \sigma_k, \ldots$ the *speculating suffix* of π. We call k the *speculative split point* of π.

By construction, we have spec $= 0$ in $\pi^{[0..k-1]}$. Therefore, letting fence$_i$ = true for any instruction $i \in \pi^{[0..k-1]}$ has no effect on the transitions in $\pi^{[0..k-1]}$. However, since spec > 0 in π^k, letting fence$_i$ = true for any $i \in \pi^k$ makes π an unfeasible path. In order to formalize this intuition we use the following definition and lemma.

Definition 3. Let P be a program and $\overline{M} = \langle \hat{X}, \widehat{Init}, \overline{Tr} \rangle$ its speculative transition system with assertions. If for $i \in P$ it holds that $\widehat{Init} \models \neg$fence$_i$, then adding a fence to i results in a new transition system $\overline{M}_i = \langle \hat{X}, \widehat{Init}_i, \overline{Tr} \rangle$ where $\widehat{Init}_i := \widehat{Init}[\neg$fence$_i \leftarrow$ fence$_i]$.

Here, $\widehat{Init}[\neg$fence$_i \leftarrow$ fence$_i]$ denotes the substitution of $\neg$fence$_i$ with fence$_i$. After substitution it holds that $\widehat{Init}_i \models$ fence$_i$, and initialization for all other variables is unchanged. Thus, the initial value of fence$_i$ is the only difference between $\overline{M}_i$ and $\overline{M}$.

Lemma 4. *Let* $\pi = \sigma_0, \ldots, \sigma_N$ *be an execution of* $\overline{M}$ *such that* $\sigma_N \models$ *Bad and* $k \leq N$ *be its speculative split point. For any instruction* $i \in P$ *such that* $i \in \pi^k$ *(see Remark 1),* $\widehat{Init} \models \neg$fence$_i$. *Moreover,* π *is not an execution of* $\overline{M}_i$ *and for every execution* $\hat{\pi} = \hat{\sigma}_0, \ldots, \hat{\sigma}_m$ *of* $\overline{M}_i$, $\hat{\pi}\pi^k$ *is not an execution of* $\overline{M}_i$.

Proof. Let $i \in \pi^k$ and j such that $(\sigma_{j-1}, \sigma'_j) \models \hat{\tau}_i$ (see Remark 1). Assume $\widehat{Init} \models$ fence$_i$. Since the values of fences never change during execution, we have $\sigma_{j-1} \models$ fence$_i$. Moreover, $\sigma_{j-1} \models$ spec > 0 and thus, $\sigma_{j-1} = \sigma_j = \sigma_N$. We have a contradiction because $\sigma_{j-1} \models$ pc $= i$. Thus, $\widehat{Init} \models \neg$fence$_i$ and π^k is not a path of $\overline{M}_i$. Therefore, no prefix $\hat{\pi}$ exists such that $\hat{\pi}\pi^k$ is an execution of $\overline{M}_i$. $\qquad\square$

By Lemma 4, it is sufficient to let fence$_i$ = true for any instruction $i \in \pi^k$ in order to make π an unfeasible path. Then, CURESPEC can be resumed and search for a different leaking execution, or prove that the added fences make P secure. This results in the following properties of our repair algorithm.

4.2 Properties of CURESPEC (Algorithm 2)

As noted earlier, CURESPEC is *parametrized* by the set VInst of possibly vulnerable instructions. While in this paper we focus on instructions that are vulnerable to SPECTRE-PHT, CURESPEC can detect other kinds of instructions that are executed under speculation, and as a result, repair such instances where execution under speculation may lead to unwanted transient behavior.

Theorem 1. *Let P be a program and $\overline{M} = \langle \hat{X}, \hat{Init}, \overline{Tr} \rangle$ its speculative transition system with assertions. Then, on input $\langle \overline{M}, Bad \rangle$, CURESPEC terminates and returns a list $\mathcal{L}$ such that for $\overline{M}_s = \langle \hat{X}, \hat{Init}_s, \overline{Tr} \rangle$, where $\forall i \in \mathcal{L} \cdot \hat{Init}_s \models$ fence$_i$, it holds that $\langle \overline{M}_s, Bad \rangle$ is SAFE (as witnessed by the final invariant F_∞).*

Proof. The **SpecLeak** and **AddFence** rules are applicable at most $|\mathcal{F}|$ times each (Lemma 4). Thus, termination follows from the termination of Algorithm 1. After the final application of rule **AddFence**, CURESPEC analyzes $\overline{M}_s$ and constructs an inductive invariant F_∞ showing that $\langle \overline{M}_s, Bad \rangle$ is SAFE. $\quad\square$

Lemma 5. *Let P be a program, M its transition system and $\overline{M}$ its speculative transition system with assertions. For $i \in P$, if $Init \models \neg$fence$_i$ then $M \leq_{sim} \overline{M}_i \leq_{sim} \overline{M}$.*

Proof. Let π be an execution of M. Since it does not involve speculative execution there exists a corresponding execution in $\overline{M}_i$ because fences only affect speculative executions. This shows $M \leq_{sim} \overline{M}_i$. Moreover, the additional fence in $\overline{M}_i$ only removes valid executions from $\overline{M}$. So, $\overline{M}_i \leq_{sim} \overline{M}$ holds, too. $\quad\square$

By Lemma 5, over-approximations computed with respect to $\overline{M}$ are also over-approximations with respect to M. This allows CURESPEC to reuse information between different repair iterations. While rule **AddFence** resets $\mathcal{Q}$, REACH, and π, it does not reset the current level N, the trace $F_0, \ldots, F_N$, and the invariant F_∞, where F_∞ and F_j over-approximate the states that are reachable (in up to j steps). Note that while **AddFence** resets REACH, in practice parts of REACH can be retained even after a fence is added. Intuitively, the repair loop resembles a Counterexample Guided Abstraction Refinement (CEGAR) loop [14].

This *incremental* way of using Algorithm 1 within CURESPEC makes repair already much more efficient than a *non-incremental* version that completely re-starts verification after every modification (see Sect. 5). However, allowing the algorithm to add fences after every instruction in P may still be inefficient, as there are (unnecessarily) many possibilities, and the repair may add many unnecessary fences. Therefore, in the following we describe two heuristics for optimizing the way fences are added.

4.3   Heuristics and Optimizations

Fence Placement Options. As described in Sect. 3, speculation can only start at conditional instructions and the information leak itself happens at an instruction $i \in \text{VInst}$. Thus, we can restrict the set of instructions for which we introduce fence variables fence$_i$ to one of the following: (i) after each conditional instruction $i \in C$, in both branches, or (ii) before every instruction $i \in \text{VInst}$. In the following, option (i) will be called the *after-branch*, and (ii) will be called *before-memory*. In both cases correctness according to Theorem 1 is preserved.

Fence Activation. When a leaking execution π is found, there might be multiple positions for a fence that would prevent it. We employ a heuristic that

---

activates a fence variable as close as possible to the bad state. In case that fence placement option (i) is used, this means that π is traversed backwards from the bad state until a conditional instruction is reached, and a fence is activated in the branch that appears in π. Under option (ii), the instruction $i \in$ VInst that is causing the leak in π is the last instruction in π, and we activate the fence right before i. In both cases, this not only removes the given leaking execution, but also other executions where speculation starts before the newly added fence (cp. Lemma 4).

5   Implementation and Evaluation

Implementation. We implemented CureSpec[5] in the SeaHorn verification framework [20]. This gives us direct access to an LLVM front-end. We compile each benchmark (see below) with Clang[6] 10.0.0 to LLVM using optimization level -O2. The speculative execution semantics is added *after* these compilation passes, which is important because they might introduce new vulnerabilities.

The modified program is encoded into Horn rules [6] and then passed to Z3 [28] version 4.10.2, SeaHorn's internal solver. Upon termination, our tool can output the inductive invariant of the repaired program, together with its speculative semantics, in SMT-LIB format, such that this certificate can be checked independently by any SMT solver that supports the LIA theory.

Evaluation: Benchmarks. We evaluated CureSpec on four sets of benchmarks. The first set consists of Kocher's 15 test cases[7], which are simple code snippets vulnerable to Spectre-PHT attacks. To show that CureSpec can also handle complex programs from a domain that handles sensitive data, we have selected representative and non-trivial (measured in the number of LLVM instructions and conditional instructions, see Table 1 for details) benchmarks from OpenSSL 3.0.0[8] and the HACL* [37] cryptographic library.

A first set of OpenSSL benchmarks includes two versions of AES block encryptions: (i) aes_encrypt, which uses lookup tables and (ii) aes_encrypt_ct, a constant-time version. Both of these only encrypt a single AES block. We also include both AES encryptions in cipher block chaining mode (aes_cbc_encrypt and aes_cbc_encrypt_ct, respectively), which encrypt arbitrarily many blocks, resulting in significantly more complex and challenging benchmarks. The second set of SSL benchmarks includes functions used for the multiplication, squaring, and exponentiation of arbitrary-size integers (bn_mul_part, bn_sqr_part, and bn_exp_part, respectively). Since the full versions of these include function calls to complex subprocedures (with an additional 18900 LLVM-instructions and more than 4100 branches), we abstract these called functions by uninterpreted functions. Finally, the HACL* benchmarks include the following cryptographic primitives: (i) Curve25519_64_ecdh, the ECDH key agreement using

---

[5] https://github.com/joachimbard/seahorn/tree/b3f52fefa1.

[6] https://clang.llvm.org/.

[7] https://www.paulkocher.com/doc/MicrosoftCompilerSpectreMitigation.html.

[8] https://github.com/openssl/openssl/tree/openssl-3.0.0.

Curve25519 [4], (ii) the stream cipher `Chacha20_encrypt`, (iii) a message authentication code using the Poly1305 hash family [3] (`Poly1305_32_mac`).

Evaluation: Repair Performance. We present in detail the repair performance of CureSpec based on the strong threat model (see Sect. 3), i.e., VInst consists of all memory accesses in the analyzed program[9], and on the semantics with unbounded speculation. For each program, we evaluated the performance for all combinations of the following options: (a) incremental or non-incremental repair, (b) fence placement at every instruction, or according to one of the heuristics (after-branch or before-memory) from Sect. 4.3. All experiments were executed on an Intel® Xeon® Gold 6244 CPU @ 3.60 GHz with 251 GiB of main memory.

Table 1. Evaluation results on Kocher's test cases, OpenSSL, and HACL* benchmarks. Test cases not in the table have the similar values as test2. Columns $\#_i$, $\#_b$, and $\#_m$ give the number of instructions, conditional instructions and memory instructions, respectively, which is also the maximal number of possible fences for the respective placement option (every-inst, after-branch, before-memory). The number of inserted fences is shown in the $\#_f$ columns. Presented times are in seconds, averaged over 3 runs, with a timeout of 2 h. RSS is the average maximum resident set size in GiB.

Benchmark	baseline				after-branch				before-memory			
	$\#_i$	$\#_f$	time	RSS	$\#_b$	$\#_f$	time	RSS	$\#_m$	$\#_f$	time	RSS
test1	14	5	0.4	0.12	2	1	0.1	0.07	2	1	0.1	0.06
test2 (and others)	18	7	0.7	0.15	2	1	0.1	0.07	4	1	0.1	0.07
test5	20	1	0.3	0.16	4	2	0.2	0.08	2	1	0.1	0.07
test7	23	11	1.5	0.17	4	2	0.2	0.07	6	2	0.2	0.08
test9	21	10	1.2	0.16	2	1	0.1	0.06	5	1	0.1	0.08
test10	20	15	1.4	0.15	4	2	0.2	0.07	4	2	0.1	0.07
test12	20	9	0.9	0.15	2	1	0.1	0.06	4	1	0.1	0.07
test15	19	8	0.9	0.15	2	1	0.1	0.07	5	1	0.1	0.08
aes_encrypt_ct	566		timeout		4	3	0.8	0.09	38	3	1.7	0.32
aes_encrypt	476	116	3712.1	11.04	4	3	0.7	0.10	97	3	4.7	0.98
aes_cbc_encrypt_ct	1302		timeout		40	15	100.7	0.65	102	10	80.6	1.41
aes_cbc_encrypt	1122		timeout		40	15	774.5	1.51	220	10	135.1	4.34
bn_mul_part	104	69	48.7	0.72	24	13	2.6	0.18	19	7	1.3	0.16
bn_sqr_part	161	71	109.5	1.05	24	13	3.4	0.19	30	9	2.5	0.22
bn_exp_part	307	139	1939.3	5.60	74	29	57.4	0.82	49	18	21.8	0.61
Chacha20_encrypt	3552		timeout		6	3	8.0	0.33	117	2	8.4	1.90
Poly1305_32_mac	483	93	1143.3	4.76	6	4	2.2	0.12	90	4	6.4	0.72
Curve25519_64_ecdh	351		timeout		8	3	3115.3	1.92	44		timeout	

---

[9] We did not analyze the benchmarks with respect to the classical Spectre model, since that requires manual annotations of the code to determine variables that the attacker can control, which require a deep understanding of the code to be analyzed.

Table 1 summarizes the most important results, comparing the **baseline** (non-incremental, every-inst fence placement without fence activation heuristic) to two options that both use incrementality and the heuristic for fence activation, as well as either **after-branch** or **before-memory** for fence placement. We observe that the latter two perform better than the baseline option, with a significant difference both in repair time and number of activated fences, even on the test cases (except for test5, which is solved with a single fence even without heuristics). Notably, all of the benchmarks that time out in the baseline version can be solved by at least one of the other versions, often in under 2 min.

Table 2. Impact of inserted fences on performance. The time for the non-fixed program is the combined runtime (for 50 random seeds), and the impact columns show the increase in runtime compared to the non-fixed program, for programs repaired with after-branch and before-memory placements, respectively.

Benchmark	input size [MiB]	non-fixed time [s]	after-branch impact	before-memory impact
aes_cbc_encrypt_ct	16	64.49	1.91%	2.26%
aes_cbc_encrypt	16	5.84	53.06%	51.55%
Chacha20_encrypt	64	9.86	6.69%	7.57%
Poly1305_32_mac	256	10.07	67.89%	49.73%

While our heuristics make a big difference when compared against the baseline, a comparison between the two fence placement heuristics does not show a clear winner. The before-memory heuristic results in the smallest number of activated fences for all benchmarks, but the difference is usually not big. On the other hand, for the HACL* benchmarks (and some others), this heuristic needs more time than the after-branch heuristic, and even times out for `Curve25519_64_ecdh`, while after-branch solves all of our benchmarks.

Furthermore, we observe that the fence placement heuristics (after-branch or before-memory) reduce the repair time by 81.8% or 97.5%, respectively. Incrementality, when considered over all parameter settings, reduces the repair time on average by 10.3%, but on the practically relevant settings (with fence placement heuristics) it reduces repair time by 21.5%.[10]

The results for a bounded speculation window are comparable, except that this seems to be significantly more challenging for CureSpec: repair times increase, in some cases drastically, and `aes_cbc_encrypt` times out regardless of the selected options. Note however that CureSpec currently does not implement any optimizations that are specific to the bounded speculation mode.

---

[10] In a few cases, the incremental approach actually performs worse than the non-incremental approach. We conjecture that this is a similar effect as can be observed in SAT or SMT solvers, where sometimes it is more efficient to re-start exploration from scratch than continue on a verification run that has already taken a lot of time.

Overall, our results show that CURESPEC can repair complex code such as the OpenSSL and HACL* examples by inserting only a few fences in the right places. Note that for the OpenSSL and HACL* functions, the fences inserted by our repair point to *possible* vulnerabilities, but they might not correspond to actual attacks because of over-approximations in the strong threat model. I.e., CURESPEC may add fences that are not strictly necessary to secure the program.

Evaluation: Performance of Repaired Programs. We evaluate the performance impact of inserted fences on an Intel Core™ i7-8565U CPU @1.80 GHz by comparing the runtimes of *non-fixed* programs to those of repaired programs obtained with after-branch and before-memory fence placements, respectively. To get meaningful results we only use benchmarks that have a non-negligible runtime and use 50 random seeds[11] to generate input data.[12]

Results are summarized in Table 2. The concrete numbers are hardware-specific and can differ greatly for other CPU hardware. However, we can observe some trends, which we discuss in the following. Note that the performance impact is negligible on the constant-time version of AES in cipher block chaining mode, while we have big performance impact on the non-constant time version, even though the number of instructions is similar, and the number of branches and added fences (using a given heuristic) is the same. Moreover, in the Chacha20-implementation, the impact of the three fences from the after-branch heuristic is smaller than the impact of the two fences from the before-memory heuristic.

In summary, the number of fences does not seem to have a strong correlation with the performance impact, which seems to depend more on *where* the fences are added, and on properties of P that are not reflected in the number of instructions or branches. For example, a fence that is placed on an error-handling branch will have much less of an impact on the overall performance than a fence that is placed on a branch that is regularly used in non-erroneous executions of the program. Therefore, we think that optimal placement of fences with respect to their performance impact will be an important direction of future research.

Limitations. As mentioned earlier, we focus on SPECTRE-PHT and do not consider other variants like SPECTRE-BTB, -RSB, and -STL [10]. Moreover, since we extend SEAHORN, its limitations carry over to our approach. In particular, our analysis happens on the LLVM-IR level and thus, compiler passes that transform it into binary code might introduce new vulnerabilities. For example, `select` instructions might be transformed into branches (see e.g. [16]). Note that CURESPEC directly supports dynamic memory allocation, i.e., vulnerable instructions that access dynamically allocated memory can and should be included into `VInst` just like accesses to statically allocated memory.

---

[11] Obtained from https://www.random.org/.

[12] Even though CURESPEC with bounded speculation window inserts fewer fences, their impact on performance is very similar. Therefore, we only give a single table.

6   Related Work

A good overview of existing research on detecting speculative information leaks is given in [12]. To the best of our knowledge, Blade [32] is the only existing tool that combines automatic repair with formal correctness guarantees against Spectre leaks. Blade implements a type-based approach, where the typing rules construct a dataflow graph between expressions in order to detect potential information leaks, similar to taint tracking. While Blade supports automatic repair and comes with a formal security guarantee, it suffers from the usual drawbacks of type-based approaches: the typing rules assume a fixed threat model, and any change to the type system requires to manually prove correctness of the resulting type system. In contrast, our approach is easily extensible, as it is guaranteed to provide a secure implementation whenever it is provided a set of vulnerable instructions.

Apart from Blade, all existing approaches that provide formal guarantees against Spectre attacks require the program to be repaired manually. This includes the approach by Cauligi et al. [11], which explicitly models the reorder buffer and the processor pipeline, and KLEESpectre [33], which explicitly models the cache, potentially achieving a higher precision than our over-approximating approach. Similarly, the technique developed by Cheang et al. [13] as well as the SPECTECTOR technique [18,19] are based on extensions of standard notions like observational determinism to speculative execution semantics, and check for these precisely. Moreover, Haunted RelSE [15] extends symbolic execution to reason about standard and speculative executions at the same time. In [27] speculative execution and attacker capabilities are axiomatically modeled in the CAT language for weak memory models, allowing for easy adaption to new Spectre variants. However, it requires to unroll the program and thus has the drawback of not handling unbounded loops/recursion. Recent work on speculative load hardening (SLH) introduced a type system for providing formal security guarantees in this setting [31], and compared different variants of SLH based on their performance penalty and their level of security [36].

On the other hand, there are approaches that automatically repair a given program, but cannot give a formal security guarantee. This includes SpecFuzz [29], which uses fuzzing to detect and repair out-of-bounds array accesses under speculation, as well as oo7 [34], which detects and repairs Spectre leaks by static analysis and a taint tracking approach on the binary level. However, it cannot give a security guarantee since its binary-level analysis may be incomplete.

Another line of work that resembles our approach is the automatic insertion of fences in weak memory models [1,8,26]. In contrast to these approaches, our algorithm is tightly coupled with the model checker, and does not use it as a black-box. CURESPEC allows SPACER to maintain most of its state when discovering a counterexample, and to resume its operation after adding a fence.

7   Conclusions

We present CURESPEC, an automatic repair algorithm for information leaks that are due to speculative execution. It is implemented in the SEAHORN verification framework and can handle C programs. When CURESPEC detects a leak, it repairs it by inserting a fence. This procedure is executed iteratively until the program is proved secure. To this end, CURESPEC uses the model checking algorithm PDR incrementally, maintaining PDR's state between different iterations. This allows CURESPEC to handle realistic programs, as shown by the experimental evaluation on various C functions from OpenSSL and HACL*. CURESPEC also returns an inductive invariant that enables a simple correctness check of the repair in any SMT solver.

Data-Availability Statement. The evaluated artifact containing the implementation and the benchmarks is publicly available at https://doi.org/10.5281/zenodo.8348711 [2].

Acknowledgments. We would like to thank the anonymous reviewers and our shepherd Lesly-Ann Daniel for their comments and suggestions helping to improve this paper.

References

1. Abdulla, P.A., Atig, M.F., Chen, Y.-F., Leonardsson, C., Rezine, A.: Counterexample guided fence insertion under TSO. In: Flanagan, C., König, B. (eds.) TACAS 2012. LNCS, vol. 7214, pp. 204–219. Springer, Heidelberg (2012). https://doi.org/10.1007/978-3-642-28756-5_15
2. Bard, J., Jacobs, S., Vizel, Y.: Artifact of automatic and incremental repair for speculative information leaks. https://doi.org/10.5281/zenodo.8348711
3. Bernstein, D.J.: The Poly1305-AES message-authentication code. In: Gilbert, H., Handschuh, H. (eds.) FSE 2005. LNCS, vol. 3557, pp. 32–49. Springer, Heidelberg (2005). https://doi.org/10.1007/11502760_3
4. Bernstein, D.J.: Curve25519: new Diffie-Hellman speed records. In: Yung, M., Dodis, Y., Kiayias, A., Malkin, T. (eds.) PKC 2006. LNCS, vol. 3958, pp. 207–228. Springer, Heidelberg (2006). https://doi.org/10.1007/11745853_14
5. Bhattacharyya, A., et al.: Smotherspectre: exploiting speculative execution through port contention. In: CCS, pp. 785–800. ACM (2019)
6. Bjørner, N., McMillan, K., Rybalchenko, A.: On solving universally quantified horn clauses. In: Logozzo, F., Fähndrich, M. (eds.) SAS 2013. LNCS, vol. 7935, pp. 105–125. Springer, Heidelberg (2013). https://doi.org/10.1007/978-3-642-38856-9_8
7. Bloem, R., Jacobs, S., Vizel, Y.: Efficient information-flow verification under speculative execution. In: Chen, Y.-F., Cheng, C.-H., Esparza, J. (eds.) ATVA 2019. LNCS, vol. 11781, pp. 499–514. Springer, Cham (2019). https://doi.org/10.1007/978-3-030-31784-3_29
8. Bouajjani, A., Derevenetc, E., Meyer, R.: Checking and enforcing robustness against TSO. In: Felleisen, M., Gardner, P. (eds.) ESOP 2013. LNCS, vol. 7792, pp. 533–553. Springer, Heidelberg (2013). https://doi.org/10.1007/978-3-642-37036-6_29

9. Bradley, A.R.: SAT-based model checking without unrolling. In: Jhala, R., Schmidt, D. (eds.) VMCAI 2011. LNCS, vol. 6538, pp. 70–87. Springer, Heidelberg (2011). https://doi.org/10.1007/978-3-642-18275-4_7

10. Canella, C., et al.: A systematic evaluation of transient execution attacks and defenses. In: USENIX Security Symposium, pp. 249–266 (2019)

11. Cauligi, S., et al.: Constant-time foundations for the new spectre era. In: PLDI, pp. 913–926. ACM (2020)

12. Cauligi, S., Disselkoen, C., Moghimi, D., Barthe, G., Stefan, D.: Sok: practical foundations for software spectre defenses. In: SP, pp. 666–680. IEEE (2022)

13. Cheang, K., Rasmussen, C., Seshia, S.A., Subramanyan, P.: A formal approach to secure speculation. In: CSF, pp. 288–303. IEEE (2019)

14. Clarke, E., Grumberg, O., Jha, S., Lu, Y., Veith, H.: Counterexample-guided abstraction refinement. In: Emerson, E.A., Sistla, A.P. (eds.) CAV 2000. LNCS, vol. 1855, pp. 154–169. Springer, Heidelberg (2000). https://doi.org/10.1007/10722167_15

15. Daniel, L., Bardin, S., Rezk, T.: Hunting the haunter - efficient relational symbolic execution for spectre with haunted relse. In: 28th Annual Network and Distributed System Security Symposium, NDSS 2021, virtually, 21–25 February 2021. The Internet Society (2021). https://www.ndss-symposium.org/ndss-paper/hunting-the-haunter-efficient-relational-symbolic-execution-for-spectre-with-haunted-relse/

16. Daniel, L., Bardin, S., Rezk, T.: Binsec/Rel: symbolic binary analyzer for security with applications to constant-time and secret-erasure. ACM Trans. Priv. Secur. 26(2), 11:1–11:42 (2023)

17. Denning, D.E., Denning, P.J.: Certification of programs for secure information flow. Commun. ACM 20(7), 504–513 (1977)

18. Fabian, X., Guarnieri, M., Patrignani, M.: Automatic detection of speculative execution combinations. In: CCS, pp. 965–978. ACM (2022)

19. Guarnieri, M., Köpf, B., Morales, J.F., Reineke, J., Sánchez, A.: Spectector: principled detection of speculative information flows. In: IEEE Symposium on Security and Privacy, pp. 1–19. IEEE (2020)

20. Gurfinkel, A., Kahsai, T., Komuravelli, A., Navas, J.A.: The SeaHorn verification framework. In: Kroening, D., Păsăreanu, C.S. (eds.) CAV 2015. LNCS, vol. 9206, pp. 343–361. Springer, Cham (2015). https://doi.org/10.1007/978-3-319-21690-4_20

21. Hoder, K., Bjørner, N.: Generalized property directed reachability. In: Cimatti, A., Sebastiani, R. (eds.) SAT 2012. LNCS, vol. 7317, pp. 157–171. Springer, Heidelberg (2012). https://doi.org/10.1007/978-3-642-31612-8_13

22. Khasawneh, K.N., Koruyeh, E.M., Song, C., Evtyushkin, D., Ponomarev, D., Abu-Ghazaleh, N.B.: Safespec: banishing the spectre of a meltdown with leakage-free speculation. In: DAC, p. 60. ACM (2019)

23. Kiriansky, V., Lebedev, I.A., Amarasinghe, S.P., Devadas, S., Emer, J.S.: DAWG: a defense against cache timing attacks in speculative execution processors. In: MICRO, pp. 974–987. IEEE Computer Society (2018)

24. Kocher, P., et al.: Spectre attacks: exploiting speculative execution. In: IEEE Symposium on Security and Privacy, pp. 1–19. IEEE (2019)

25. Komuravelli, A., Gurfinkel, A., Chaki, S.: SMT-based model checking for recursive programs. Formal Methods Syst. Des. 48(3), 175–205 (2016)

26. Kuperstein, M., Vechev, M.T., Yahav, E.: Automatic inference of memory fences. In: Bloem, R., Sharygina, N. (eds.) Proceedings of 10th International Conference on

Formal Methods in Computer-Aided Design, FMCAD 2010, Lugano, Switzerland, 20–23 October, pp. 111–119. IEEE (2010)

27. de León, H.P., Kinder, J.: Cats vs. spectre: an axiomatic approach to modeling speculative execution attacks. In: 43rd IEEE Symposium on Security and Privacy, SP 2022, San Francisco, CA, USA, 22–26 May 2022, pp. 235–248. IEEE (2022). https://doi.org/10.1109/SP46214.2022.9833774

28. de Moura, L., Bjørner, N.: Z3: an efficient SMT solver. In: Ramakrishnan, C.R., Rehof, J. (eds.) TACAS 2008. LNCS, vol. 4963, pp. 337–340. Springer, Heidelberg (2008). https://doi.org/10.1007/978-3-540-78800-3_24

29. Oleksenko, O., Trach, B., Silberstein, M., Fetzer, C.: Specfuzz: bringing spectre-type vulnerabilities to the surface. In: USENIX Security Symposium, pp. 1481–1498. USENIX Association (2020)

30. Schwarz, M., Schwarzl, M., Lipp, M., Masters, J., Gruss, D.: NetSpectre: read arbitrary memory over network. In: Sako, K., Schneider, S., Ryan, P.Y.A. (eds.) ESORICS 2019. LNCS, vol. 11735, pp. 279–299. Springer, Cham (2019). https://doi.org/10.1007/978-3-030-29959-0_14

31. Shivakumar, B.A., et al.: Typing high-speed cryptography against spectre v1. In: SP, pp. 1094–1111. IEEE (2023)

32. Vassena, M., et al.: Automatically eliminating speculative leaks from cryptographic code with blade. Proc. ACM Program. Lang. 5(POPL), 1–30 (2021)

33. Wang, G., Chattopadhyay, S., Biswas, A.K., Mitra, T., Roychoudhury, A.: Kleespectre: detecting information leakage through speculative cache attacks via symbolic execution. ACM Trans. Softw. Eng. Methodol. 29(3), 14:1–14:31 (2020)

34. Wang, G., Chattopadhyay, S., Gotovchits, I., Mitra, T., Roychoudhury, A.: oo7: Low-overhead defense against spectre attacks via program analysis. IEEE Trans. Softw. Eng. 47(11), 2504–2519 (2021)

35. Yan, M., Choi, J., Skarlatos, D., Morrison, A., Fletcher, C.W., Torrellas, J.: Invisispec: making speculative execution invisible in the cache hierarchy (corrigendum). In: MICRO, p. 1076. ACM (2019)

36. Zhang, Z., Barthe, G., Chuengsatiansup, C., Schwabe, P., Yarom, Y.: Ultimate SLH: taking speculative load hardening to the next level. In: USENIX Security Symposium, pp. 7125–7142. USENIX Association (2023)

37. Zinzindohoué, J.K., Bhargavan, K., Protzenko, J., Beurdouche, B.: Hacl*: a verified modern cryptographic library. In: Thuraisingham, B., Evans, D., Malkin, T., Xu, D. (eds.) Proceedings of the 2017 ACM SIGSAC Conference on Computer and Communications Security, CCS 2017, Dallas, TX, USA, 30 October–03 November 2017, pp. 1789–1806. ACM (2017). https://doi.org/10.1145/3133956.3134043

Sound Abstract Nonexploitability Analysis

Francesco Parolini[✉][iD] and Antoine Miné[iD]

Sorbonne Université, CNRS, LIP6, 75005 Paris, France
{francesco.parolini,antoine.mine}@lip6.fr

Abstract. Runtime errors that can be triggered by an attacker are sensibly more dangerous than others, as they not only result in program failure, but can also be exploited and lead to security breaches such as Denial-of-Service attacks or remote code execution. Proving the absence of exploitable runtime errors is challenging, as it involves combining classic techniques for safety with novel security analyses. While numerous approaches to statically detect runtime errors have been proposed, they lack the ability to classify program failures as potentially exploitable or not. In this paper, we bridge the gap between traditional safety properties and security hyperproperties by putting forward a novel definition of *nonexploitability*, which we leverage to propose a sound static analysis by abstract interpretation to prove the absence of exploitable runtime errors. While false alarms can occur, if our analysis determines that a program is *nonexploitable*, then there is a strong mathematical guarantee that it is *impossible* for an attacker to trigger a runtime error. Furthermore, our analysis reduces the noise generated from false positives by classifying each warning as security-critical or not. We implemented the first nonexploitability analyzer for a subset of C, and we evaluated it on a set of 77 real-world programs taken from the GNU Coreutils package that are long up to $4,188$ lines of code. Our analysis was able to *prove* that more than 70% of the runtime errors previously reported ($3,498$ over $4,715$) cannot be triggered by an attacker.

Keywords: Security and Privacy · Static Analysis · Abstract Interpretation

1 Introduction

Program failures that can be triggered by a malicious user are sensibly more dangerous than others, as they can lead to security breaches. Attackers can exploit well-known runtime errors, such as index out-of-bounds and double free, to perform dangerous attacks including Denial-of-Service (DoS) attacks or remote code execution. Numerous companies identified such exploitable vulnerabilities in their systems, including Meta [3], Apple [2], and Google [4]. Microsoft recently published a report showing that consistently over 20 years, around 70% of the security breaches that have been reported in their systems are due to exploitable

R. Dimitrova et al. (Eds.): VMCAI 2024, LNCS 14500, pp. 314–337, 2024.
https://doi.org/10.1007/978-3-031-50521-8_15

memory corruption [8]. As it is difficult to identify program errors with manual inspection, static analysis is an invaluable tool to automatically detect them.

While sound static analyzers can report *all* possible runtime errors, including the exploitable ones, they often raise a high number of false positives. If the noise generated by the false alarms is elevated, the report of the analyzer quickly becomes unintelligible, and it is then difficult to identify the true exploitable runtime errors. In order to filter out the warnings that do not concern security issues, it is necessary to combine a traditional analysis for safety properties with a security analysis for hyperproperties [19].

In this paper, we bridge the gap between classic safety and security. We first formalize *nonexploitability* as a hyperproperty, and then we propose an alternative characterization based on *semantically tainted* (i.e. user-controlled) variables. We leverage such a characterization to put forward a sound analysis by abstract interpretation [23] that can prove the absence of exploitable runtime errors. Our analysis has the capability to classify each warning by its threat level (security-related or not), which makes the report of the analyzer more intelligible.

We leverage an underlying abstract value domain to infer numeric invariants, which we pair with a *semantic taint analysis* that tracks the set of user-controlled variables. Combining the two is necessary in order to infer an overapproximation of the exploitable runtime errors. By taking advantage of the semantic information inferred by the abstract numeric domain, our taint analysis achieves enhanced precision compared to traditional methods. Furthermore, our framework can handle programming language features that are essential to analyze real-world programs, such as nondeterminism and runtime user input.

We implemented and evaluated the first analyzer for nonexploitability in the MOPSA [35] static analysis platform. The analysis targets a large subset of C and it is *fully automatic*. We analyzed 77 real-world programs, each up to $4,188$ lines long, taken from the GNU Coreutils package, to which we added $13,261$ test cases taken from the Juliet test suite developed by NIST [7]. We found that our tool can *prove* that more than 70% of the warnings previously raised by the analyzer ($3,498$ over $4,715$) cannot be triggered by an attacker, while incurring a performance overhead of less than 16%.

In this paper, we claim the following contributions:

- We introduce a novel property, *nonexploitability*, and we give its semantic characterization as a hyperproperty.
- We put forward an alternative characterization of nonexploitability in terms of *semantically tainted* (i.e. user-controlled) variables.
- We introduce a new practical, modular analysis that combines a traditional value analysis with a taint analysis to prove nonexploitabilty.
- We implement our analysis and evaluate it on a large set of real-world C programs.

2   Motivation

Figure 1 represents an exploitable program where a buffer overflow can occur depending on the user's input. A malicious user can take advantage of this type

```
1   #include <stdio.h>
2   #include <string.h>
3
4   void use_input(const char* input) {
5       char dest[10];
6       strcpy(dest, input);
7   }
8
9   void main() {
10      char buff[100];
11      fgets(buff, sizeof(buff), stdin);
12      use_input(buff);
13  }
```

Fig. 1. C program with exploitable buffer overflow.

of vulnerability to execute sophisticated, dangerous attacks. There are numerous examples of well-known attacks that exploit runtime errors: among them we find the Morris Worm [45], Code Red [17], SQL Slammer [51], and Heartbleed [26].

While techniques such as testing and human inspection by security experts are useful to detect (exploitable) runtime errors, the only option to rule out their existence is through formal methods. In particular, abstract interpretation [23] has been effective in proving the absence of program failures in real-time avionics software [21]. While analyses by abstract interpretation are *sound*, they can often raise false positives. If the noise generated by the false alarms is too high, the analyzer quickly becomes unusable.

Reducing the number of false alarms is usually achieved by employing more precise abstract domains. This paper takes an orthogonal approach to the problem, reducing the number of alarms by reporting the subset of possible runtime errors that can be triggered by an attacker. As these errors are comparatively more dangerous than the others, the report of the analyzer becomes more intelligible, enhancing the usefulness of sound static analysis tools. Girol et al. make the same observation, leveraging the concept of *robust reachability* to identify errors that are relatively more dangerous [28]. A bug is *robustly reachable* if there exists a user input for which the bug is always reached, regardless of the value of the uncontrolled input. The main difference with our concept of exploitability, is that we require the user input to be actually used in triggering the bug, while strong reachability does not (see Sect. 8 for a detailed comparison).

Taint analysis is a popular technique used in computer security to track the flow of untrusted data within a program, and we leverage this method to prove nonexploitability. While taint analyzers often rely on heuristics to track the flow of unsafe data [11], our approach is *semantic*, namely grounded in a definition based on the formal semantics of programs. While formal methods techniques extensively studied the verification of security properties such as *noninterference* [20,29,30], the exploitability analysis is, to the best of our knowledge, uninvestigated (see Sect. 8 for a comparison). This paper bridges the gap between classic safety properties analysis and security hyperproperties [19] in order to rule out the existence of exploitable runtime errors from a software system. Our approach is closely related to [22], which proposes a technique that

$$P := S \qquad \text{(Programs)}$$
$$S := \texttt{skip} \mid \texttt{x} \leftarrow \texttt{input}() \mid \texttt{x} \leftarrow \texttt{rand}() \mid \texttt{x} \leftarrow \texttt{A} \qquad \text{(Statements)}$$
$$\mid \texttt{S}; \texttt{S} \mid \texttt{if (B) S else S} \mid \texttt{while (B) S}$$
$$\texttt{A} := n \mid \texttt{x} \mid \texttt{A} \diamond \texttt{A} \; (\diamond \in \{+, -, *, / \}) \qquad \text{(Arithmetic Expressions)}$$
$$\texttt{B} := \texttt{tt} \mid \texttt{ff} \mid \texttt{A} < \texttt{A} \mid \neg\texttt{B} \mid \texttt{B} \diamond \texttt{B} \; (\diamond \in \{ \&\&, || \}) \qquad \text{(Boolean Expressions)}$$

Fig. 2. Syntax of the WHILE language.

can prove noninterference using the abstract interpretation framework. Nevertheless, our technique supports features such as nondeterminism and dynamic user input that are not considered in [22]. Furthermore, we leverage the values of the variables to enhance the precision of our analysis, which is an extension proposed but not implemented in [22].

3 Syntax and Concrete Semantics

In this section, we define a *reachability semantics* that computes the set of reachable program states. As we are interested in program errors, the semantics also collects the set of error states. The finite set of program variables is denoted as $\mathbb{V}$, and in Fig. 2 we present the syntax of the WHILE language that we consider.

Expressions are deterministic, and nondeterminism is isolated in the language in specific statements (`rand`, `input`) to simplify the presentation. We define the set of *program memories* as $\mathbb{M} \triangleq \mathbb{V} \rightarrow \mathbb{Z}$. The value $\frac{1}{2}$ represents a runtime error, and $\mathbb{Z}_{\frac{1}{2}} \triangleq \mathbb{Z} \cup \{ \frac{1}{2} \}$. The *arithmetic evaluation* $\mathcal{A}[\![\texttt{A}]\!] : \mathbb{M} \rightarrow \mathbb{Z}_{\frac{1}{2}}$ definition is straightforward: it results in an error if there is a division by zero. For illustration purposes, we consider only runtime failures arising from divisions by zero, but our implementation supports all classic C arithmetic and memory errors. The set $\mathbb{B}$ is $\{ \texttt{tt}, \texttt{ff} \}$, and $\mathbb{B}_{\frac{1}{2}} \triangleq \mathbb{B} \cup \{ \frac{1}{2} \}$. The *boolean evaluation* $\mathcal{B}[\![\texttt{B}]\!] : \mathbb{M} \rightarrow \mathbb{B}_{\frac{1}{2}}$ results in a runtime error if the arithmetic evaluation results in a runtime error. The definitions of $\mathcal{A}[\![\texttt{A}]\!]$ and $\mathcal{B}[\![\texttt{B}]\!]$ are standard, and we do not report them.

The *program states* are triplets $(m, i, r) \in \mathbb{M} \times \mathbb{Z}^{\omega} \times \mathbb{Z}^{\omega} \triangleq \mathbb{S}$. The first element is the program memory, the second is the unbounded sequence of inputs provided by the user, and the third is the unbounded sequence of random numbers. We explicitly represent states in which a runtime error occurred by setting a special *return* variable to 1. All error-free states have the return variable, denoted as `ret`, set to 0. Programs cannot read nor write explicitly to `ret`, as it cannot syntactically appear in statements. Our semantics relies on pairs of initial-reachable states. A set of initial-reachable states is a relation $\mathcal{R} \in \wp(\mathbb{S} \times \mathbb{S}) \triangleq \mathbb{D}$.

In this section, we will define the *reachability semantics* of statements $\mathbb{S}[\![\texttt{S}]\!] : (\mathbb{D} \times \mathbb{D}) \rightarrow (\mathbb{D} \times \mathbb{D})$ by induction. The first element in the input pair is the set of pre-post states that reach the current statement without encountering an error, while the second is the set of pre-post states that previously resulted in an error. $\mathbb{S}[\![\texttt{S}]\!](\mathcal{R}, \mathcal{E})$ outputs both the reachable and the error states after executing S. The set of initial states is $\mathcal{I} \triangleq \{ ((m, i, r), (m, i, r)) \mid (m, i, r) \in \mathbb{S}, m[\texttt{ret}] = 0 \}$.

We define the semantics of programs $\mathcal{S}[\![P]\!] \in \mathbb{D}$ by merging the reachable states at the end of the program with those that resulted in an error. Let P:=S.

$$\mathcal{S}[\![P]\!] \triangleq \text{let } (\mathcal{R}, \mathcal{E}) = \mathcal{S}[\![S]\!](\mathcal{I}, \emptyset) \text{ in } \mathcal{R} \cup \mathcal{E}$$

We now define by structural induction the reachability semantics of statements. As the definitions for skip and $S_1 ; S_2$ are standard, we do not report them. For the input read statement, we update the memory by assigning the first number in the infinite input sequence to the assigned variable, and then we shift the input sequence. The operator $\circ$ denotes the relation composition, while hd and tl respectively extract the head and the tail of a sequence.

Input read statement $(S := x \leftarrow \text{input}())$.

$$\mathcal{S}[\![S]\!](\mathcal{R}, \mathcal{E}) \triangleq (\mathcal{R} \circ \{\, ((m, i, r), (m[x \leftarrow \text{hd}(i)], \text{tl}(i), r) \mid (m, i, r) \in \mathbb{S}) \,\}, \mathcal{E})$$

The random read statement is similar to the input read statement, but uses the infinite sequence of random numbers.

Random read statement $(S := x \leftarrow \text{rand}())$.

$$\mathcal{S}[\![S]\!](\mathcal{R}, \mathcal{E}) \triangleq (\mathcal{R} \circ \{\, ((m, i, r), (m[x \leftarrow \text{hd}(r)], i, \text{tl}(r)) \mid (m, i, r) \in \mathbb{S}) \,\}, \mathcal{E})$$

Assignments can result in errors, which in our semantics are represented as states where ret is 1. If a runtime error occurs, the program sets ret to 1 and adds the state to the second element of the output pair. Error states are collected throughout the execution, and are propagated at the end of the program even in case of non-termination. Note, however, that non-termination is not considered to be an error. We define $\text{ok}[\![A]\!] : \mathbb{D} \to \mathbb{D}$ and $\text{err}[\![A]\!] : \mathbb{D} \to \mathbb{D}$ to collect respectively regular and error states in the evaluation of A.

$$\text{ok}[\![A]\!]\mathcal{R} \triangleq \mathcal{R} \circ \{\, ((m, i, r), (m[x \leftarrow \mathcal{A}[\![A]\!]m], i, r)) \mid (m, i, r) \in \mathbb{S}, \mathcal{A}[\![A]\!]m \neq \lightning \,\}$$

$$\text{err}[\![A]\!]\mathcal{R} \triangleq \mathcal{R} \circ \{\, ((m, i, r), (m[\text{ret} \leftarrow 1], i, r)) \mid (m, i, r) \in \mathbb{S}, \mathcal{A}[\![A]\!]m = \lightning \,\}$$

Assignment statement $(S := x \leftarrow A)$

$$\mathcal{S}[\![S]\!](\mathcal{R}, \mathcal{E}) \triangleq (\text{ok}[\![A]\!]\mathcal{R}, \mathcal{E} \cup \text{err}[\![A]\!]\mathcal{R})$$

We define $\text{test}[\![B]\!] : \mathbb{D} \to \mathbb{D}$ to filter states according to a boolean condition B: $\text{test}[\![B]\!]\mathcal{R} \triangleq \mathcal{R} \circ \{((m, i, r), (m, i, r)) \mid (m, i, r) \in \mathbb{S}, \mathcal{B}[\![B]\!]m = \text{tt}\}$. We abuse the notation, and we use $\text{err}[\![B]\!]$ for the boolean evaluation of errors.

If statement $(S := \texttt{if (B) } S_t \texttt{ else } S_e)$

$$\mathcal{S}[\![S]\!](\mathcal{R}, \mathcal{E}) \triangleq \text{let } (\mathcal{R}_t, \mathcal{E}_t) = \mathcal{S}[\![S_t]\!](\text{test}[\![B]\!]\mathcal{R}, \mathcal{E}) \text{ in}$$
$$\text{let } (\mathcal{R}_e, \mathcal{E}_e) = \mathcal{S}[\![S_e]\!](\text{test}[\![\neg B]\!]\mathcal{R}, \mathcal{E}) \text{ in}$$
$$(\mathcal{R}_t \cup \mathcal{R}_e, \mathcal{E}_t \cup \mathcal{E}_e \cup \text{err}[\![B]\!]\mathcal{R})$$

The semantics of while statements is a classic fixpoint definition. The operator $\dot{\cup}$ denotes the point-wise set union on pairs. For the rest of the paper, we denote the point-wise lifting of operator $\diamond$ to tuples as $\dot{\diamond}$.

While statement $(S := \texttt{while (B) } S_b)$

$$\mathcal{S}[\![S]\!](\mathcal{R}, \mathcal{E}) \triangleq \text{let } (\mathcal{R}_f, \mathcal{E}_f) = \text{lfp } F \text{ in } (\text{test}[\![\neg B]\!]\mathcal{R}_f, \mathcal{E}_f)$$
$$\text{where } F(\mathcal{R}_1, \mathcal{E}_1) \triangleq (\mathcal{R}, \mathcal{E}) \dot{\cup} \mathcal{S}[\![\texttt{if (B) } S_b \texttt{ else skip}]\!](\mathcal{R}_1, \mathcal{E}_1)$$

---

4   Nonexploitability

In this section, we first give a formal definition of *nonexploitability* as a hyperproperty [19]. Then, we put forward an alternative characterization based on semantically tainted (i.e. user-controlled) variables, which we leverage to introduce a sound, effective analysis for nonexploitability.

Nonexploitability formalizes the idea that by modifying only the user input at the beginning of a program, it is not possible to change whether the program results in a runtime error or not. Since we designed our concrete semantics to explicitly represent runtime errors as states with the return variable set to 1, we use program memories to differentiate erroneous states from regular ones.

Definition 1. (Nonexploitability).

$$\mathcal{NE} \triangleq \{\, \mathcal{R} \in \mathbb{D} \mid \forall((m_0, i_0, r_0), (m_1, i_1, r_1)), ((m'_0, i'_0, r'_0), (m'_1, i'_1, r'_1)) \in \mathcal{R} :$$
$$m_0 = m'_0, i_0 \neq i'_0, r_0 = r'_0 \implies m_1[\texttt{ret}] = m'_1[\texttt{ret}] \,\}$$

Example 1. Accordingly to our definition, the following program is exploitable: $x \leftarrow \texttt{input()}; \ 1/x$. This is because if we consider two initial states, one in which the first element of the input sequence is zero, and the other in which it is not, we observe that the value of $\texttt{ret}$ changes. Conversely, the program $x \leftarrow \texttt{rand()}; \ 1/x$ is *nonexploitable*: even if there is a possible division by zero, once we fix the sequence of random numbers, changing the user input does not result in modifying the value of $\texttt{ret}$. If we did not compare pairs of initial states with the *same* sequence of random numbers, the program would be exploitable, even if the user input is never read.

Example 2. (Comparison with robust reachability [28]). A bug is *robustly reachable* if there exists a user input for which the bug is always reached, regardless of the value of the random input. Consider a program that always results in a division by zero: 1/0. The program is nonexploitable: for any possible user input, the value of ret will always be 1. Conversely, the error is robustly reachable, as it is trivially reached for *any* user input. This highlights an important difference between the two concepts: nonexploitability requires the user input to be *effectively used* in triggering program errors, while robust reachability does not.

In what follows, we show that nonexploitability can be expressed in terms of *semantically tainted variables*. Intuitively, a variable is tainted if an attacker can control its value. Taint analysis [36] is a well-known technique in computer security to track the variables that are controlled by external users. However, many existing approaches use heuristics and syntactic formulations of the problem, which may be both imprecise and unsound. In contrast, we rely on a *semantic* approach, which is grounded in the formal semantics of programs. The following hyperproperty captures the set of semantics where the value of a variable x depends on the user's input, i.e. x is tainted. We compare pairs of executions that in the initial states differ only for the user's input, but then result in different values for x. The definition formalizes the intuition that x is tainted if, by modifying only the user input, it is possible to change the value of x.

Definition 2. (Taint). Let $x \in \mathbb{V}$.

$$\mathscr{T}(x) \triangleq \{ \mathscr{R} \in \mathbb{D} \mid \exists((m_0, i_0, r_0), (m_1, i_1, r_1)), ((m_0', i_0', r_0'), (m_1', i_1', r_1')) \in \mathscr{R} :$$
$$m_0 = m_0', i_0 \neq i_0', r_0 = r_0' : m_1[x] \neq m_1'[x] \}$$

We define abstraction and concretization functions for $\wp(\mathbb{V})$.

$$\alpha_t(\mathscr{R}) \triangleq \{ x \in \mathbb{V} \mid \mathscr{R} \subseteq \mathscr{T}(x) \} \qquad \gamma_t(\mathcal{T}) \triangleq \bigcap_{x \in \mathcal{T}} \mathscr{T}(x)$$

As it turns out, there is a Galois connection between $\wp(\mathbb{D})$ and $\wp(\mathbb{V})$ defined by α_t and γ_t. The order for the abstract domain $\wp(\mathbb{V})$ is $\supseteq$ because if we consider more relations, we obtain fewer tainted variables common to *all* of these relations. Notice that this is different from observing that larger relations present more tainted variables, which will be discussed later in this section. A variable x is tainted in a program P if $x \in \alpha_t(\{\mathcal{S}[\![P]\!]\})$.

Example 3. (Implicit flows). If statements can generate *implicit flows* [25], namely dependencies that arise from the program control flow. Consider the following: x ← input(); if (x==0) y ← 1 else y ← 2. Depending on the user's input, y can be either 1 or 2, and accordingly to our semantic characterization of tainted variables, y is tainted. Taint analyzers (e.g. [5,6,11,53]) often ignore implicit flows, considering only *explicit flows* (i.e. when tainting is propagated through assignments only), which is unsound in our framework.

If the user cannot control the value of ret, then they cannot control whether there is a runtime error, i.e. the program is nonexploitable. This is the fundamental observation used in the following alternative characterization of $\mathscr{NE}$.

$$\mathscr{R} \in \mathscr{NE} \iff \text{ret} \notin \alpha_t(\{\mathscr{R}\}) \tag{1}$$

Equation (1) is significant because it shows that nonexploitability can be verified with a taint analysis. In contrast to classic taint analyses, simply tracking the set of user-controlled variables is not sufficient, as to infer whether `ret` is tainted we also need to detect runtime errors. In fact, `ret` does not syntactically appear in programs, and its value changes only when program failures occur. To determine when this happens, is it important to consider the *values* of the variables. Without semantic information about the values of the variables, every expression with a division should be considered dangerous in order to be sound, and this would result in an unacceptable loss of precision. In Sect. 6 we put forward a sound analysis by abstract interpretation that can prove programs to be nonexploitable by combining a classic value analysis with a taint analysis. The former detects program locations that potentially present runtime errors, while the latter determines whether the user can trigger those errors.

Hyperproperties verification is challenging for analyses based on abstract interpretation, because not every hyperproperty is *subset-closed* [19]: by computing an overapproximation $\mathcal{R}_1$ of $\mathcal{R}_0$, the fact that $\mathcal{R}_1$ respects an hyperproperty does not, in the general case, imply that $\mathcal{R}_0$ respects the hyperproperty. To overcome this problem, many works rely on *hypersemantics* [12,37–39,56]: the concrete semantics of a program is a set of sets of states, in contrast to a classic set of states. The main disadvantage of hypersemantics is that *hyperdomains* [37–39] are incompatible with regular abstract domains: the former abstract hypersemantics, while the latter abstract regular semantics.

In this paper, we rely on the standard abstract interpretation framework. In the rest of this section, we show that an overapproximation of the concrete semantics is sufficient to prove nonexploitability. A significant benefit of using the standard framework is that we can combine a taint analysis with any existing over-approximating value domain, which leads to a modular design. Furthermore, enhancing the precision of the numeric analysis improves the precision of the taint analysis as well. Observe that, as discussed in this section, in our context, it is important to rely on a classic safety analysis (and hence, on regular abstract numeric domains) to identify expressions that potentially present runtime errors.

We observe that larger semantics have more tainted variables. This holds due to the existential quantifier in Definition 2. Let $\mathcal{R}_0, \mathcal{R}_1 \in \mathbb{D}$.

$$\mathcal{R}_0 \subseteq \mathcal{R}_1 \implies \alpha_t(\{\mathcal{R}_0\}) \subseteq \alpha_t(\{\mathcal{R}_1\}) \tag{2}$$

By using this result, we observe that if `ret` is not tainted in $\mathcal{R}_1$, it cannot be tainted in $\mathcal{R}_0$. This implies that if $\mathcal{R}_1$ is nonexploitable, then $\mathcal{R}_0$ is nonexploitable, namely $\mathcal{N\!E}$ is subset-closed.

Theorem 1. ($\mathcal{N\!E}$ is subset-closed). *Let* $\mathcal{R}_0, \mathcal{R}_1 \in \mathbb{D}$.

$$(\mathcal{R}_0 \subseteq \mathcal{R}_1 \text{ and } \mathcal{R}_1 \in \mathcal{N\!E}) \implies \mathcal{R}_0 \in \mathcal{N\!E}$$

Theorem 1 is significant because it implies that by overapproximating the semantics of a program, we can still prove that it is nonexploitable. This justifies why the standard abstract interpretation framework is sufficient, and allows using the large library of existing abstract value domains. The theorem formalizes the intuition that if it is not possible for an attacker to trigger any runtime error, by further reducing the semantics of the program–and hence the capabilities of the attacker–he is still not able to make the program fail.

5   Taint Concrete Semantics

In this section, we define the non-computable concrete taint semantics that we overapproximate in Sect. 6. The semantics associates the reachable states with the set of semantically tainted variables using the abstraction function α_t. As the semantics is not *structural* (i.e. defined by induction on the program syntax), we also develop a structural equivalent definition. This is necessary in order to overapproximate the concrete taint semantics with an inductive and effectively computable abstract semantics.

We first define the reachability taint semantics of statements $\mathcal{S}_t[\![S]\!] : (\mathbb{D} \times \mathbb{D}) \to (\mathbb{D} \times \mathbb{D} \times \wp(\mathbb{V}))$. This semantics associates each statement with its set of truly tainted variables by relying on α_t.

$$\mathcal{S}_t[\![S]\!](\mathcal{R}, \mathcal{E}) \triangleq \text{let } (\mathcal{R}_1, \mathcal{E}_1) = \mathcal{S}[\![S]\!](\mathcal{R}, \mathcal{E}) \text{ in } (\mathcal{R}_1, \mathcal{E}_1, \alpha_t(\{\mathcal{R}_1\}))$$

We then define the reachability taint semantics for programs $\mathcal{S}_t[\![P]\!] \in \mathbb{D} \times \wp(\mathbb{V})$. As regular and erroneous states are merged at the end of programs, we use α_t to obtain the tainted variables in the union. Let P:=S.

$$\mathcal{S}_t[\![P]\!] \triangleq \text{let } (\mathcal{R}, \mathcal{E}, \mathcal{T}) = \mathcal{S}_t[\![S]\!](\mathcal{I}, \emptyset) \text{ in } (\mathcal{R} \cup \mathcal{E}, \mathcal{T} \cup \alpha_t(\{\mathcal{R} \cup \mathcal{E}\}))$$

Observe that only at the end of the program ret can become tainted: regular and erroneous states are partitioned in the semantics for statements, so that ret is always constant (0 for the normal executions and 1 for the others). The program P is then nonexploitable iff ret is not tainted in $\mathcal{S}_t[\![P]\!]$.

Example 4. The statement x ← rand() can taint x if there are two executions in which the sequence of random numbers is out-of-sync due to a user action. This is because in the definition of $\mathcal{T}$ we compare pairs of execution with the *same* sequence of random numbers. Consider the following program:

x ← input(); if (x != 0) { y ← rand() }; z ← rand()

The user can control whether z is assigned to the first or the second number in the random sequence. If we fix as random sequence $1, 2, \dots$, we can observe that z can be either 1 or 2 at the end of the program, depending on the user's input.

Observe that this behaviour is relevant in scenarios where the attacker has partial knowledge about the uncontrolled random input. For instance, consider the program in Fig. 3, where the application first reads a character from standard input. If the character is a, the program reads the first pseudo-random number, and then it assigns z to rand(). As the sequence of random numbers has not been initialized, it does not change and could be predicted across different executions.

```
1   void main() {
2      if (getchar() == 'a') rand();
3      int z = rand();
4   }
```

Fig. 3. C program that reads pseudo-random numbers.

The user can make the program assign z to the first or second number in the sequence, being able to influence the assigned value. Another relevant case is when a program reads a file with unmodifiable but public content. If an attacker can control which bytes are read, then they can influence the execution of the program without even modifying the contents of the file.

The fact that random read statements can potentially taint the assigned variable directly derives from our semantic definition of $\mathcal{T}$. By changing the definition of $\mathcal{T}$, it is possible to choose whether random read statements can taint assigned variables. We make the choice to use random read statements as potential sources of tainted data because, in a context in which an attacker has (partial) knowledge about the unmodifiable pseudo-random input, such statements can be exploited to influence the execution of the program. While it would be possible to support the classic model where the attacker has no knowledge about the random input, this is less interesting in a context where security is considered fundamental.

As we want to overapproximate the concrete taint semantics by induction on the program structure, we give a structural equivalent definition of $\mathcal{S}_t[\![S]\!]$. The non-computable semantics $\hat{\mathcal{S}}_t[\![S]\!] : (\mathbb{D} \times \mathbb{D} \times \wp(\mathbb{V})) \rightarrow (\mathbb{D} \times \mathbb{D} \times \wp(\mathbb{V}))$ inductively collects the truly tainted variables. The semantics takes as additional input parameter the set of previously tainted variables, which are used to infer the set of tainted variables after the execution of the statement. Due to space constraints, here we present only the definition for assignments.

Assignment statement (S := x ← A)

$\hat{\mathcal{S}}_t[\![S]\!](\mathcal{R}, \mathcal{E}, \mathcal{T}) \triangleq$

    let $(\mathcal{R}_1, \mathcal{E}_1) = \mathcal{S}[\![S]\!](\mathcal{R}, \mathcal{E})$ in

    let $\mathcal{T}_1 = \{ y \in \mathcal{T} \mid y \neq x \} \cup$

        $\{ x \mid \exists ((m_0, i_0, r_0), (m_1, i_1, r_1)), ((m'_0, i'_0, r'_0), (m'_1, i'_1, r'_1)) \in \mathcal{R} :$

            $m_0 = m'_0, i_0 \neq i'_0, r_0 = r'_0 : \xi \neq \mathcal{A}[\![A]\!]m_1 \neq \mathcal{A}[\![A]\!]m'_1 \neq \xi \}$ in

    $(\mathcal{R}_1, \mathcal{E}_1, \mathcal{T}_1)$

We define the concrete inductive taint semantics for program P:=S as follows.

$$\hat{\mathcal{S}}_t[\![P]\!] \triangleq \text{let } (\mathcal{R}, \mathcal{E}, \mathcal{T}) = \hat{\mathcal{S}}_t[\![S]\!](\mathcal{I}, \emptyset, \emptyset) \text{ in } (\mathcal{R} \cup \mathcal{E}, \mathcal{T} \cup \alpha_t(\{\mathcal{R} \cup \mathcal{E}\}))$$

The following result formalizes that $\hat{\mathcal{S}}_t[\![P]\!]$ and $\mathcal{S}_t[\![P]\!]$ are equivalent.

> **Theorem 2. (Correctness of $\hat{S}_t[\![P]\!]$).** $\hat{S}_t[\![P]\!] \doteq S_t[\![P]\!]$

6 Taint Abstract Semantics

In this section, we introduce a computable sound overapproximation of the taint semantics presented in Sect. 5. This abstraction of the concrete non-computable semantics is parametric in the underlying abstract domain used to overapproximate the values of the variables. In contrast to traditional techniques, we leverage numeric invariants to improve the precision of the taint analysis.

Let $\mathbb{D}^\sharp$ be the abstract domain used to overapproximate $\mathbb{D}$, and $\gamma_d : \mathbb{D}^\sharp \to \mathbb{D}$ be the concretization function.[1] The domain $\mathbb{D}^\sharp$ is equipped with partial order $\sqsubseteq_d^\sharp$ and abstract join $\sqcup_d^\sharp$, while $\bot_d^\sharp$ is the bottom element. We assume $S_d^\sharp[\![S]\!] :$ $(\mathbb{D}^\sharp \times \mathbb{D}^\sharp) \to (\mathbb{D}^\sharp \times \mathbb{D}^\sharp)$ given by the numeric domain to be a sound computable abstraction of $S[\![S]\!]$: $\forall \mathcal{R}^\sharp, \mathcal{E}^\sharp \in \mathbb{D}^\sharp : S[\![S]\!](\dot{\gamma}_d(\mathcal{R}^\sharp, \mathcal{E}^\sharp)) \dot{\sqsubseteq} \dot{\gamma}_d(S_d^\sharp[\![S]\!](\mathcal{R}^\sharp, \mathcal{E}^\sharp))$. The abstract value domain also exposes the abstract functions $\mathtt{test}^\sharp[\![B]\!]$ and $\mathtt{err}^\sharp[\![B]\!]$ to overapproximate the concrete ones.

In the rest of the section, we structurally define the *abstract taint semantics* $S_t^\sharp[\![S]\!] : (\mathbb{D}^\sharp \times \mathbb{D}^\sharp \times \wp(\mathbb{V})) \to (\mathbb{D}^\sharp \times \mathbb{D}^\sharp \times \wp(\mathbb{V}))$. The semantics collects an overapproximation of the reachable states, the error states, and the tainted variables. The concretization function $\gamma : (\mathbb{D}^\sharp \times \mathbb{D}^\sharp \times \wp(\mathbb{V})) \to (\mathbb{D} \times \mathbb{D} \times \wp(\mathbb{V}))$ is defined as $\gamma(\mathcal{R}^\sharp, \mathcal{E}^\sharp, \mathcal{T}^\sharp) \triangleq (\gamma_d(\mathcal{R}^\sharp), \gamma_d(\mathcal{E}^\sharp), \mathcal{T}^\sharp)$. The soundness criterion states that the abstract semantics exhibits more tainted variables than those in the concrete semantics $\hat{S}_t[\![S]\!]$. Let $\mathcal{R}^\sharp, \mathcal{E}^\sharp \in \mathbb{D}^\sharp, \mathcal{T}^\sharp \in \wp(\mathbb{V})$.

$$\hat{S}_t[\![S]\!](\gamma(\mathcal{R}^\sharp, \mathcal{E}^\sharp, \mathcal{T}^\sharp)) \dot{\sqsubseteq} \gamma(S_t^\sharp[\![S]\!](\mathcal{R}^\sharp, \mathcal{E}^\sharp, \mathcal{T}^\sharp)) \tag{3}$$

In our abstract semantics, we taint $\mathtt{ret}$ every time there is a *possible* runtime error due to user input. This ensures that if $\mathtt{ret}$ is untainted in $S_t^\sharp[\![S]\!]$, it will be untainted at the end of the program, i.e. the program is nonexploitable. Let $\mathtt{P}{:=}\mathtt{S}$, and let $\mathcal{T}^\sharp \in \mathbb{D}^\sharp$ be an overapproximation of the set of initial states, namely $\mathcal{T} \subseteq \gamma_d(\mathcal{T}^\sharp)$. Let $(\mathcal{R}^\sharp, \mathcal{E}^\sharp, \mathcal{T}^\sharp) = S_t^\sharp[\![S]\!](\mathcal{T}^\sharp, \bot_d^\sharp, \emptyset)$.

$$\mathtt{ret} \notin \mathcal{T}^\sharp \implies S[\![P]\!] \in \mathscr{N\!E} \tag{4}$$

In the rest of this section, we define by structural induction $S_t^\sharp[\![S]\!]$. The abstract semantics collects an overapproximation of the tainted variables, and specifically taints $\mathtt{ret}$ whenever a runtime error potentially caused by the user occurs. We will take advantage of the helper function $\mathtt{taint}^\sharp[\![A]\!] : (\mathbb{D}^\sharp \times \wp(\mathbb{V})) \to \mathbb{B}$ that returns $\mathtt{ff}$ only if the result of the evaluation of A definitely does not depend on tainted variables. Standard value-insensitive taint analyses ignore

---

[1] While the concrete semantics is defined as a set of input-output relations to express nonexploitability, in the numeric abstraction it is possible to use numeric domains that abstract sets of states by abstracting only the image of the relations, and then consider each possible state as initial in the concretization.

the values of the variables and simply return tt if a tainted variable *syntactically* appears in A. This is sound, but imprecise. For instance, consider the program x ← input(); y ← x; z ← x-y. The user cannot control the value of z, as it is always 0. By using a relational abstract domain such as polyhedra or octagons [40], it is possible to determine that z is constant, and therefore that it is not tainted. The actual definition of $\mathtt{taint}^\sharp[\![A]\!]$ depends on the underlying abstract value domain, presenting numerous opportunities to improve the function's precision.

The abstract semantics for skip and $S_1; S_2$ are standard, and we do not report them. As variables read from user input are the main sources of tainted data, we always taint variables read from input statements.

Input read statement $(S := x \leftarrow \mathtt{input}())$

$$\mathcal{S}_t^\sharp[\![S]\!](\mathcal{R}^\sharp, \mathcal{E}^\sharp, \mathcal{T}^\sharp) \triangleq \text{let } (\mathcal{R}_1^\sharp, \mathcal{E}_1^\sharp) = \mathcal{S}_d^\sharp[\![S]\!](\mathcal{R}^\sharp, \mathcal{E}^\sharp) \text{ in}$$
$$(\mathcal{R}_1^\sharp, \mathcal{E}_1^\sharp, \mathcal{T}^\sharp \cup \{x\})$$

As observed in Sect. 5 (see Example 4), random read statements can taint the assigned variable in case the user controls the position of the value which is read in the random input sequence. For the abstract semantics, it would be sound to always taint the assigned variable. Nevertheless, this is too coarse, and we propose an abstraction that improves the precision. The idea is to represent the sequence of random numbers as a queue: programs read from it at index i, and then increment i. In this model, x ← rand() is syntactically substituted with x ← rand[i]; i ← i+1. We assume that the abstract semantics $\mathcal{S}_d^\sharp[\![S]\!]$ can handle reading from the queue. The special index variable i is then handled by the numeric domain as any other variable. We taint the result of x ← rand() only if i is tainted: this happens when the user can control which number is read from the random sequence.

Random read statement $(S := x \leftarrow \mathtt{rand}())$

$$\mathcal{S}_t^\sharp[\![S]\!](\mathcal{R}^\sharp, \mathcal{E}^\sharp, \mathcal{T}^\sharp) \triangleq \text{let } (\mathcal{R}_1^\sharp, \mathcal{E}_1^\sharp) = \mathcal{S}_d^\sharp[\![x \leftarrow \mathtt{rand}[i]; i \leftarrow i+1]\!](\mathcal{R}^\sharp, \mathcal{E}^\sharp) \text{ in}$$
$$\text{let } \mathcal{T}_1^\sharp = \{y \in \mathcal{T}^\sharp \mid y \neq x\} \cup \{x \mid \mathtt{taint}^\sharp[\![i]\!](\mathcal{R}^\sharp, \mathcal{T}^\sharp)\} \text{ in}$$
$$(\mathcal{R}_1^\sharp, \mathcal{E}_1^\sharp, \mathcal{T}_1^\sharp)$$

Assignments can present runtime errors, so that we need to taint ret in case the user can trigger a program failure. To determine if there is an exploitable runtime error in the evaluation of an expression, we rely on the function $\mathtt{exploit}^\sharp[\![A]\!] : (\mathbb{D}^\sharp \times \wp(\mathbb{V})) \to \mathbb{B}$. The function returns tt if there is a possible runtime error when evaluating A, and such an error can be triggered by the user. We assume the existence of a function $\mathtt{zero}^\sharp[\![A]\!] : \mathbb{D}^\sharp \to \mathbb{B}$, which is provided by the numeric domain and returns tt if the evaluation of A is possibly zero. Let $x \in \mathbb{V}$. We define $\mathtt{exploit}^\sharp[\![x]\!]$ and $\mathtt{exploit}^\sharp[\![n]\!]$ as ff, while for binary expressions we need to consider the values of the variables.

$$\mathtt{exploit}^\sharp[\![A_1 \diamond A_2]\!](\mathcal{R}^\sharp, \mathcal{T}^\sharp) \triangleq \begin{cases} \mathtt{tt} & \text{if } \diamond = /, \mathtt{zero}^\sharp[\![A_2]\!]\mathcal{R}^\sharp, \mathtt{taint}^\sharp[\![A_2]\!](\mathcal{R}^\sharp, \mathcal{T}^\sharp) \\ \mathtt{tt} & \text{if } \mathtt{exploit}^\sharp[\![A_1]\!](\mathcal{R}^\sharp, \mathcal{T}^\sharp) \text{ or } \mathtt{exploit}^\sharp[\![A_2]\!](\mathcal{R}^\sharp, \mathcal{T}^\sharp) \\ \mathtt{ff} & \text{otherwise} \end{cases}$$

Assignment statement $(S := x \leftarrow A)$

$\mathcal{S}_t^\sharp[\![S]\!](\mathcal{R}^\sharp, \mathcal{E}^\sharp, \mathcal{T}^\sharp) \triangleq$

  let $(\mathcal{R}_1^\sharp, \mathcal{E}_1^\sharp) = \mathcal{S}_d^\sharp[\![S]\!](\mathcal{R}^\sharp, \mathcal{E}^\sharp)$ in

  let $\mathcal{T}_1^\sharp = \{\, y \in \mathcal{T}^\sharp \mid y \neq x \,\} \cup$
    $\{\, x \mid \mathtt{taint}^\sharp[\![A]\!](\mathcal{R}^\sharp, \mathcal{T}^\sharp) \,\} \cup \{\, \mathtt{ret} \mid \mathtt{exploit}^\sharp[\![A]\!](\mathcal{R}^\sharp, \mathcal{T}^\sharp) \,\}$ in
  $(\mathcal{R}_1^\sharp, \mathcal{E}_1^\sharp, \mathcal{T}_1^\sharp)$

As discussed in Sect. 4 (see Example 3), if statements can generate *implicit flows* [25], namely dependencies that originate from the program control flow. When an attacker can control which branch of an if statement is executed, and in that branch a variable is assigned, then the variable could be tainted.

The set of variables that become tainted as a result of a tainted condition is traditionally overapproximated (when conditions are handled at all) with the variables that *syntactically* appear in the assignments of the branches. This is a coarse overapproximation, and we can improve this result by using the values of the variables. For instance, consider the program $x \leftarrow y$; if $(y < x) \{ z \leftarrow 10 \}$. The assignment is never executed, and a relational analysis can deduce that z is never assigned. The traditional syntactic approach is not sufficient to infer this information. We rely on the function $\mathtt{assigned}^\sharp[\![S]\!] : \mathbb{D}^\sharp \to \wp(\mathbb{V})$ that returns an overapproximation of the set of variables that are *semantically* assigned when executing S. If there is a state in the concretization of the abstract input $\mathcal{R}^\sharp$ in which a variable x changes value during the execution of S, then $x \in \mathtt{assigned}^\sharp[\![S]\!]\mathcal{R}^\sharp$. Observe that in case an exploitable runtime error occurs, $\mathtt{assigned}^\sharp[\![S]\!]\mathcal{R}^\sharp$ includes $\mathtt{ret}$, which does not syntactically appear in the program. A straightforward implementation can run the regular value analysis and inductively collect the variables that are assigned. While doing this, the function discards unreachable code and assignments that do not modify the state, such as $x \leftarrow 0$ when x is already 0, being effectively more precise than a syntactic approach. We define the following function to compute the set of variables that are tainted due to implicit flows.

$$\mathtt{diff}^\sharp[\![\mathtt{if} \ (B) \ S_t \ \mathtt{else} \ S_e]\!](\mathcal{R}^\sharp, \mathcal{T}^\sharp) \triangleq$$
$$\{\, x \in \mathtt{assigned}^\sharp[\![\mathtt{if} \ (B) \ S_t \ \mathtt{else} \ S_e]\!]\mathcal{R}^\sharp \mid \mathtt{taint}^\sharp[\![B]\!](\mathcal{R}^\sharp, \mathcal{T}^\sharp) \,\}$$

Tainted variables can also become untainted due to conditionals. For instance, the variable x is not tainted inside of the then branch in the following program: $x \leftarrow \mathtt{input}()$; if $(x == 0) \{ \ldots \}$. The reason is that x equals zero when entering

```
x ← input()
y ← 1
if (x==0) {
    z ← rand()
    if (z==0) { 1/x }
    if (z==1) { y ← z }
}
w ← rand()
```

```
x ← input()
if (x <= 0) { x ← 1 }
while (tt) {
    1 / x
    x ← rand()
}
```

(a)   (b)

Fig. 4. Programs that read values from the user and the random queue.

the first branch, and constants are by definition not controlled by the user. Classic methods ignore this, and do not filter tainted variables after conditionals. This is sound, but we can again achieve better precision by taking into account the values of the variables. We define the function $\mathbf{refine}^{\sharp}[\![B]\!] : (\mathbb{D}^{\sharp} \times \wp(\mathbb{V})) \to \wp(\mathbb{V})$ as $\mathbf{refine}^{\sharp}[\![B]\!](\mathcal{R}^{\sharp}, \mathcal{J}^{\sharp}) \triangleq \mathcal{J}^{\sharp} \backslash \mathbf{const}^{\sharp}(\mathbf{test}^{\sharp}[\![B]\!]\mathcal{R}^{\sharp})$, where $\mathbf{const}^{\sharp}$ returns the set of constant variables in the abstract state in its argument. The function $\mathbf{refine}^{\sharp}[\![B]\!]$ filters out the variables that are definitely constant after the execution of the test B, improving the precision of the analysis. We can now give the definition of the abstract semantics for if statements.

If statement $(S := \text{if } (B) \, S_t \text{ else } S_e)$

$\mathcal{S}^{\sharp}_t[\![S]\!](\mathcal{R}^{\sharp}, \mathcal{E}^{\sharp}, \mathcal{J}^{\sharp}) \triangleq$

    let $(\mathcal{R}^{\sharp}_t, \mathcal{E}^{\sharp}_t, \mathcal{J}^{\sharp}_t) = \mathcal{S}^{\sharp}_t[\![S_t]\!](\mathbf{test}^{\sharp}[\![B]\!]\mathcal{R}^{\sharp}, \mathcal{E}^{\sharp}, \mathbf{refine}^{\sharp}[\![B]\!](\mathcal{R}^{\sharp}, \mathcal{J}^{\sharp}))$ in

    let $(\mathcal{R}^{\sharp}_e, \mathcal{E}^{\sharp}_e, \mathcal{J}^{\sharp}_e) = \mathcal{S}^{\sharp}_t[\![S_e]\!](\mathbf{test}^{\sharp}[\![\neg B]\!]\mathcal{R}^{\sharp}, \mathcal{E}^{\sharp}, \mathbf{refine}^{\sharp}[\![\neg B]\!](\mathcal{R}^{\sharp}, \mathcal{J}^{\sharp}))$ in

    let $\mathcal{J}^{\sharp}_{te} = \mathbf{diff}^{\sharp}[\![\text{if } (B) \, S_t \text{ else } S_e]\!](\mathcal{R}^{\sharp}, \mathcal{J}^{\sharp})$ in

    $(\mathcal{R}^{\sharp}_t \cup^{\sharp}_d \mathcal{R}^{\sharp}_e, \mathcal{E}^{\sharp}_t \cup^{\sharp}_d \mathcal{E}^{\sharp}_e \cup^{\sharp}_d \mathbf{err}^{\sharp}[\![B]\!]\mathcal{R}^{\sharp}, \mathcal{J}^{\sharp}_t \cup \mathcal{J}^{\sharp}_e \cup \mathcal{J}^{\sharp}_{te})$

Example 5. The program in Fig. 4a demonstrates various ways in which our analysis differs from other taint analyses. First, we can infer that the program is exploitable: if the user inputs zero, then there is the possibility, depending on the sequence of random numbers, that a runtime error is triggered. The value analysis is important to infer that x is zero when performing the division, so that we can deduce that there is a division by zero. Second, we can use the semantic information inferred by the numeric domain to deduce that y is not tainted. Even if y is assigned inside of a branch that depends on the user's input, the variable y does not change, as it is still 1 after the execution of the statement. An interval analysis is sufficient to deduce this. Third, we can infer that the

variable w is tainted: depending on user input, it is assigned either to the first
or the second value in the sequence of random numbers.

Further precision improvements can be implemented. For instance, consider
the program if (x < 10) {y ← 0} else {y ← 1}. If the abstract value domain
can determine that before the execution of the statement the value of x is less
than 10, the statement is semantically equivalent to y ← 0. This implies that,
even if x is tainted, the user cannot control the value of y. When the analysis
can infer that one of the two branches is never executed, the if statement can be
substituted with the other branch, ignoring the implicit flows that are generated
by the condition, and improving again the precision.

The abstract semantics for while statements is a classic limit computation
that relies on the widening operator ∇ to guarantee convergence in a finite
number of iterations. As the number of variables is finite, $\wp(\mathbb{V})$ has finite height,
so that the widening operator for $\wp(\mathbb{V})$ is simply the set union. The operator
$(\mathcal{R}_1^\sharp, \mathcal{E}_1^\sharp, \mathcal{T}_1^\sharp) \,\dot\cup^\sharp\, (\mathcal{R}_2^\sharp, \mathcal{E}_2^\sharp, \mathcal{T}_2^\sharp)$ denotes $(\mathcal{R}_1^\sharp \cup_d^\sharp \mathcal{R}_2^\sharp, \mathcal{E}_1^\sharp \cup_d^\sharp \mathcal{E}_2^\sharp, \mathcal{T}_1^\sharp \cup \mathcal{T}_2^\sharp)$, and the operator
$(\mathcal{R}_1^\sharp, \mathcal{E}_1^\sharp, \mathcal{T}_1^\sharp) \,\dot\nabla\, (\mathcal{R}_2^\sharp, \mathcal{E}_2^\sharp, \mathcal{T}_2^\sharp)$ denotes $(\mathcal{R}_1^\sharp \nabla \mathcal{R}_2^\sharp, \mathcal{E}_1^\sharp \nabla \mathcal{E}_2^\sharp, \mathcal{T}_1^\sharp \cup \mathcal{T}_2^\sharp)$.

While statement (S := while (B) S_b)

$$\mathcal{S}_t^\sharp[\![S]\!](\mathcal{R}^\sharp, \mathcal{E}^\sharp, \mathcal{T}^\sharp) \triangleq \text{let } (\mathcal{R}_f^\sharp, \mathcal{E}_f^\sharp, \mathcal{T}_f^\sharp) = \lim F^n(\bot_d^\sharp, \bot_d^\sharp, \emptyset) \text{ in}$$

$$(\text{test}^\sharp[\![\neg B]\!]\mathcal{R}_f^\sharp, \mathcal{E}_f^\sharp, \text{refine}^\sharp[\![\neg B]\!](\mathcal{R}_f^\sharp, \mathcal{T}_f^\sharp))$$

where

$$F(\mathcal{R}_1^\sharp, \mathcal{E}_1^\sharp, \mathcal{T}_1^\sharp) \triangleq \text{let } (\mathcal{R}_2^\sharp, \mathcal{E}_2^\sharp, \mathcal{T}_2^\sharp) = \mathcal{S}_t^\sharp[\![\text{if (B) } S_b \text{ else skip}]\!](\mathcal{R}_1^\sharp, \mathcal{E}_1^\sharp, \mathcal{T}_1^\sharp) \text{ in}$$

$$(\mathcal{R}_1^\sharp, \mathcal{E}_1^\sharp, \mathcal{T}_1^\sharp) \,\dot\nabla\, ((\mathcal{R}^\sharp, \mathcal{E}^\sharp, \mathcal{T}^\sharp) \,\dot\cup^\sharp\, (\mathcal{R}_2^\sharp, \mathcal{E}_2^\sharp, \mathcal{T}_2^\sharp))$$

Example 6. In principle, it is possible to first run a value analysis, and then use
the inferred numeric invariants in a taint analysis to prove nonexploitability. Nevertheless, as shown by the program in Fig. 4b, executing the two together achieves
strictly superior precision by leveraging the reduction between the domains. The
invariant inferred at statement 1/x entails that x can be zero, so that there is a
potential runtime error. Furthermore, a taint analysis infers that x is tainted at
the same program location. By combining these information, the division by zero
is exploitable. However, if we execute the value and the taint analyses together,
we can observe that it is never true *at the same time* that x is 0 and tainted,
so that the program failure cannot be triggered by an attacker. Our framework
runs the two analyses together, and is thus able to prove that the program
is nonexploitable. The reduction between the two domains can also refine the
taint information using the value information, and this can improve subsequent
results.

7   Experimental Evaluation

Implementation. We propose MOPSA-NEXP, the *first* analyzer dedicated to nonexploitability. We implemented our analysis for a large subset of C in the MOPSA framework [35], which is a modular platform to build static analyzers based on abstract interpretation. MOPSA offers an extensive collection of ready-to-use abstract domains for analyzing C and Python, providing the flexibility to tune the tradeoff between precision and performance. MOPSA is implemented in 120,000 lines of OCaml code, and our exploitability analysis accounts for around 10,000 of them. Thanks to MOPSA's modular design, we were able to use most of the C analysis with minimal modifications.

In our implementation, we maintain taint information at the level of *memory blocks*, i.e. we perform a *field-insensitive* taint analysis. While this can result in a loss of precision, the implementation is simple and efficient. Proposing an enhanced field-sensitive taint analysis for C is out of the scope of this paper, and it is left as future work. As MOPSA performs dynamic expression rewriting to encourage a design based on layered semantics, to retrieve sources of tainted data, during the analysis we have to consider the expressions' rewriting history.

Our analysis can detect a wide variety of runtime errors, including double free, index-out-of-bounds, and null pointer dereference. While the formal presentation in this article, for the sake of simplicity, only supports division-by-zero errors, it was trivial to adapt our analysis to identify different types of failures. In the report of the analyzer each warning is classified as possibly exploitable or not, and we infer a sound overapproximation of both the regular runtime errors and the exploitable ones. All the warnings that are not labelled as exploitable are thus *proved* to be nonexploitable. If the analyzer does not report *any* exploitable warning, then this is a proof that the program is nonexploitable.

The functions that read data from the user are part of the C standard library. They include, for instance, `getchar`, `scanf`, and `recv`. MOPSA provides a stub modeling language to specify the behaviour of library functions [46]. We have extended this language to support the fact that some functions generate tainted data, and then we annotated our stubs for the C standard library to take into account the taint information. This model makes it trivial to update the list of dangerous sources, and the user of the analyzer does not have to annotate the source code to run the exploitability analysis, which is *fully automatic*.

Performance and Precision Evaluation. To assess the usefulness of our tool, we have analyzed real-world C programs from the GNU Coreutils package, which is a collection of command-line utilities. The test suite is composed of 77 programs that are long up to $4,188$ lines of code. To them, we added a large set of short C programs taken from the Juliet test suite developed by NIST [7]. These programs contain examples of various runtime errors that can trigger well-known security vulnerabilities. In fact, Juliet is based on the CVE database [1], which enumerates vulnerabilities and focusses on security. The tested runtime errors include double frees, index out-of-bounds, and null pointer dereferences. The test

Table 1. Evaluation results.

Test suite	Domain	Analyzer	Alarms	Time
Coreutils	Intervals	Mopsa	4,715	1:17:06
		Mopsa-Nexp	1,217	1:28:42
	Octagons	Mopsa	4,673	2:22:29
		Mopsa-Nexp	1,209	2:43:06
	Polyhedra	Mopsa	4,651	2:12:21
		Mopsa-Nexp	1,193	2:30:44
Juliet	Intervals	Mopsa	49,957	11:32:24
		Mopsa-Nexp	13,906	11:48:51
	Octagons	Mopsa	48,256	13:15:29
		Mopsa-Nexp	13,631	13:41:47
	Polyhedra	Mopsa	48,256	12:54:21
		Mopsa-Nexp	13,631	13:21:26

cases are specifically designed to assess the precision of static analysis tools, and use a large set of features from the C standard. For Juliet, we considered $13,261$ different test cases that amount to a total of $2,861,980$ lines of code. Each test case comes with two versions: one that triggers a runtime failure, and one where the error is fixed. We run our analysis on both versions. An artifact to reproduce our experimental evaluation is available on Zenodo [47].

We compare the performance and number of alarms between Mopsa-Nexp and Mopsa. The analyses are parametric in the underlying abstract numeric domain, and we consider intervals, octagons [40], and polyhedra. Observe that to compare only the number of alarms raised by the two analyzers it is not necessary to run both tools, as Mopsa-Nexp can report all warnings raised by Mopsa. Notice that while the ground truth about the errors provided with Juliet can be used to evaluate the precision of a classic safety analysis, this is not the case for nonexploitability. In fact, the benchmarks categorize the test cases as either dangerous or not, but they do not include any information about whether an attacker can trigger the errors. We ran our experiments on a server with 128GB of RAM, with 48 Intel Xeon CPUs E5-2650 v4 @ 2.20GHz and Ubuntu 18.04.5 LTS. In Table 1 we report the results of our experiments.

For Coreutils, in the case of intervals, our analysis was able to *prove* that $3,498$ over $4,715$ runtime errors previously reported by the analyzer cannot be triggered by an attacker. For octagons and polyhedra, our analysis proved that respectively 3,464 and 3,458 potential runtime errors over 4,673 and 4,651 are not exploitable. Overall, this results in filtering out 74.13%-74.35% of the warnings. We found similar results for Juliet, where Mopsa-Nexp was able to prove that 71.75%-72.16% of the warnings are not exploitable. For Coreutils, Mopsa-Nexp raises 1,193 to $1,217$ warnings, which are those that can be potentially triggered by an attacker. The user of the analyzer could prioritize those alarms over the regular ones, as they are comparatively more dangerous.

The exploitability analysis incurs a performance overhead ranging from 13.89% to 15.05% for Coreutils and 2.4% to 3.5% for Juliet. During the analysis we consider expressions' rewriting history to preserve taint information, and this history is sensibly larger in real-world programs, which justifies the performance overhead difference between Coreutils and Juliet. Observe that we found octagons to be less efficient than polyhedra. This is due to the fact that MOPSA relies on the APRON [34] library, which uses a sparse representation for polyhedra, and can be very efficient if the number of variables is low and there are few constraints. As octagons use a dense representation, even if their algorithmic complexity is better, they are slightly slower for our case.

Discussion. We observed that MOPSA-NEXP is able to consistently filter out more than 70% of the warnings raised by the regular analyzer, while imposing low performance overhead. The Juliet test cases show that MOPSA-NEXP can handle almost the whole C specification, while the Coreutils experiments confirm that our analysis is effective even for real-world programs. The significant advantage of being able to classify each warning as security-critical or not outweighs the reasonable performance cost overhead. Observe that the alarms raised by MOPSA-NEXP are a subset of those reported by MOPSA. This implies that the exploitability analysis is, in the worst case, as precise as the regular analysis.

While it would be desirable to determine how many truly exploitable alarms are raised by MOPSA-NEXP, this cannot be done automatically. In fact, there is no ground truth that classifies program errors as nonexploitable or not, so that human inspection is the only option. In future work, we would like to conduct such an inspection.

8   Related Work

Secure Information Flow. In [25] the authors propose the first mechanism to verify the secure flow of information in a program, namely checking that a program cannot cause supposedly nonconfidential results to depend on confidential input data. Their formulation of the problem is based on the syntax of a program, and does not take into account its semantics. The concept of secure information flow is related to *noninterference* [20,29,30], which is a semantic definition. A program is *noninterferent* if its public output data does not depend on private input data. Checking that a program cannot cause nonconfidential results to depend on confidential input data has been widely studied through type systems [9,14,32,43,44,49,50,52,57–59]. Because these works perform only syntactic checks without taking into account semantic information, the results are generally very imprecise. In contrast, our approach tracks the flow of data generated by the user through a semantic taint analysis, which achieves enhanced precision by leveraging an overapproximation of the values of the variables.

Noninterference and nonexploitabilty are closely related: nonexploitability can be seen as a type of noninterference where the only public output variable is `ret`. Nevertheless, we do not rely on the static partitioning of variables into public and private, as our definition supports dynamic user input reads. Our framework can be used to prove noninterference: it is sufficient to read all private input variables at the beginning of the program, and then verify that the public output variables are not tainted. On the contrary, traditional methods to prove noninterference cannot prove nonexploitability, as they do not take the values of the variables into account.

Hyperproperties Verification. Clarkson and Schneider [19] put forward the framework of *hyperproperties*, namely program properties that relate different sets of executions. Hyperproperties are able to express security policies, such as secure information flow. K-hypersafety properties [19] can be verified with traditional techniques for safety properties on the k-times self-composed system [16,54], even though this can be computationally expensive [10]. HyperLTL and HyperCTL/CTL* [18,27] define extensions of temporal logic able to quantify over multiple traces to address the verification of hyperproperties.

Noninterference Verification by Abstract Interpretation. Cousot [22] put forward a semantic definition of dependencies in the abstract interpretation framework. He proposes a sound analysis of dependencies, capable of proving noninterference. Similarly to us, he does not rely on hypersemantics, using standard abstract interpretation techniques. Nevertheless, the abstract dependency semantics is not structural (i.e. defined by induction on the program syntax), as it does not take the values of variables into account. The author proposes leveraging the values of the variables to give a structural definition of the semantics, and this paper attempts to implement such an extension. Since his definition of the dependency semantics does not take into account the values of the variables, it is not possible to define an analysis that leverages numeric abstract domains to enhance the precision of the dependency analysis. Another significant difference is that the dependencies are relative to the *initial values* of variables. Our analysis computes *dynamic tainting*, which is the dependency of a variable from *any* input statement (including those within conditionals and loops), so that it generalizes the dependency analysis from the beginning of the program.

There are numerous papers that use an alternative version of the abstract interpretation framework based on *hypersemantics* [12,37–39,56], where the concrete domain is a set of sets of states, rather than a set of states. This is to overcome the difficulties related to the fact that not every hyperproperty is subset-closed, and classic overapproximation techniques seem to fail. However, as argued in this paper and [22], this is not the case for standard noninterference and nonexploitability. Relying on the classic abstract interpretation framework allows using the large library of existing abstract domains, and leveraging the semantic information inferred by such domains is not only essential for nonexploitabilty, but also enhances the precision of the taint analysis. Another approach to noninterference verification is introduced in [55], where the authors combine abstract interpretation with symbolic execution to define a sound analysis.

Errors Classification. In [28] the authors put forward the concept of *robust reachability*. A runtime error is robustly reachable if a controlled input can make it so the bug is reached whatever the value of uncontrolled input. The authors use symbolic execution and bounded model checking techniques to find robustly reachable bugs. Similarly to this paper, [28] classifies runtime failures by their dangerousness and filters out less interesting alarms that do not concern security issues. Nevertheless, the concept of robustly reachable runtime error is different from nonexploitability: a bug is considered robustly reachable even if it is triggered *for all* possible user input, while such an error is not exploitable according to our formal definition of exploitability. In fact, we require the user input to be actually involved in triggering an error to consider a program exploitable.

Other techniques relying on probability theory to differentiate classes of bugs have been proposed. They include *probabilistic model checking* [13,31], *probabilistic abstract interpretation* [24,41,42,48], *quantitative robust reachability* [15], and *quantitative information flow analysis* [33]. An interesting extension of this paper would be to use ideas from these approaches to put forward a *quantitative* exploitability analysis to classify more finely the level of threat caused by alarms.

9 Conclusions

In this paper, we introduced the novel definition of nonexploitability, which we leveraged to put forward a sound analysis by abstract interpretation. The framework supports constructs that are essential to analyze real-world programs, such as nondeterminism and dynamic user input reads. Our analysis performs a semantic taint analysis that achieves superior precision through a modular reduction with existing numeric abstract domains. The theoretical framework bridges the gap between traditional safety properties and security hyperproperties, and our analysis can rule out the existence of exploitable runtime errors in programs.

We implemented our analysis in the MOPSA-NEXP tool, the first analyzer dedicated to nonexploitability. The tool is fully automatic, and to assess its effectiveness, we evaluated it on a large set of real-world C programs. The analyzer can consistently *prove* that more than 70% of the previously raised warnings cannot be triggered by an attacker, all while incurring less than 16% performance overhead. While usually the number of false positives is lowered by increasing the precision of the abstract domains, we take an orthogonal approach by reporting only the alarms that can be triggered by an attacker. By leveraging the fundamental observation that security-related warnings are more dangerous than the others, our technique dramatically reduces the noise generated by false alarms, enhancing the usefulness of the analyzer.

In future work, we would like to extend our analysis to prove the absence of other classes of exploitable bugs. A promising path forward is to leverage probability theory to perform a *quantitative* exploitability analysis capable of further reducing the number of alarms. Another interesting extension of this paper is to adapt our framework to rule out the existence of exploitable liveness errors, such as exploitable deadlocks in multithreaded programs.

Acknowledgments. We would like to thank the anonymous reviewers for their comments. This work was supported by the SECURVAL project. The SECUREVAL project was funded by the "France 2030' government investment plan managed by the French National Research Agency, under the reference ANR-22-PECY-0005.

References

1. Common vulnerabilities and exposures (CVE) database. https://cve.mitre.org/. Accessed 30 Aug 2023
2. CVE-2019-8745. Available from NIST, CVE-ID CVE-2019-8745. https://nvd.nist.gov/vuln/detail/CVE-2019-8745. Accessed 30 Aug 2023
3. CVE-2022-36934. Available from NIST, CVE-ID CVE-2022-36934. https://nvd.nist.gov/vuln/detail/CVE-2022-36934. Accessed 30 Aug 2023
4. CVE-2022-4135. Available from NIST, CVE-ID CVE-2022-4135. https://nvd.nist.gov/vuln/detail/CVE-2022-4135 Accessed 30 Aug 2023
5. The Infer static analyzer. https://fbinfer.com/
6. The Pysa static analyzer. https://engineering.fb.com/2020/08/07/security/pysa/
7. Juliet C/C++ test suite (2017). https://samate.nist.gov/SARD/test-suites/112. Accessed 30 Aug 2023
8. Microsoft: a proactive approach to more secure code (2019). https://msrc.microsoft.com/blog/2019/07/a-proactive-approach-to-more-secure-code/. Accessed 30 Aug 2023
9. Agat, J.: Transforming out timing leaks. In: Principles of Programming Languages, POPL, pp. 40–53. ACM (2000). https://doi.org/10.1145/325694.325702
10. Antonopoulos, T., Gazzillo, P., Hicks, M., Koskinen, E., Terauchi, T., Wei, S.: Decomposition instead of self-composition for proving the absence of timing channels. In: Conference on Programming Language Design and Implementation, PLDI, pp. 362–375. ACM (2017). https://doi.org/10.1145/3062341.3062378
11. Arzt, S., et al.: FlowDroid: precise context, flow, field, object-sensitive and lifecycle-aware taint analysis for android apps. In: Programming Language Design and Implementation, PLDI, pp. 259–269. ACM (2014). https://doi.org/10.1145/2594291.2594299
12. Assaf, M., Naumann, D.A., Signoles, J., Totel, E., Tronel, F.: Hypercollecting semantics and its application to static analysis of information flow. In: Principles of Programming Languages, POPL (2017). https://doi.org/10.1145/3009837.3009889
13. Aziz, A., Sanwal, K., Singhal, V., Brayton, R.: Verifying continuous time Markov chains. In: Alur, R., Henzinger, T.A. (eds.) CAV 1996. LNCS, vol. 1102, pp. 269–276. Springer, Heidelberg (1996). https://doi.org/10.1007/3-540-61474-5_75
14. Banerjee, A., Naumann, D.A.: Secure information flow and pointer confinement in a Java-like language. In: Computer Security Foundations Workshop CSFW, pp. 253. IEEE Computer Society (2002). https://doi.org/10.1109/CSFW.2002.1021820
15. Bardin, S., Girol, G.: A quantitative flavour of robust reachability. CoRR abs/2212.05244 (2022). 10.48550/arXiv. 2212.05244

16. Barthe, G., D'Argenio, P.R., Rezk, T.: Secure information flow by self-composition. Math. Struct. Comput. Sci. **21**(6), 1207–1252 (2011). https://doi.org/10.1017/S0960129511000193

17. Berghel, H.: The code red worm. Commun. ACM **44**(12), 15–19 (2001). https://doi.org/10.1145/501317.501328

18. Clarkson, M.R., Finkbeiner, B., Koleini, M., Micinski, K.K., Rabe, M.N., Sánchez, C.: Temporal logics for hyperproperties. In: Abadi, M., Kremer, S. (eds.) POST 2014. LNCS, vol. 8414, pp. 265–284. Springer, Heidelberg (2014). https://doi.org/10.1007/978-3-642-54792-8_15

19. Clarkson, M.R., Schneider, F.B.: Hyperproperties. In: 21st IEEE Computer Security Foundations Symposium, pp. 51–65 (2008)

20. Cohen, E.S.: Information transmission in computational systems. In: Symposium on Operating System Principles, SOSP, pp. 133–139. ACM (1977). https://doi.org/10.1145/800214.806556

21. Cousot, P., et al.: The ASTREÉ analyzer. In: Sagiv, M. (ed.) ESOP 2005. LNCS, vol. 3444, pp. 21–30. Springer, Heidelberg (2005). https://doi.org/10.1007/978-3-540-31987-0_3

22. Cousot, P.: Abstract semantic dependency. In: Chang, B.-Y.E. (ed.) SAS 2019. LNCS, vol. 11822, pp. 389–410. Springer, Cham (2019). https://doi.org/10.1007/978-3-030-32304-2_19

23. Cousot, P., Cousot, R.: Abstract Interpretation: A Unified Lattice Model for Static Analysis of Programs by Construction or Approximation of Fixpoints. Principles of Programming Languages, POPL (1977)

24. Cousot, P., Monerau, M.: Probabilistic abstract interpretation. In: Seidl, H. (ed.) ESOP 2012. LNCS, vol. 7211, pp. 169–193. Springer, Heidelberg (2012). https://doi.org/10.1007/978-3-642-28869-2_9

25. Denning, D.E., Denning, P.J.: Certification of programs for secure information flow. Commun. ACM **20**(7), 504–513 (1977). https://doi.org/10.1145/359636.359712

26. Durumeric, Z., et al.: The matter of heartbleed. In: Internet Measurement Conference, IMC, pp. 475–488. ACM (2014). https://doi.org/10.1145/2663716.2663755

27. Finkbeiner, B., Rabe, M.N., Sánchez, C.: Algorithms for model checking HyperLTL and HyperCTL*. In: Kroening, D., Păsăreanu, C.S. (eds.) CAV 2015. LNCS, vol. 9206, pp. 30–48. Springer, Cham (2015). https://doi.org/10.1007/978-3-319-21690-4_3

28. Girol, G., Farinier, B., Bardin, S.: Not all bugs are created equal, but robust reachability can tell the difference. In: Silva, A., Leino, K.R.M. (eds.) CAV 2021. LNCS, vol. 12759, pp. 669–693. Springer, Cham (2021). https://doi.org/10.1007/978-3-030-81685-8_32

29. Goguen, J.A., Meseguer, J.: Security policies and security models. In: Security and Privacy, pp. 11–20. IEEE Computer Society (1982). https://doi.org/10.1109/SP.1982.10014

30. Goguen, J.A., Meseguer, J.: Unwinding and inference control. In: Security and Privacy, pp. 75–87. IEEE Computer Society (1984). https://doi.org/10.1109/SP.1984.10019

31. Hansson, H., Jonsson, B.: A logic for reasoning about time and reliability. Formal Aspects Comput. **6**(5), 512–535 (1994). https://doi.org/10.1007/BF01211866

32. Heintze, N., Riecke, J.G.: The slam calculus: programming with secrecy and integrity. In: Principles of Programming Languages, POPL, pp. 365–377. ACM (1998). https://doi.org/10.1145/268946.268976

33. Heusser, J., Malacaria, P.: Quantifying information leaks in software. In: Annual Computer Security Applications Conference, ACSAC, pp. 261–269. ACM (2010). https://doi.org/10.1145/1920261.1920300
34. Jeannet, B., Miné, A.: APRON: a library of numerical abstract domains for static analysis. In: Bouajjani, A., Maler, O. (eds.) CAV 2009. LNCS, vol. 5643, pp. 661–667. Springer, Heidelberg (2009). https://doi.org/10.1007/978-3-642-02658-4_52
35. Journault, M., Miné, A., Monat, R., Ouadjaout, A.: Combinations of reusable abstract domains for a multilingual static analyzer. In: Chakraborty, S., Navas, J.A. (eds.) VSTTE 2019. LNCS, vol. 12031, pp. 1–18. Springer, Cham (2020). https://doi.org/10.1007/978-3-030-41600-3_1
36. Li, L., et al.: Static analysis of android apps: a systematic literature review. Inf. Softw. Technol. 88, 67–95 (2017). https://doi.org/10.1016/j.infsof.2017.04.001
37. Mastroeni, I., Pasqua, M.: Hyperhierarchy of semantics - a formal framework for hyperproperties verification. In: Static Analysis Symposium, SAS. vol. 10422, pp. 232–252 (2017). https://doi.org/10.1007/978-3-319-66706-5_12
38. Mastroeni, I., Pasqua, M.: Verifying bounded subset-closed hyperproperties. In: Static Analysis Symposium, SAS. vol. 11002, pp. 263–283 (2018). https://doi.org/10.1007/978-3-319-99725-4_17
39. Mastroeni, I., Pasqua, M.: Statically analyzing information flows: an abstract interpretation-based hyperanalysis for non-interference. In: Symposium on Applied Computing, SAC, pp. 2215–2223 (2019). https://doi.org/10.1145/3297280.3297498
40. Miné, A.: The octagon abstract domain. High. Order Symbolic Comput. (HOSC) 19(1), 31–100 (2006). https://doi.org/10.1007/s10990-006-8609-1, http://www-apr.lip6.fr/mine/publi/article-mine-HOSC06.pdf
41. Monniaux, D.: Abstract interpretation of probabilistic semantics. In: Palsberg, J. (ed.) SAS 2000. LNCS, vol. 1824, pp. 322–339. Springer, Heidelberg (2000). https://doi.org/10.1007/978-3-540-45099-3_17
42. Monniaux, D.: An abstract analysis of the probabilistic termination of programs. In: Cousot, P. (ed.) SAS 2001. LNCS, vol. 2126, pp. 111–126. Springer, Heidelberg (2001). https://doi.org/10.1007/3-540-47764-0_7
43. Myers, A.C., Liskov, B.: A decentralized model for information flow control. In: Symposium on Operating System Principles, SOSP, pp. 129–142. ACM (1997). https://doi.org/10.1145/268998.266669
44. Ørbæk, P., Palsberg, J.: Trust in the lambda-calculus. J. Funct. Program. 7(6), 557–591 (1997). https://doi.org/10.1017/s0956796897002906
45. Orman, H.K.: The Morris worm: a fifteen-year perspective. IEEE Secur. Priv. 1(5), 35–43 (2003). https://doi.org/10.1109/MSECP.2003.1236233
46. Ouadjaout, A., Miné, A.: A library modeling language for the static analysis of C programs. In: Pichardie, D., Sighireanu, M. (eds.) SAS 2020. LNCS, vol. 12389, pp. 223–247. Springer, Cham (2020). https://doi.org/10.1007/978-3-030-65474-0_11
47. Parolini, F., Miné, A.: Sound Abstract Nonexploitability Analysis Artifact (2023). https://doi.org/10.5281/zenodo.8334112
48. Di Pierro, A., Wiklicky, H.: Probabilistic abstract interpretation: from trace semantics to DTMC's and linear regression. In: Probst, C.W., Hankin, C., Hansen, R.R. (eds.) Semantics, Logics, and Calculi. LNCS, vol. 9560, pp. 111–139. Springer, Cham (2016). https://doi.org/10.1007/978-3-319-27810-0_6
49. Pottier, F., Simonet, V.: Information flow inference for ML. ACM Trans. Program. Lang. Syst. 25(1), 117–158 (2003). https://doi.org/10.1145/596980.596983
50. Sabelfeld, A., Sands, D.: Probabilistic noninterference for multi-threaded programs. In: Computer Security Foundations Workshop, CSFW, pp. 200–214. IEEE Computer Society (2000). https://doi.org/10.1109/CSFW.2000.856937

51. Schultz, E., Mellander, J., Peterson, D.: The MS-SQL slammer worm. Netw. Secur. **2003**(3), 10–14 (2003). https://doi.org/10.1016/S1353-4858(03)00310-6
52. Smith, G., Volpano, D.M.: Secure information flow in a multi-threaded imperative language. In: Principles of Programming Languages, POPL, pp. 355–364. ACM (1998). https://doi.org/10.1145/268946.268975
53. Spoto, F., et al.: Static identification of injection attacks in Java. ACM Trans. Program. Lang. Syst. **41**(3), 18:1–18:58 (2019). https://doi.org/10.1145/3332371
54. Terauchi, T., Aiken, A.: Secure information flow as a safety problem. In: Hankin, C., Siveroni, I. (eds.) SAS 2005. LNCS, vol. 3672, pp. 352–367. Springer, Heidelberg (2005). https://doi.org/10.1007/11547662_24
55. Tiraboschi, I., Rezk, T., Rival, X.: Sound symbolic execution via abstract interpretation and its application to security. In: Verification, Model Checking, and Abstract Interpretation, VMCAI. LNCS, vol. 13881, pp. 267–295. Springer (2023). https://doi.org/10.1007/978-3-031-24950-1_13
56. Urban, C., Müller, P.: An abstract interpretation framework for input data usage. In: European Symposium on Programming, ESOP. vol. 10801, pp. 683–710 (2018). https://doi.org/10.1007/978-3-319-89884-1_24
57. Volpano, D.M., Irvine, C.E., Smith, G.: A sound type system for secure flow analysis. J. Comput. Secur. **4**(2/3), 167–188 (1996). https://doi.org/10.3233/JCS-1996-42-304
58. Volpano, D.M., Smith, G.: Probabilistic noninterference in a concurrent language. J. Comput. Secur. **7**(1), 231–253 (1999). https://doi.org/10.3233/jcs-1999-72-305
59. Zdancewic, S., Myers, A.C.: Secure information flow and CPS. In: Sands, D. (ed.) ESOP 2001. LNCS, vol. 2028, pp. 46–61. Springer, Heidelberg (2001). https://doi.org/10.1007/3-540-45309-1_4

Author Index

Printed in the United States
by Baker & Taylor Publisher Services